ADVANCE PRAISE FOR

Liberal Forces in Twentieth Century Yugoslavia

"In this book I met a number of people of whom and whose names I have heard. I found the memoirs very instructive because it showed me the other side of the political upbringing from which I benefited. Vladislav Bevc has done a very fine job editing it, and I commend Peter Lang for publishing it."

Aloysius Cardinal Ambrožič, Archbishop of Toronto

"The book is beautiful—the life story of the author is very rich and dramatic. The book will enrich our library and Sokol archive and will meaningfully complete the mosaic of history of the world Sokol movement."

Vladimir Dostal, Chairman of the International Department, Czech Sokol Accociation, Tyrš Hall, Prague, Czech Republic

Liberal Forces in Twentieth Century Yugoslavia

Studies in Modern European History

Frank J. Coppa
General Editor

Vol. 58

PETER LANG
New York • Washington, D.C./Baltimore • Bern
Frankfurt am Main • Berlin • Brussels • Vienna • Oxford

Liberal Forces in Twentieth Century Yugoslavia

Memoirs of Ladislav Bevc

Edited by
Vladislav Bevc

PETER LANG
New York • Washington, D.C./Baltimore • Bern
Frankfurt am Main • Berlin • Brussels • Vienna • Oxford

Library of Congress Cataloging-in-Publication Data

Bevc, Ladislav, 1890–1988.
Liberal forces in twentieth century Yugoslavia: memoirs of Ladislav Bevc /
Ladislav Bevc; edited by Vladislav Bevc.
p. cm. — (Studies in modern European history; v. 58)
Includes bibliographical references and index.
1. Bevc, Ladislav, 1890–1988. 2. Civil engineers—Yugoslavia—Biography.
3. Civil engineers—United States—Biography. 4. Slovenia—History—20th century—
Sources. 5. Yugoslavia—Politics and government—20th century.
I. Bevc, Vladislav. II. Title. III. Title: Liberal forces in 20th century Yugoslavia.
TA140.B49A3 949.73'02092—dc22 [B] 2007009787
ISBN 978-1-4331-0008-6
ISSN 0893-6897

Bibliographic information published by **Die Deutsche Bibliothek.**
Die Deutsche Bibliothek lists this publication in the "Deutsche
Nationalbibliografie"; detailed bibliographic data is available
on the Internet at http://dnb.ddb.de/.

Layout and typesetting by
Jutro d.o.o., P.O. Box 4986, SI 1001, Ljubljana

And is it his task to open [the Pearly Gates]
even to the souls of Slovenian liberals?

Ivan Tavčar: 4000

Acknowledgments

Thanks are due to my sister Breda and her husband Dr. Richard George both of whom read with great care the English translation of the manuscript and made numerous and well placed improvements to the text.

The idea for including the photographs in this book, as well as in the earlier Slovenian edition, originated with Stane Kodrič and his staff at the Publishing House Jutro in Ljubljana, Slovenia, who designed the layout, executed the typesetting, and gave unstintingly of their time and advice in carrying out the project.

The Editor

And is it his task to open The Pearly Gates!

Acknowledgments

Contents

Refugee Camps

List of Abbreviations

AMG	Allied Military Government
AVNOJ	Antifašističko Veče Narodnog Oslobodjenja Jugoslavije — Antifascist Council for People's Liberation of Yugoslavia
IRO	International Refugee Organization
KNOJ	Korpusi Narodnog Oslobodjenja Jugoslavije — Corps of People's Liberation of Yugoslavia
NCWC	National Catholic Welfare Conference
ORJUNA	Organizacija Jugoslovenskih Nacionalista — Organization of Yugoslav Nationalists
OSS	Office of Strategic Services
OZNA	Organizacija zaštite naroda — Organization for People's Protection
SNB	Czechoslovak political police
UNRRA	United Nations Rehabilitation and Relief Administration
CBE	Companion of the British Empire

Introduction

Ladislav Bevc wrote these memoirs in California — in another world from Yugoslavia politically and geographically. Knowing well the strange ways of Slovenia's destiny we need wonder what had taken him so far from his native land. The savage upheavals of Yugoslavian and Slovenian politics during World War II were reasons enough for distancing oneself from that part of the world. Having mentioned "Yugoslav "politics I must hasten to make it clear that the protagonists of that policy were not mainly the Serbs, as is so often thought, — but rather the half-Croatian and half-Slovenian Josip Broz, alias Tito, and his "intellectual" Slovenian accomplice Edward Kardelj.

As a Slovenian liberal Ladislav Bevc was active between the two wars as a leading member of the Slav gymnastic and patriotic organization *Sokol* as well as other liberal political groups within the *Yugoslav Nationalist Party.* His standing as the presiding elder of his Sokol association and as a respected citizen of Ljubljana shows that he was one of the more prominent liberals. Professionally he was the senior engineering counselor for the regional government just as the first automobile highways were being built in Slovenia.

Today when independence of Slovenia or rather Slovenia as an integral part of the European Union, is considered great progress it is necessary to clarify the enthusiasm for Yugoslavia which once existed between the two wars and even prior to World War I. Slovenians were facing the threat of being partitioned and divided between Italy and Austria. At the end of World War I the Italians acquired a good third of the Slovenian national territory and assiduously applied themselves to its total Italianization. Austrian Germans already firmly established in the towns of Lower Styria attempted to act in a similar way. Even in Ljubljana German sympathizers were not lacking. In some places Slovenians resorted to arms in order to defend themselves as in Maribor where General Rudolf Maister (1874-1934) successfully repelled the German nationalist *Abwehrkampf,* but they usually pinned their hopes on collaboration with the Croats and Serbs. It should not be forgotten that tiny Serbia heroically and, at first, successfully resisted the superior power of Austria's Germans. Eventually the Serbian army, with the support of the Entente liberated all its territory and penetrated as far west as the Alps. The Serbs were recognized as allies who could help liberating the splintered parts of the Austro-Hungarian Empire populated by the Slovenians, Croats and Serbs living in Austria. One must realize that all these nationalities were constantly pressured by arrogant Austrian Germans and Hungarians who conceitedly imagined that they could advance their "culture" if they managed to alienate the Slav and Rumanian barbarians from their own national heritage.

Such were the conditions under which Yugoslavia was created by universal agreement of its constituent nations. Bogumil Vošnjak, a Slovenian scholar and politician and Ante Trumbič, the Croatian Mayor of Split, who were members of the émigré Yugoslav Committee in New York, traveled during the war to the Adriatic Island of Corfu where the Serbian government had taken refuge under the protection of the allied navies. There they with Serbian politicians wrote the *Corfu Declaration* which stated that the Slovenians, Croats, and Serbs wanted to live together in one nation. This historical decision is now all but forgotten and even intellectuals in the West entertain the absurd notion that Yugoslavia was an invention of Josip Broz, a former Austrian noncommissioned officer and communist conspirator.

Divergent tendencies soon arose in the new South Slavic nation. Among the Croats the Radić brothers became the most powerful political leaders and not only organized a boycott of the Belgrade Parliament thus making it possible to adopt a centralist constitution but also flirted with Leninism. Their party was called *Croatian Republican Peasant's Party* (*Hrvatska republikanska seljačka stranka.*) The problems of the new state were compounded by the communist assassination of the Minister of the Interior, Drašković, after which the Radić brothers changed their mind and entered into a coalition government with the Serbian radicals. There was every indication that harmony would endure until a Serbian chauvinist with personal grievances shot at the Croatian deputies in the Parliament and triggered a disaster.

The reaction of King Alexander, known as an energetic leader already during World War I, was at first moderate. He appointed Anton Korošec, a Slovenian priest and politician, as Prime Minister but then he changed his mind, assumed dictatorial powers, made General Živković prime minister. Some politicians, such as the successor of the Radićes Vlatko Maček, were imprisoned while others, including Korošec, were interned.

It was a difficult situation, to say the least, but before hastily condemning Yugoslavs as savages one should recall that at much the same time the Austrian authorities shelled the suburbs of Vienna populated by Social Democrats and the Nazis assassinated the Austrian dictator Dolfuss. Moreover, in Janka Pusta the Hungarians at Mussolini's request established an Ustashi training camp where terrorist attacks on Yugoslavia were organized.

All of this led Slovenian liberals to join the Yugoslav Nationalist Party (*Jugoslovanska nacionalna stranka — JNS)* which supported the king's dictatorship although their colleagues from the former Independent Democratic Party (*Samostalna*

Demokratska Stranka — SDS), the so-called Pribičevićes, opposed it and their leader Pribičević had to seek refuge in Prague.

Macedonian extremists and Ustashi assassinated King Alexander when on a state visit to France. The regency led by the king's cousin, Prince Paul, from the first tried to find a rapprochement with the Croats. Then a new party came on the scene, The Yugoslav Radical Union (*Jugoslovanska radikalna zajednica — JRZ),* which included the Slovenian People's Party (*Slovenska ljudska stranka — SLS*). The Serbian opposition, however, did not join. Negotiations with the Croats ultimately produced the Maček - Cvetković agreement which represented the best possible solution for Yugoslavia. It should be noted that the supporters of Pribičević, mainly Serbs who lived in Croatia were then already aligned with Maček in his peasant-democratic coalition. It was then no surprise that Albert Kramer, the leader of the Slovenian part of the Yugoslav Nationalist Party, was already engaged in talks with Maček. From 1935 on the Yugoslav Nationalist Party supported the United Opposition.

These developments could not but affect the role of the Yugoslav Sokol Association of the Kingdom of Yugoslavia, a gymnast patriotic organization, to which Ladislav Bevc was devoting much of his life work. The *Orel (Eagles)*, the gymnastics organization of the Slovenian People's Party, was resurrected and became active as the Slovenian Youth *(Slovenski fantje).* Knowing that they could never attract many supporters if they came out in the open, the communists began to infiltrate other organizations in order to spread their ideology under various disguises. At that time judge Jože Rus began his attempts to subvert the Sokol organization and pave the way for certain Sokols to cooperate with the communists. He was successful to some extent although later the communists at Pugled submerged the identity of the Sokol apostates and other communist fellow-travelers in their ranks. Ladislav Bevc, presiding elder of the Sokol I Association at Tabor in Ljubljana, was prominent in rooting out the communists from his Sokol association. This resistance to communist takeovers was similar to the fight against the communists that Ronald Reagan led in Hollywood at about the same time. The Hollywood actor eventually ascended to the most powerful political office in the world and is credited by some with causing the demise of the Soviet Union. Ladislav Bevc, having incurred a mortal enmity of the communists, had to leave Yugoslavia and to continue his political activities as an émigré in the United States.

The agreement with the Croats brought enormous, widely shared relief on the international level and Maček, who became the vice-president of the Yugoslav government, received a majestic reception on his arrival in Belgrade. With World War II in Europe already underway the new government strove to act calmly and deliberately. That was indicated particularly after the German invasion of France and Hitler's good relations with his fellow dictator in Russia.

In the last years prior to World War II Maček's party was the dominant party in Croatia. In Serbia The United opposition of the Radicals, Democrats and Farmers was in power; and in Slovenia The Slovenian People's Party under the leadership of Anton Korošec dominated. Korošec died in the fall of 1940 to the great detriment of the Yugoslav politics as he was undoubtedly one of the ablest and most highly respected Yugoslav politicians. It is interesting that after the agreement with Maček he declined to grant to Slovenia the same status as Croatia. He was succeeded by Kulovec who was unfortunately killed by a German bomb in Belgrade.

Prior to World War II the communists were most successful in Slovenia. They induced Edward Kocbek and his Christian Socialists to become a leftist opposition within the Slovenian People's Party. There was an Old and Young split in the liberal parties. The Young regarded the Old as inept and opposed them in everything. The dissidents of New Yugoslavia were the first to leave and eventually joined Črtomir Nagode, a politician who favored alignment with the Soviet Union. At the same time the Young Yugoslavia (*Omladina JNS*), a liberal youth organization headed by the highly regarded law professor Stojan Bajič. Notwithstanding subsequent attempts to renew cooperation especially after the expulsion of Nagode's Ancient Rights (*Stara Pravda*) from the so-called Liberation Front the relations between the various factions of the liberals were never fully restored in the homeland. It was only in exile in Riccione, Italy in 1946 that the representatives of the liberal parties agreed to join in the Slovenian Democratic Party (*Slovenska demokratska stranka*) with Ladislav Bevc as its chairman.

This was the situation just prior to World War II and the German invasion of Yugoslavia in April 1941. Hitler had been pressuring Prince Paul that Yugoslavia join the tripartite pact allying Yugoslavia with the Axis powers. Paul eventually agreed and signed the pact on March 25, 1941. Within two days the military reacted with a *coup d'etat* and removed Paul as regent. On April 6 the Germans invaded Yugoslavia.

In this turmoil Ladislav Bevc had to make crucial decisions. His convictions and sense of duty led him to remain a loyal supporter of the king and the king's political supporters. Unlike others he recognized the ability of certain older politicians and statesmen. With fellow liberals Bevc helped organize a clandestine military organization, the Sokol Legion, to combat the enemies of freedom. They encountered serious difficulties because of the intrigues of Jože Rus and the enormous power of the communists.

As long as the Nazis and the Soviets were allies the communists agitated against Franco-British imperialists and for "peace," that is to say, for communism, because they hold that peace is impossible until communism is established everywhere. When Hitler attacked the Soviet Union the slogan changed to "war of national liberation."

They insisted that the liberation must be conducted under communist leadership and that the next phase would be the elimination of the class enemies.

Clever slogans and distrust of the elders caused total confusion among the population, particularly in Ljubljana. The farmers in the province were, however, harder to persuade. Because of their recalcitrance the communists began accusing them of cooperating with the enemy and proceeded to systematically kill their leaders. Occupation authorities were no less confused: their first step was to intern all Yugoslav officers who were in fact opposed to communism. The situation was grave because the same confusion and political disorientation existed throughout Yugoslavia. With the Germans came the Ustashi, Croatian political extremists, whose mass murders of Serbs in Croatia and Bosnia played right into the hands of the communists filling partisan detachments with Serbs fleeing into the woods. Independent Democratic Party (*SDS*) politicians like Sava Kosanović and Srdjan Budisavljević who should have been allies of the liberals in Slovenia also joined the communists.

Where General Dragoljub Mihajlović operated in Serbia clashes between his forces and Tito's partisans soon broke out. Cooperation between the two groups was completely impossible since, according to Milovan Djilas, the communists were from the start determined to crush the former Yugoslav officers hiding in the mountains. Instructions sent to Slovenia by Mihajlović tended to aggravate problems rather than smooth them over. With their tactics and propaganda the communists gained control. They went so far as to proclaim at the end of 1941 they proclaimed the class struggle and began to kill right and left. The carnage in Montenegro was so appalling that the entire land turned against them. Their Italian occupiers were less of a threat than the communists and even protected intended victims. Something similar also happened in Slovenia.

These developments had fatal consequences for Yugoslav relations with the western allies. Following the German breakthrough in France the British found themselves facing the Germans and their Italian allies alone. They were concerned that except for the Greeks no one would come to their aid. After the *coup d'etat* in Yugoslavia, the Germans cleared the Balkans of potential enemies again leaving the British alone, but when the Nazis attacked the Soviet Union the British were relieved and readily abandoned their reservations about communism.

Even Churchill set aside his antipathy toward the Comintern while Roosevelt was more concerned about Churchill's imperialism than Stalin's communism. Stalin's partition of Poland and his attack on Finland were forgotten. Because Mihajlović moved with caution to avoid reprisals against the population, the Allies threw their support to Tito with his ruthless revolutionary tactics.

Although the Allies used to be anti-communist they did not understand what communism really was. It is well known that when Fitzroy MacLean tried to warn

Churchill that Yugoslavia would be taken over by the communists, the latter retorted that since neither he nor MacLean intended to live in Yugoslavia after the war who ruled it was of no concern to them. Churchill even partly believed Tito's assurances that he was not opposed to democracy. The Allies appeared to have no idea that the communists were lying. They did not know that the communists offered to cooperate with the Germans if they were allowed to annihilate Mihajlović's *chetniks*. In 1944 Tito was much more interested in using the arms supplied by the Western powers against domestic opponents rather than against the Germans. Tito and his cohorts always considered the Americans and the British to be their class enemies.

The Yugoslav government of the Air Force General Simović who organized the *coup d'etat* went into exile with enormous prestige but his government's rigid positions eroded that prestige. Although Serbian outrage over the Ustashi's mass murders of Serbs in Croatia and Bosnia was justified, the efforts of many Serbian politicians to ascribe guilt to the entire Croatian nation and to revoke the Croatian agreement caused an irreparable rift in the government in exile. As a consequence all its influence was lost. Still worse were the consequences of intrigues concerning Mihajlović and the inability of the democratic politicians in London to counter the clever maneuvering of Tito's communists. In brief, the Yugoslav government in exile was unable to clarify the situation at home to its estranged allies.

The politicians in Slovenia always expected that the London government would assist them through intervention by the Western Allies. But the allies themselves did not understand Yugoslavia and the Yugoslav government in exile did not have close ties with the allied governments. Even if they had been more resourceful they could hardly find arguments by which to persuade Churchill that his plan of reconciling Tito and the monarchy were starry-eyed nonsense.

Thus the communists controlled the events to the very end while the liberals and the democrats suffered one defeat after another. No blame attaches to the liberal politicians like Ladislav Bevc who did all they could. Events were carried forward on a wave of total ignorance of the communists and their well published tactics. When the Western powers finally realized what was going on it was too late. In case of Yugoslavia the family quarrels between Tito and Stalin and Tito's expulsion from the resurrected Comintern compounded the confusion.

Men of principle were fortunate if they managed to reach the West and there begin a new life as did Ladislav Bevc in California. The writer visited him and his wife in the vine growing Napa Valley north of San Francisco. We could only regret the misfortune that had befallen our homeland; we could not think of anything that anyone of us could have done differently to alter the outcome.

Glasgow, March 2006 Ljubo Sirc, CBE

Early Years

My Home and Family

My papers say that I was born on August 8, 1890 in Škocjan in Dolenjska, parish Mokronog, Land Carniola. However, when I needed a new birth certificate during World War II, the rector wrote me that I was actually born on June 8 and that I was accordingly two months older than I thought.

Close to Straža near Mokronog is a small settlement called Bevc. My family no doubt originated there although there are in Slovenia at least five other places with the same name. At the Technical University in Vienna a professor told me that Bevc is an ancient Slavic name, it might have meant a carpenter — a man who hammers things together, perhaps even a blacksmith. Straža is the name given to places where centuries ago people watched for bonfires on the neighboring hills which were lit as a signal that the Turks were again on a rampage and that their raid was imminent. At Straža, too, a bonfire was ready to be lit so that people could take a timely refuge either in the castles of the feudal lord or, more often, in the woods, subterranean caves or in churches fortified with a wall around the cemetery.

In his History of Parish Šentrupert, author Steklasa[1] writes that in 1977 one of the keyholders to the church at Vesela Gora was Jakob Bevc of Straža near Mokronog. The keyholders were a supervisory board which, jointly with the local nobility, ensured that the accounts of the parish were properly managed.

My father Josip Bevc was son of farmer Janez Bevc and his wife Marija Berk, of Mokronog, house No. 25. My mother was Marija Gorenc, daughter of quarter[2] farmer Janez Gorenc and his wife Marija Možina of Mokronog, house No. 34. Both my parents came from large families. I had six uncles and aunts on my father's side and seven uncles and aunts on my mother's side.

My parents had nine children. Two boys, Rudolf and Emil, died as infants, a girl, Angela, died when she was two years old. I thus had two elder sisters, Josephine (Pepa) and Marie (Mici) and three younger brothers Josip (Pepi), Albin and Viljem (Vilko).

---

[1] Ivan Steklasa: *Zgodovina Župnije Šentrupert na Dolenjskem* [*History of Parish Šentrupert in Dolenjska*], Ljubljana, 1913, p.215. My other relatives mentioned in the book are: Josip Bevk(?), p. 147; Jera Gorenc, p. 192; Matija Gorenc, p.215; Anton Možina, Helena Gorenc, Martin Belc, p. 221; Berg (Berk?), p. 333.

[2] Quarter farms result when a farm is divided in four parts, usually of very modest size.

At first my father worked as gardener at the nearby castle at Mirna. The castle was owned by Count Barbo, an Austrian nobleman of Italian extraction. When the emperor began to build his bureaucracy, the mayor suggested that my father take a civil service examination. He did very well and was appointed assessor — tax collector. His job required him to travel a great deal to make assessments. In most places he

Ancestral homestead of the Bevc family at Straža near Šentrupert

was received with hospitality and treated to tasting home grown wine as people generally thought it was a good idea to be on good terms with the tax authorities. Once, however, he came to a remote settlement where the primitive inhabitants threw rocks at him, thinking perhaps that they would avoid taxation in this way. After that incident my father would pack a shining revolver which he got to keep after his military service in Bosnia where Austria had sent an expeditionary corps under pretext that the area had to be pacified. I have long kept that revolver in my study along with my modern officer's sidearm. When the Italians, in 1941, occupied our part of Yugoslavia[3] they demanded that all weapons be surrendered to them. At that time I had still not returned from the front and my wife did not know where I kept the weapons. Finally our son Vlado told her where they were because once when he was rifling through the library he discovered where I kept the weapons and used to examine them when there was no one at home to interfere with his curiosity.

My father had a deep baritone voice and was a good singer. He also liked to joke and tell funny stories. I remember a story he was telling about Count Barbo. The

[3] In April of 1941, Germany invaded Yugoslavia. Italy, Germany's ally, occupied the Slovenian capital Ljubljana and all the territory south of Sava River and to the East of Ljubljana.

count, apparently an eccentric, was in the habit of waiting in the evening with his gun in front of the castle for the deer to show up at the edge of the forest quite some distance away. When the deer showed up, the count would fire at him. If asked whether he scored a hit, he would reply: "Oh no, that one was too far away..." Once, when my father and I we were walking along the road, we encountered the mayor who apparently had had a few drinks too many at the local pub so that he was staggering all over the road. Embarrassed he approached us and said: "You know, a log fell on my feet." From then on we always said about drunks that a log had fallen on their feet.

Politically my father was a staunch Slovenian and a liberal. He was promoting his party all the time and was thus a thorn in the side of the clericals who took every opportunity to get back at him and had him transferred many times. Thus he served in Žiri, Litija, Zalog, Vače and, finally, in Semič.

My sister Josephine was sent to learn sewing in a convent in Ljubljana. The nuns in the convent were very strict and the place was the last thing an impressionable young girl needed. She escaped from that institution several times. A friend of hers emigrated to America and after a while wrote her from Colorado that she had found a suitable bridegroom for her. In those days it took three weeks for the mail to come from America to Yugoslavia: by train across two continents and by boat across the ocean. Today, when airplanes can fly across the continent and the ocean in eight hours, it still takes three weeks for the mail to arrive from America to Yugoslavia. My sister in fact decided to go to America. When she arrived to Pueblo, Colorado, a long line of young men with horse drawn carriages was waiting for her at the railroad station. Some of these young men may have owned their carriages and some may have rented them, there was "Rent-a-carriage" already at that time. My sister accepted the ride from Martin Mufich (Muhvić in his native Croatian). A few weeks later she and Martin were married. I have not seen my sister for thirty five years but we kept in touch by mail all the time except, of course, during the two wars. I only saw her again in 1949 when on her and her family's invitation I came to California. Her youngest son Warren (Buddy) signed the required affidavits stating I would not become a public charge in the United States and sent me the ticket.

We were in Vače when a great fire erupted. It was caused by children, my younger brother Albin among them, who played with matches and set fire to a barn from where the fire spread and destroyed almost the entire town. My father hardly got my mother out of the house: he found her in the bedroom where she was gathering up linen and other clothing she hoped to save. One of the houses was saved by owners who were spreading wet blankets and sheets on the roof and poured water all around it. When it became apparent that that house would not catch fire after all, a resentful man threw a burning ember in its direction shouting: "Will that damn thing not catch fire?"

Schooling

I attended the elementary school in Žiri and Vače. In Žiri I learned crocheting as this was the domestic industry there but I did not retain that skill. Because I did well in school they sent me to the Real Gymnasium School (Realka — intermediate college preparatory school) in Ljubljana. At that time my family lived in Semič. The Real Gymnasium was a German bastion as all instruction, with the exception

Preparatory school at Vegova Street in 1902; the author is first from the right in the first row

of the Slovenian language, was in German. Because of that we had to learn many things by rote, particularly in the lower classes. Thus I have not retained as much as I would want to of subjects such as physics and chemistry, which were taught in German. Every day at the main recess at ten in the morning when we were released to the courtyard the din and shouting that ensued was so loud that it could be heard at the farmer's market, a considerable distance away. People there would note that the "real-foxes" were out for recess.

My life was typical for a student from the country. The lodging was with a landlady who offered room and board to several students. My schoolfellows would sometimes play practical jokes on the landlady. For example, she would require us to say our prayers gathered around the table. As she was hard of hearing, our colleague Kolbe would, instead of praying, recite with a dead pan expression: "Look at the dragon how she is watching us..." Naturally, we could not control our laughter and the landlady would become infuriated, saying: "Out with you rascals — Kolbe is the only decent fellow among you!" Kolbe was quite a naughty fellow. One of his pranks was stepping in front of people on the sidewalk and then moving to the same side as the blocked person wanted to take to get around him. Once I went with him to a pastry shop where he spent all his money on cream puffs. When his money was gone he asked me to lend him some which I did. After he finished off the additional cream puffs purchased with the money I loaned him he said: "Don't think I shall ever repay you."

For a time I was getting lunch at the home of Franja Tavčar, wife of the Slovenian writer, lawyer, and liberal politician Ivan Tavčar. The arrangement was for ten students to come for lunch every day. The student guests at Tavčar's home were from families of liberal political orientation. Students from conservative families, on the other hand, were supported by the rather extensive foundations operated by the Catholic Church, such as, for instance, the Saint Aloysius Home.

I remember the catechist Dr. Miha Opeka, a very strict man who had studied in Rome of which he was very proud and later wrote a book about it.[4] Reverend Opeka used to stand in front of the church taking down the names of students who were late for the school's Sunday service. Once I, too, was late and Dr. Opeka duly made a notation of it in the class records book.[5] When the professor in charge of the class called me to account for this, I pointed out that the service had not yet begun as Dr. Opeka, who was to officiate at the mass, was still in front of the church taking down names of the tardy and so I could not have missed any part of the service. The professor accepted my defense and did not reprimand me, annotating the book to that effect. That elicited a sour faced smile from the catechist when he checked on how the matter was disposed.

In the upper division I initially had difficulty with descriptive geometry but I worked hard at it and one day I saw the light. From then on any solution of a problem in descriptive geometry seemed to come naturally to me. In fact, many years later I taught the subject as lecturer at the Technical University in Ljubljana.

---

[4] Dr. Mihael Opeka: *Iz mojih rimskih let [Out of my Roman Years]*, Misjonska tiskarna, Ljubljana, Groblje-Domžale, 1935.

[5] Each class had a records book which was a daily log of instruction, and a record of absences and students' disciplinary infractions. One faculty member was assigned to supervise the administration of each class and deal with tardiness, absences, and other disciplinary matters.

Going home for vacations I would take a train to Novo Mesto and then, if possible, catch a horse drawn carriage across the Gorjanci range to Semič. Once I got a ride with doctor Derganc who was a prominent physician in the area. That was a very luxurious trip. In Semič, my father who came to collect me, immediately tried to discuss politics with the doctor but when he became critical of the clergy, Derganc asked him to stop. Not even a doctor was supposed to listen to criticism of sacred cows.

When I was sixteen I thus once went home after a finished term at school. My father waited for me at the bridge in Novo Mesto. He did not look well, he had been ill for some time but we never found out what his ailment was. I showed him my report card which was excellent and he read it with great care. He died the same year. As he lay dying my mother and two sisters were at his bedside. When he appeared to have expired my sisters burst into tears whereupon my father regained consciousness and his agony continued for another ten hours. He was buried in Semič. During World War II, when I could not travel there, a soldier of the National Guard[6] was buried in his grave. I had great difficulties and expenses to have the dead soldier removed and buried elsewhere, the rector demanded a new funeral with all the services that go with the interment. After the end of World War II, our friend Štrumbelj in Semič, who used to take care of my father's grave, was impoverished because of communist expropriations and could no longer help in this way. He was unable to pay the annual rent for the cemetery plot which is required in Yugoslavia and the authorities probably took it over and assigned it to somebody else. Prior to his departure for the United States my son went to see Štrumbelj who was ill and disabled and could only describe the location where the grave might be found but the youth could not find it in the rather neglected cemetery.

After my father died my mother moved to Ljubljana where she lived at Valvazor Square at a rather meager pension she was receiving. I helped somewhat by tutoring. As an upper division student I could start tutoring students who needed help. The schools in Austria were very demanding and an education received in those schools meant something: it opened the way to professional or government employment. Parents were thus willing to pay a tutor when their children needed one. In this way I could help support the family. During my last two years at the Real Gymnasium I was tutoring Nino Perhauz with whom we became good friends. I was staying with the Perhauz family who were very friendly; they even invited me to spend a summertime vacation with them at Duino on the Adriatic coast. I enjoyed the rugged coast and spent countless hours climbing the boulders on the shore. Mr. Perhauz owned a travel agency and sold steamship tickets for overseas travel to America. The business was booming and he was quite prosperous. Perhauz's elder

[6] *National Guard = Domobranci*

daughters were courted by army officers who were regular guests at lunch. Thus I had the opportunity to observe their polished demeanor and learned a great deal about urban social life and habits. When World War I broke out, many of these young officers were among the first who died at the front. The Austrian army was at that time still unfamiliar with the trench warfare and its soldiers were running without cover about the battlefield where they could be readily mowed down by enemy fire.

Another friend of mine at the Real Gymnasium was Hugo Delcott, scion of an old Ljubljana's noble family. We spent a great deal of time studying together. Unfortunately, he committed suicide for reasons nobody knew.[7]

I decided to go to Vienna and study civil engineering at the Technological University (Techische Hochschule Wien) there which had much greater renown than the somewhat closer university in Graz.

Events of September 1908 in Ljubljana

On September 18 a protest rally was held in Ljubljana because of the unheard of attack of Germans and Germanophiles on the convention of members of the *Cyril and Methodius Association*[8] in Ptuj. Franja Tavčar, wife of liberal politician Ivan Tavčar told the attendees that the Germans in Ptuj crowded upon her, verbally abused her and even stepped on her skirts. Her and similar other reports enraged the attendees who walked out on the streets and loudly protested. Soon the crowd grew to a number which could not be controlled by the police. All this lasted until the shots rung out on September 20 following which people began to take down bilingual Slovenian-German signs. I saw Jew Goldstein who at Tranča helped the crowd to toss his own commercial sign into the Ljubljanica River.

On that fatal day the police split the demonstrators into two groups, one in the direction of Saint Nicholas (Šenklavž) the other in the direction of the courthouse. In the latter group were my friend Šuflaj and I. In front of the courthouse mounted dragons amused themselves by chasing people around. The dragons' spirits were kept high by the Cossler brewery which provided free beer for them. I recall that these events reached a crisis on September 20th. 1908 when the military were called out. This was the regiment of "canaries" — called so because of the yellow tabs they wore — who were Germans from upper Styria, billeted

[7] Many years later, after World War II, when my wife and children were evicted from our apartment by the communists, they found temporary lodging in the Delcott villa at Strossmajer Street 3 in the Poljane district.

[8] *Saint Cyril and Methodius Society* was promoting construction and staffing of Slovenian schools on the ethnically endangered territories of Slovenia.

in the Belgian barracks. At the command the soldiers fired into the crowd. Two innocent young men Ivan Adamič and Rudolf Lunder were fatally wounded as martyrs for the Slovenian national cause. When the shots were fired I was near the Prešeren memorial. A reporter from "Slovenec" thought that people were throwing firecrackers. Funeral of the two victims of Austro-Hungarian savagery was held two days later. Over twenty thousand people attended the funeral of the two victims of Teutonic barbarism.

At that tragic event all the hatred Slovenians felt for the Hapsburg dynasty came to light. At that time there used to be a bust of the emperor in front of the courthouse with a kneeling fairy in front of him tendering him a wreath of laurels. Ljubljana's pranksters added to this a derisive object which the authorities naturally readily removed. They could not, however, find the prankster so that he could be shot for insulting the "illustrious emperor."

Ivan Hribar was in his capacity of the mayor also the prefect of police. The governor was later defending the use of the Austro-Hungarian military forces stating that Hribar requested his assistance because the metropolitan police could not control the enormous crowd of the demonstrators. Hribar denied this. When the city council reelected Hribar for the sixth time the Austrian emperor refused to approve the election. The government dissolved the city council and put its own commissioner in its place.

At that time I was eighteen. In July I was graduated from the Real Gymnasium with a maturity certificate and was getting ready to leave for studies at the Technological University in Vienna. I left for Vienna in October 1908. In my room I kept on the table a portrait of Adamič and Lunder. In my lapel I wore a memorial badge with some kind of a macabre motif which was noted by my German schoolfellows who pointed at it and whispered at each other "Schau" — (See!...).

Studies in Vienna

Thus I decided to study civil engineering at the Technological University in Vienna. Scholarships were unattainable but I managed to secure a loan of 200 Krone a year from the Radogoj Student's Supportive Association the president of which was Ivan Hribar. The loan was paid to to me in monthly installments of 20 Krone under condition that I present to the Association proof of timely completion of all examinations and colloquia. At that time Slovenians considered that supporting their future intelligentsia was a good investment. At the end of World War I when Austrian currency was for all practical purposes devalued, I could easily have paid off my loan but I paid it off with the still good currency because I was grateful that

Technological University in Vienna (1910)

they made it possible for me to get my education. Before I left for Vienna my mother wanted me to get a nice, dark green, formal suit which I then wore at all examinations and other formal occasions.

At first I lived in Schonbrunerstrasse, then in Mariahilferstrasse, and during my last years in Margarethenstrasse. My quarters, or rather my room, was so small that I had to do all technical drawings keeping the drawing board on my knees. I got my meals in a student restaurant ran by volunteer ladies. The ladies were very kindly disposed toward the students and often provided us with a free second meal at lunch or dinner.

Because I was always current with my examinations and received good grades my tuition was waived. Once I almost had to pay the tuition, which was substantial, because a pesky clerk in the Registrar's office refused to accept a document submitted with my application for waiver on grounds that in that document my marital status was not entered in German. I wrote to the mayor of my birthplace requesting to enter into that form the German expression "ledig" instead of the Slovenian term "samski," that is, "single" which was the only word that had to be entered on the bilingual form. But the mayor again sent me the requested document filled out in the Slovenian language. I had to find in Vienna an authorized translator who would attest to the accuracy of the translation. Finally I found someone who advertised that he was translating documents from all the languages of the world but the person had never heard of the Slovenian language. It was quite a task to persuade him, but as a sum of some 100 Krone was involved I persisted while he was looking into his dictionaries. Finally he acceded to my request and confirmed that the translation, of the one and only word in question, was accurate, pressed his stamps on the document and collected his fee. Triumphantly I took the verified document to the clerk at the university who could do nothing but place a big fat question mark in red ink next to the translated word on the Slovenian source document thus expressing his doubt whether the translation agreed with the original.

Higher mathematics and calculus was taught by professor Gustav Herglotz who spent most of the time discussing the limits and abstract concepts of higher mathematics which are usually of only secondary interest to engineers. After a while, however, Herglotz was called to a famous German university and the German Burschen gave him a festive sendoff. I do not know if this was the same Herglotz who in 1904 discovered a method for complex integration of the wave equation before the theory of relativity was published by Einstein. Herglotz was replaced by Emmanuel

Czuber who, after he found out how little subject material had been covered by his predecessor, stated that we would have thenceforth step up to a much faster express train. Czuber had written several excellent textbooks on higher mathematics and calculus. Once some pranksters drew on the blackboard a barrel with a head, hands and legs because his name, Czuber, meant a barrel. The professor on entering the lecture hall and seeing the drawing, remarked: "Meine Herren, Sie irren sich, das Fass ist umgeben mit Reifen, ich aber bin umgeben mit Unreifen."[9] Many years later I found textbooks for advanced mathematics in America similar to those written by Czuber and sent them to my son in Yugoslavia who was studying calculus on his own. My copies were, of course, confiscated by the communists.

At my examination in descriptive geometry professor Emil Muller was fascinated with my performance, the efforts of my years at the Real Gymnasium were thus well rewarded. But as everywhere many things at the Technological University depended also on luck. Once we had to prepare extensive land surveying drawings. The assistant professor assigned each student a sector which he had to draw. As my turn came, the assistant asked me: "Well, what do you want me to give you?" For one of the sectors I happened to know that a good drawing already existed in the student archive. I therefore pointed to the adjoining sector and said: "Perhaps you can give me this one?" The assistant grinned and said: "You would like that, wouldn't you? Nothing doing — you'll do the sector next to it." Thus I got exactly what I wanted. I could simply copy the existing drawing in the archive in a few days otherwise I would have to work for a few weeks.

Although all structures at that time were of steel the reinforced concrete had also started to come in use. The professor who lectured on structures, however, was none too enthusiastic with the novelty and, instead of telling us to get acquainted with reinforced concrete, mentioned only that reinforced concrete was some kind of a novelty which started to crumble when he kicked it. Later, during World War II , I helped my friend civil engineer Ciril Juvan edit the Slovenian edition of his book on design of reinforced concrete structures.[10] Professor Johann Brick who taught bridge design and construction told us that he was in the habit of swimming in every river over which he had designed a bridge and that he had thus swam in every major river of Europe. Once he was called to the rescue at the construction of a bridge. The engineer who specified the length of a section had mistakenly used the distance between the piers instead of the distance between their centers. Thus the fabricated section was too short. Brik readily redesigned the section by adding appropriate latticework. At the formal opening of that bridge, the imperial minister of public works in reviewing the staff mustered out for reception stopped

---

[9] "Gentlemen, you are mistaken, a barrel is encircled by hoops (*Reifen*), but I am encircled by immature (*Unreifen*)." — A word play in German.

[10] Ciril Juvan: *Dimenzioniranje železobetonskih prerezov*, Ljudska Tiskarna, Ljubljana 1942.

in front of the unfortunate engineer and contemptuously remarked: "So that's the individual."

Professor Eduard Doležal, known for his tables of logarithms,[11] taught descriptive geometry, and professor Dokulil lectured on the technique of terrains. Once on a field trip we stopped for lunch at which the students began to sing: "Alle Schalen

Members of Slovenian Engineering Students in Vienna (1912); author as president is in the middle of the second row, his friend of many years Stanko Dimnik is second from the right in the second row

brechen, alle Schalen brechen nur die Dole-Schale nicht! Alle Lilien weichen, alle Lilien weichen nur die Doku-Lilie nicht!"[12] The two professors took this in good humor.

My friend Ivan Vurnik, later professor of architecture at the University in Ljubljana, one fine day disappeared from the lectures and the mess hall. The lady at the cash register in the restaurant where we took our meals, gave me an address where I might find him. I went to that address, knocked at the door and my friend opened them. He introduced me to his wife, an artist, whom he had just married.

[11] Stampfer-Doležal: *Sechstellige logaritmisch-trigonometrische Taffeln*, Karl Gerolds Sohn, Wien, 1909.

[12] All cups (*Schalen*) break, but Dole-cup does not, all lillies (*Lilien*) fade, only Doku-lilly does not! Word play on the names of the professors.

Slovenian university students in Vienna organized our own association in 1910. I was president for three years. Serbs and Croats also had their own separate associations which were formed before ours, the Bulgars founded their association after we founded our own. The government commissioner did not conceal his hostility to our organizations, he would even correct commas in the text of our bylaws. When all four societies were established, we agreed to found "Yugoslavia" — federation of associations of engineering students in Vienna. That federation was founded just before the war between Serbia and Bulgaria and it fell apart when the war between the two countries broke out. At our meetings, which had to be held indoors so that the German Burschen could not annoy us, we would sing Onam, onamo (Over

Certificate admitting Slovenian Engineering Students in Vienna to the status of voting membership in the Cyril Methodius Association (1913)

there) and inspired each other with thoughts of Yugoslavia. Once, when Peter I. Karadjordjević was called to the Serbian throne, he was traveling through Vienna and a reception was held in his honor. But our Serbian colleagues did not invite us to the reception.

Stanko Dimnik and I had the idea to embark on a campaign selling matches to collect funds for the Cyril Methodius Association which was building Slovenian school in ethnically endangered places like Trieste, Gorizia, Volkermarkt and the like. We ordered large boxes of matches from Šarabon in Ljubljana and pushed the matches on our colleagues. Thus we managed to achieve a net profit of 200 Goldinars.

In recognition of our contribution the Cyril Methodius Association conferred upon the Slovenian Technology Students in Vienna the status of a sustaining member with voting privileges at the Association's meetings. The certificate attesting to this is still on the wall of my library.

In Vienna I had the opportunity to attend the performances of the Imperial Opera. All operas were sung in German as all librettos were translated. On the other hand in the United States all operas are sung in the original language with the exception of "The Bartered Bride" by Smetana which is sung in German instead of the original Czech. In Vienna, at that time, it was not allowed to interrupt a singer by applause as is now the habit everywhere including, as I hear, even in Vienna.

Automobiles were rather scarce at that time and there were but a few of them in Vienna. Because of noise the imperial court eschewed motor cars and continued to use horse drawn carriages. Once I was walking along a street and, being somewhat absent minded, did not notice that I was followed by a motor car in which probably rode some dignitary. The chauffeur tooted his horn impatiently and gave me a dirty look because I did not get out of his way fast enough. Fortunately that driver has not yet acquired the driving habits of Californians and that dirty look was the end of the incident.

During the vacations I was twice employed at the City Engineering Office in Ljubljana.

Government Service in the Slovenian Littoral

In December of 1913, after five years of studies, I was graduated as a professional civil engineer. Under the laws then in force I was still considered a minor. Therefore my mother had me officially declared of age. That used to be called "buying the age." In Vienna I made acquaintance with Valentin Žun, a high official in the Ministry of Finance. At my request he was kind enough to recommend me to Member of Parliament Dr. Ribar, who recommended me to prince Hohenlohe, Imperial Governor of the Slovenian Littoral, as a candidate for an engineering position. I had also applied for an engineering position at the Engineering Directorate in Ljubljana but my application was refused by return mail. The Germans did not want to appoint Slovenians in Slovenia because they intended to Germanize the land just as they have done with Carinthia which is today totally Germanized. There was nothing left for me but wait for an opening somewhere.

At the end of April 1914, during the time when I was walking about Rožnik, a telegram was brought to my home inquiring if I was still interested in a position in the Littoral. I immediately went to Trieste to present myself to governor Hohenlohe. I was assigned as a civil engineering practitioner at the Engineering Directorate for the regulation of the Isonzo River and its affluents. I was excused from taking an examination in the Italian language because I had taken an elective course in the language at the University in Vienna.

I made official trips to the places along the Isonzo River, through Bovec and the delta as well as the region near Vipava as far as Dornberk. All the works were being carried out by the Engineering Directorate with its own budget; our supervisory authority consisted of representatives of various ministries (Buildings, Railways, and Agriculture), representatives of the autonomous Land Administration of Gorizia as well as representatives of various municipalities which contributed funds to cover expenses.

The supervisory authority would meet from time to time, examine the accounts and issue instructions for further work. Director of the Engineering Directorate engineer Jakob Hassler, a fanatical German, would repeatedly start arguments in the office on the subject of Carinthia being a German land. The only parts of Carinthia that he would concede of still being Slovenian were Jezersko and the Mežica valley. His fanaticism can be best illustrated by the incident when a Slovenian filing arrived from Vipava valley requesting that the banks of the Vipava River be protected by suitable levees. Hassler handed me the filing to be translated into German, then gave me his response to be translated from German into Slovenian. Shortly after I handed in my translation he rushed to my work area and excitedly complained that I had translated also the wording on the stamp with the file number. He was upset because I translated the word "Zahl" (number) with "Št." while the official language was German and not Slovenian and therefore such translations were inadmissible.

My mentor at work was civil engineer Dobrkovski, on loan to us from the Tyrolean Land Government. Dobrkovski was a Pole married to an Italian from South Tyrol. It was thus with him that I was going around on business trips. Each engineer had a bicycle provided and maintained by the mechanical shop of the Engineering Directorate. In the beginning of June Dobrkovski and I headed for Dornberk with our bicycles. We took a train from Gradišče to Gorizia where we sat on our bicycles and headed toward the government railway station. The street with rails of the local streetcar company was paved between the tracks and so we rode on that pavement. Suddenly a streetcar appeared in the opposite direction and I had to avoid it in a hurry by turning onto the street. In the process my rear wheel got caught by the grooved rail and I fell on the ground with my bicycle. I picked myself up immediately and continued on my way. From the government railroad station we took the train to Prvačina, mounted our bicycles again and were in the inn at eight in the morning from where we went each to our own assignment. I took a sharp bend of the Vipava River which we were supposed to straighten. It began drizzling and my foot was hurting me in the ankle all the time. At about noon I collapsed near my transit. They took me to the inn, placed me on a wagon and drove to Gorizia to the doctor. He found that my left foot was broken in the ankle and put it in the cast. I spent three weeks in bed at Borgo Gerusaleme near the Isonzo river levee where I was lodging. After three weeks I was again taken to the doctor who replaced the cast with a new one to last another three weeks. After six weeks I was finally rid of the cast and able to walk again but I still needed a walking stick.

World War I

Conscripted into the Austrian Army

In 1908 Austria annexed Bosnia and Herzegovina to the Hapsburg Empire. Serbia protested against this and tensions between Austria and Serbia exacerbated. Following the assassination of Archduke Ferdinand these tensions lead to a military conflict. At the same time when the assassination in Sarajevo took place the engineering staff in Gradiška was being visited by a joint ministerial commission. As the tension between Serbia and Austria increased, university students who following graduation from the intermediate school completed the mandatory military service were conscripted into the military. Their studies were thus interrupted and they missed at least one year of their studies.

On July 28, 1914 Austria declared war on Serbia thereby starting World War I. The Austrian war against Serbia was opposed by the Slav nations whose nationals were victims of systematic persecution such as confinement, internment or worse. People were subjected to such treatment on the most flimsy suspicions of sympathizing with the Russians or because of some anonymous information received by the police. Although Austria was not prepared for war it chose to declare it out of sheer hatred of Serbia.

Predictions of German Emperor Wilhelm to the effect that the war would be over in the fall before the trees shed their foliage did not materialize. Because of its obsolescent military practices Austria had by that time suffered very heavy losses. Thus, for example, instead of protecting the troops with trenches, the Austrians would initially have them run over the battlefield fully exposed to devastating enemy machine gun fire. In this way fiancées of two young ladies in the Perhauz family, where I was employed as tutor during my secondary school years, both career military officers, were killed in the first few weeks of the war.

The heavy casualties had so depleted the Austrian armed forces that the government had to resort to conscription of those who were not normally subject to military service. I was called up among the first although I had been previously excused from the military service at three physicals during the time I studied in Vienna. In Gradišče, however, I was among the first who were subject to the call up and ordered to appear for the army physical. Serving on the conscription board was the municipal commissioner count Attems[13] whose duty it was to protect the interests of each conscript from his community. When the conscription board determined that I was fit to be conscripted and directed me to the *27th National Guard infantry regiment* in Ljubljana, Attems did not object, although I was the sole support of my widowed mother. He probably was happy to rid his community of a Slovenian who to him was not welcome.

I kept a fairly good diary in which I noted the more important events. Unfortunately, my diary was among the papers seized by the Gestapo in 1941, when its agents came to search my library and arrest me. Thus I can only rely on my recollection of events which are still vivid in my memory although they may sometimes not be reported in correct chronological sequence.

Cadets of Autro-Hungarian Imperial Army (1915); author is second from the right in the first row

In the confusion of the unexpected mobilization I was assigned to the infantry while technical services charged with maintenance of railroads, bridges, and construction were lacking technically qualified personnel. Following the basic training, I was sent to a three month reserve officers' training school in Breze, North Carinthia, upon the completion of which I was ordered to report back to the

[13] Family Attems originated from Castle Attems in Friuli. The family spread throughout Friuli and Austria. In 1717 Ignac Maria Attems bought the castle of Janez Wildenstein in Slovenska Bistrica establishing the Styrian branch of the family. Ignac Maria Attems bought in 17th and 18th centuries castles Starnberg, Podčetrtek, Brežice, Pilštajn and the castle in Slovenska Bistrica. Grof Attems and his spouse of Slovenska Bistrica, who were citizens of Yugoslavia, were murdered by the communists who seized their property in 1945. Their son Johannes Attems to this date (2006) did not succeed in recovering from the Slovenian government his parents' property to which he is heir. Note by editor.

Regiment which had been in the process of moving to Maribor. In Maribor I was assigned to the company in Radvanje as the different companies were billeted in the city's surroundings. Soon all the graduates of the reserve officers' school were called to appear before a commission for a qualifying examination which I passed. The following day orders were issued promoting us to the officer candidates with an assigned rank of sergeant. The name of the captain who commanded the company sounded Slovenian but I never heard a Slovenian word from him.

It was in Maribor that a strange case of mistaken identity occurred which undoubtedly played an important part in determining my destiny at least for the duration of the war.

Men continued to be called up and there was ever more personnel which was being sent to the front after a relatively short training. Training of the recruits was delegated to officer candidates. Another officer candidate and I were responsible for the training of one of the platoons. We were billeted in a warehouse with no furniture, no beds. My colleague was called Von Vorbeck, a German of noble extraction, and son of one of those officials the government used to post all over Slovenia. Von Vorbeck did not know a single Slovenian word and was not suitable for dealing with Slovenian personnel. He was a big, strong, silent type, detached and resigned. The commander, who naturally noticed that I was carrying the greater part of the work load, was favorably disposed toward me. Once, at lunch when he used to read the new directives of the higher command, he announced that the *23 National Guard Regiment Zadar*, Dalmatia, which had been moved to Karlovac, requested six Slovenian officer candidates because it needed officers who could understand its Croatian speaking recruits. I was included among those six who were to go. When, prior to my departure, I called upon the commander to take my leave, he affably inquired: "What brings you in, Von Vorbeck?« In response I rattled off the standard reporting phrase which included my name. On hearing my name the commander was at first startled, and then he muttered: "It is too late now.« When he received the request for sending officer candidates to Karlovac he tried to include in the group one who was least useful to him, that is, Von Vorbeck whose name he thought was Bevc, and to keep me whom he had mistakenly thought to be Von Vorbeck. When he came to realize that he had a case of mistaken identity on hands, it was already too late. He let me go only with regrets. I, however, welcomed the transfer in a way. One reason was that the commander of the Maribor district, a colonel, in charge of examinations of the officer training school graduates was an uncouth and brutal man: once he had actually shoved officer-candidate Flerin. Battalions were dispatched to the front in short intervals, when I was there the 12th battalion had just been marched off. The colonel had the right to promote officer candidates prior to their departure to the front but he only rarely exercised that privilege. All in all, I got to like the transfer to Karlovac.

Later I heard that Von Vorbeck had been sent to the Italian front on the Isonzo River where the ferocious fighting caused him to suffer a nervous breakdown. I do not know if he ever got well or if he could return to the front at all.

While on the train to our new post, we agreed not to take the direct route Maribor, Zidani Most, Zagreb, Karlovac but to make a detour at Zidani Most to Ljubljana and thence go via the Dolenjska railroad to Bela Krajina on to Karlovac. In this way each of us could visit his home and relatives in Ljubljana. Unfortunately, my mother was not at home because she went to see her relatives in Dolenjska searching for food which had already been rationed by that time. Thus locked out of our apartment I spent the night in the almost vacant *National Guard* barracks.

The following day, we took the train to Novo Mesto and Metlika, as agreed, and, as we were approaching Metlika, decided on the spur of the moment to make a stop over in that town. Our appearance in that provincial town aroused universal interest. The winds of war had not yet reached that town and the population was not sure what credibility could be ascribed to the reports of the heavily censored press. Six young men whose yellow tabs on their tunics must have signified a higher rank represented an opportunity for finding out something more about the war. We established ourselves in Hotel Makan where, after a long time, we again slept in comfortable beds and free of charge at that. People in the hotel's restaurant were falling over each other ordering for us all the food we wanted. Local residents were coming to the restaurant bar throughout the day and plied us with drinks, of which there were more than one could imbibe, late into the night to hear first hand accounts of the news. Any of us who managed to take a few steps out of the hotel represented an object of intense interest to the people at the windows. This went on for four days until we realized that we were five days overdue in Karlovac. A large crowd accompanied us to the station to send off the unusual visitors.

On arrival to Karlovac we promptly reported to the command headquarters of the *23rd National Guard Regiment Zadar*, or, as it used to be called by the mangled Italian word, "Zara,« expecting with some apprehension to be required to account for our delay. We were received by the commander's adjutant, a Croatian, who was very pleased to see us and did not mention our late arrival at all. The commander, on being told that we have reported, immediately came to greet us: he was Colonel Gjuro Klikić, a Croatian, small of stature, wiry, with a tinge of grey in his black hair, and blue eyes in which one could see that he was a different man than the colonel in Maribor. Colonel Klikić was reputed to be legendary strict and was universally feared by his officers all of whom were trying to defer their departure to the front for as long a time as possible. Before the war Klikič was a career military officer.

As a colonel of his regiment he had once conducted military maneuvers in Istria. In the sweltering heat of the day several men collapsed, and some of them died. Because of the pressure brought by several Members of the Parliament in Vienna who demanded that the colonel be relieved of his command, Klikič was forced to retire but when the war broke out he was reinstated to active duty. His reputation notwithstanding, I noted that he was fair, never acting out of personal animosity, and was willing to be of help whenever possible. When, during our conversation, it was brought to his attention that our rank seemed to be a matter of title only without the actual privileges Klikić promptly and in our presence directed his adjutant to include in the order of the day our promotions to full fledged officer candidates — sergeants. Our concern as how to pay for our meals in the officers' mess was thus readily resolved.

Jocif, a teacher from Kranj, and I were assigned to the company billeted in Duga Resa, 10 kilometers from Karlovac to which we promptly reported. The commander in Duga Resa was a reactivated retired first lieutenant (Oberleutnant) Beban, an Italian who, being from Dalmatia, spoke the Croatian language. Because even the base extension at Duga Resa could not absorb the influx of all the conscripts that were being called up, the commander requisitioned an elementary school building in St. Peter, three kilometers away, into which he packed the two hundred men of the Duga Resa Company. Jocif and I were put in charge of the group at St. Peter. For the more important drills we used to take the men to the Pačurkovo Brdo in Duga Resa. The men obviously appreciated that we were speaking with them in their native language. They were addressing us as "Mr Cadet" by which title they continued to address me even after I became a second lieutenant in the corps of engineers at the front. Often I would overhear conversations from which I gathered that someone had a problem and was being advised to see "Mr Cadet" about it. Once a young man from a nearby village came to see me; he told me that he was 18 years old, married, and expected to be conscripted to the army, he had talked with my men who suggested that he consult me as to whether he could join my group if he must be called up. I explained that because he was a domiciled in Croatia he could not serve in a Dalmatian regiment as the Croatians had their own National Guard.

I delegated the morning drill of the newly arrived 18-year old recruits to a Croatian noncommissioned officer who had learned the Croatian language commands on his own. I noted the happy faces of the recruits who were being trained in their native language. From Vienna came several new officer-candidates who were assigned to the main body of the regiment. Once, when we were marching along a back road to Karlovac, I noticed that one of the Viennese candidates was attempting to talk, dictionary in his hand, to a soldier who obviously did not understand him. I walked up to them and inquired of the officer-candidate what was he trying to do. He showed me an Italian dictionary and said he was trying to talk to the soldier who refused to understand him. I explained to him that there were no Italians in our company and that the men spoke and understood only their native Croatian language.

Several times I had to take the battalion to the exercising grounds in Karlovac where the drill was supervised by Colonel Klikić. On one occasion the officers in Duga Resa, probably thinking that Colonel Klikić would not show up, remained behind dumping on me the responsibility for taking some five to six hundred men to Karlovac. Having no special instructions, I deployed the men into a battalion formation and conducted the drill. Colonel Klikić, however, did show up riding a horse and stopped directly in front of my battalion to watch the drill. At the end of the exercise and before we started the next one the colonel called me and said that the procedure in which the drill was conducted was incorrect. Responding to his critique I pointed out that the execution of the drill was performed in accordance with the rules applicable to the battalion. He listened in silence and rode away without contradicting me. I do not know, however, what explanations the officers whom he found absent from the drill on that day had to make.

All the officers and officer-candidates had to report weekly at the battalion headquarters and it was on those occasions that the colonel would select those officers who were to be sent to the front. Standing next to me was reserve First Lieutenant Fuks from Trieste, a corpulent man always perspiring out of fear that he might be selected.

In the meantime the command of the Sarajevo zone announced that professional engineers who were called up for emergency military service would be promoted to the rank of lieutenants in the corps of engineers if a physical examination should establish that they could serve only in a restricted capacity without bearing the arms, that is, if they were given the so-called classification C. Everybody was suggesting that I should apply for this classification and I did. The company physician, Dr. Fabiani, was a staunch Croatian. He had several assistants among the military personnel most of whom were university students. The doctor, of course, knew about my problem and readily classified me as category C which was a prerequisite for applying, through official channels, for reclassification. At the next inspection of the battalion, Klikić came straight to me, placed his hand on my belt and very kindly inquired: "Why did you get yourself reclassified?" I replied: "For medical reasons." Those medical reasons dated back to my 16th birthday when I had a serious cold compounded by an incipient pneumonia that left a scar on the top of my left lung manifesting itself by a wheezing sound at every physical examination. The command in Sarajevo declined to accept my application on grounds that Karlovac was not in its District but Colonel Klikić endorsed my application again explaining that Karlovac belonged in the Sarajevo District because the cadres originated from Dalmatia which belonged under the Sarajevo District in its entirety.

Time went on and soon it was Jocif's turn to be sent to the front at which time Klikić promoted him to the rank of cadet. The stress of the imminent departure to the front, however, caused some ill effects to Jocif's heart and he reported for a medical examination. The doctor found a heart murmur which required medical classification B, in other words, Jocif was found incapable of combat duty and was immediately sent back to the cadre where the colonel cancelled his promotion.

When a new battalion was being formed to march off to the front, all officer candidates were instructed to select one hundred men each from what were mostly eighteen year old recruits. Each officer candidate summoned the soldiers to assemble in front of him. In front of Jocif and me appeared long lines of several hundred men all of whom wanted to be in our groups, while the officer candidates from Vienna hardly had anybody. The Viennese complained to the commander and he simply detailed one hundred men to each of us. Klikić obviously had a reason to request Slovenian officer candidates.

In October of 1915, Klikić summoned all the senior personnel who were no longer required to perform combat duty, four officers and four officer candidates to the military exercise grounds in Zagreb, where the company had moved from Karlovac. He announced the formation of four engineering companies of 250 men, each under the command of an officer assisted by an officer candidate, which would be sent to the front. Among the officer candidates were Jocif and I. I was assigned to the reserve First Lieutenant Oser, a landowner from Upper Austria, while Jocif was to serve under reserve Second Lieutenant Fuks, a merchant from Trieste. Klikić inspected the formations and when he came by me he turned to his aide-de-camp and, pointing at me, said: "These men are to be promoted to the rank of cadet on their departure."

Each engineering company was allotted fourteen horses, six horse drawn wagons with tools and equipment, and one field kitchen. It took a good month for the companies to complete their preparations for the departure to the front.

A week before the Christmas 1915, during lunchtime, a member of the battalion staff got up and announced that the engineering companies would depart to the northern front and that all four officer candidates were promoted to cadets. There was barely enough time for us to dress up our uniforms with golden trim and silver stars and to acquire short officer's swords and write home. As I already mentioned, my group of officer candidates who were transferred to the base of the *23rd National Guard Regiment Zadar*, while enroute from Maribor to Karlovac, stopped over in Ljubljana to take leave of our friends and relatives. I, however, had not found my mother at home at that time either. When, in January of 1915, my mother received the news that our departure to the front was imminent, she was inconsolable as she had bad premonitions which under the circumstances could well have been justified. Two of my younger brothers, Albin and Pepi, who were called up when the war broke out had been wounded in action and partially disabled. Because I missed my mother on my way from Maribor to Karlovac she decided to visit me in Duga Resa when I sent her my address. And indeed she was waiting for me there one day when I returned with my men from the Pečurkovo Brdo. Thus her and my wish to say our farewells in person was fulfilled. As it turned out, it was not our last farewell for after three and a half years, in November 1918, I came back from my odyssey to Russia, thence across Germany and Belgium to France from where the Administration of the Slovenian Littoral had me recalled to civil service as I was needed to work on the reconstruction of the areas devastated by war.

The winter set in, the ground was soon covered by snow. As we were assembling at the railway station snowflakes danced in the air. The men were in the forty year age group but they were always complaining that they were "getting into their senior years." From among those men I, still back in Duga Resa, selected one as my orderly who applied himself to the job with great zeal, loyalty and pride. He was from the farm country, Morich by name. Although I did not require it, he felt it his duty to come every night as I retired on the mattress spread on the floor that served as my bed to tuck me in and to share some innocuous gossip about the men while I was about to doze off. Thus Morich was standing at my side as we were preparing to board the train. Suddenly a Croatian captain under whose administrative command were all the engineering companies, appeared to hand me my first officer's pay along with the supplement for the uniform and equipment to which every newly appointed officer was entitled. He also gave me a decree on my professional qualification annotated by Klikić: "A brave cadet." Trailing behind the captain was a lady decked in a fur coat and accompanied by her daughter and husband in the uniform of one of the engineering companies. Without asking me anything, the captain simply presented that soldier to me as my orderly. He was Giovanni Milič, official of a steamship company in Trieste, who could speak Croatian and always spoke it with me. His upbringing, however, was Italian. Milič was my faithful helper until the beginning of 1917 when the emperor, on request of Milič's daughter, had him transferred from the front to Lower Styria where the working conditions were much worse, there he had to clean horses and stables. Morich was dejected because he was pushed aside but I had nothing to do with him being relieved.

The Eastern Front in Russia

Our military transport moved on across Hungary, the Carpathian Mountains and Galicia to Volinia. The general staff of the Austrian army was in Volinia's capital Vladimir Volinski,[14] which is part of the Soviet Union today. We were to be stationed east of this town. We spent the first night under the clear sky in an abandoned trench with no shelter of any kind. The following day we reached the commander of fortifications who was billeted in a village from which the population had been evacuated. The commander was assigning each company to its sector of activity which changed as needed. We worked on roads, fortifications, sanitary installations, bridges and all kinds of general construction projects. From that time on our four companies went their separate ways and had no further contact.

Movement in the countryside where we were sent on our first assignment was extremely difficult as it was everywhere in Russia where we were operating. In spring, when the thaw began, or in rain it was simply impossible for the heavier wagons to move along the roads. In fact, the roads were nothing more than certain directions across the fields and the steppes as there were no solid roadbeds nor even layers of gravel. Once in retreat through an implacable rain we simply joined the endless lines of military conveyances all of which would regularly get stuck in the mud all the way to their axles. To extricate a vehicle men would place logs in front of the wheels so that horses could pull the wagon out and move it a few paces only to sink again into the mud. We would walk alongside, soaked by the rain through to our skins. From time to time we would come to a canal filled with water which we crossed by placing a log or a thicker branch over them, securing a foothold in the middle of it, and springing across. On one such occasion I slipped and fell into the ditch with water coming up to my waist. Of course there was no place where I could change nor did I have a spare set of clothes and dry underwear. I had to let the clothes dry on my body until nightfall. Other mishaps, too, abounded. The axle of our field kitchen got bent so that the wheels assumed a camber of some 45 degrees. The kettles had to be removed and placed on the wagons, the carriage was abandoned. Every time we wanted to use the field kitchen we had to build fireplaces strong enough to support the kettles.

One of our frequent tasks was paving the roads. For this we had a special procedure. We would cut fir or spruce trees from the neatly grown groves into logs the

[14] Vladimir Volinski 50° 51' N 24° 19' E

length of which would be twice as wide as the road, split them lengthwise, place them, alternating the flat and the curved parts, across the roadway and, in absence of better material such as gravel, cover everything with earth to provide at least some stability. In some places large military sawmills operating in the middle of the woods were used to cut the logs into beams and boards. Thus, for example, the corps commander in Mikulicze[15] requested that we build a three kilometer long timber paved road from his quarters across the swamp to the railway station so that he could drive there in the staff car. In addition to roads we also were building fortifications, various sanitary, delousing, and bathing facilities, bridges and anything that was needed. Priorities were a matter of the commander's personal judgment. Soon it became the practice to assign top priority to every order or directive from the higher commands which necessarily diluted the effectiveness intended by such classifications. Thus, for example, the commander received, while dining, a telegram marked "urgent." Noting that he did not bother to open the telegram, his entourage inquired when he planned to read it only to be told that the commander was in no hurry to attend to the message, it was the sender to whom the matter was urgent.

The first place where we were sent by the commander of fortifications was Gruzyatin.[16] At the time the local population had not yet been evacuated. The town was garrisoned by a Hungarian company which considered itself to be in charge. The Hungarians were much younger than our men, their commander, second lieutenant in the reserve Niedert, was arrogant and aloof.

At one end of Gruzyatin was a one story house owned by a Jewish merchant who, among other things, was providing banking services. At the entrance of his house there was a small waiting room with no merchandise on display where one could place his order. On receiving the order the merchant would disappear behind a curtain and fetched the requested merchandise. Before evacuation, Oser went to that merchant and demanded that he sell him a cow. The merchant objected and a rather poignant scene occurred when his children were paraded before us barefoot on the cold winter morning to indicate how hard up the family was. None of this impressed Oser in the least who, commenting to me that bringing in the children was obviously a trick of the wily Jew. Oser simply handed the merchant the requisition papers which were to be presented to the Austrian authorities for payment. In this way our milk supply was secured. Our provisions and other supplies had to be brought in from the commissary far behind the lines. We would send two experienced men with a wagon to bring in the provisions; the trip would take at least three days. Every time they also brought livestock which was slaughtered. On

[15] Mikulicze 50° 46' N 24° 31' E

[16] Gruzyatin, Ukraine 51° 06' N 25° 26' E

rendering our monthly accounts we had to turn in the skin. The cow, acquired by Oser, however, remained alive for quite some time and so our monthly settlements continued to be charged by one cow's skin. It was only when we were being sent from Russia to France that our supply officer Cipin, a Czech, helped me out and traded our already barren cow for another which we could slaughter and turn in its skin to clear our account.

We had not been in Gruzyatin long when early in the morning Niedert received orders over the telephone instructing him to evacuate all civilian population to places at least a day's walk away behind the front. Niedert and his Hungarians got out of this assignment by simply directing our company to carry out the order. First Lieutenant Oser assigned me to take all those wretches far behind the front as, being unable to communicate with the natives, neither he nor the Hungarians could accomplish this task.

The people who were to be evacuated from their homes were, naturally, in despair; they had to take their old and indigents and whatever belongings they had, load them on wagons pulled by oxen, and leave. I saw an arrogant Hungarian soldier brutally shoving a man carrying a sack containing a litter of sucklings so that the piglets spilled out of the sack and ran all over the place. Niedert's men picked them up for their own feast. The departure was to take place at noon. In the houses that had to be abandoned by the evacuees we later found meals which they prepared but had no time to eat. Early in the morning a woman came to see commander Oser tearfully imploring him to have pity on her. Notwithstanding the curfew, her husband (*chelovek*) had surreptitiously left the village the night before taking grain to the mill. He would only be able to return the following night when everybody would have been gone but she was ordered to leave by herself with the children into uncertainty. Oser, of course, had no authority to defer her departure; all he could do was give her a few loaves of bread from the commissary.

The long column of evacuees set on its way at noon. It was a crisp frosty January day, the sky was crystal clear. Because the road, or what we referred to as road, was frozen, it was easier to travel along it than it would have been in rain. I stopped the caravan several times so that people, as well as the animals, could catch up and rest. It was well after nightfall when, at about eight o'clock in the evening, we arrived to the village designated as our destination where I readily found the local army headquarters. The commander was dining with his staff and, naturally, could not be bothered to come out and direct my people to their lodgings. He simply suggested that they ask the locals for shelter. Soon all the people in my care were placed with the local families. The women came to take their leave of me, causing some embarrassment when, in keeping with their habit, they gratefully kissed my hands.

My dinner was ready when I returned to the command post. To my considerable surprise the commander's aide turned out to be one of my schoolfellows from the Technical University in Vienna, it was he who recognized me first. He was a native of Lower Austria, Meyer by name. At the university he kept company with a group of some ten other Germans who used to take turns in coming early to the always crowded lecture halls reserving the better seats by placing on them notes saying "occupied." As some 300 students attended the lectures there certainly was an advantage in being able to sit in the front row. We, the Yugoslavs, also were trying to reserve some seats but we never came closer than the fifth or sixth row. We never had contact with Mayer and his group; the Germans generally ignored us and we, for our part, did not seek their company either. Thus the unexpected encounter with Mayer was indeed a surprise: In Vienna we never spoke to each other for five years but at the front he acted as if we had both been part of an old boy network. The dinner and social hour over, I returned to Gruzyatin which, unencumbered by a caravan of evacuees, I reached at about midnight. Much later, when we had been already transferred elsewhere, the official military bulletins wrote at some length about fierce fighting that went on in Gruzyatin.

One day, still in Gruzyatin, a thick folder arrived from the War Department with instructions that the attachments were to be delivered to me. Oser, who was none too good at the military gobbledygook, handed the papers to me saying: "It appears that your request for reclassification as an engineering lieutenant has been turned down." Soon after that, Oser, who was a landowner in Upper Austria, was granted the so-called harvest time furlough. He turned his command over to me and was never seen again.

A month or so later, when we already were stationed in another village, several of my "seniors" were recalled to our base at Zagreb on grounds that they were over age for front line duty. I prepared the necessary papers listing their names giving my rank with signature as cadet. Within a short time, I received from colonel Klikić's aide a copy of the battalion order of the day which mentioned that I had been promoted to the rank of engineering lieutenant; the colonel had asked him to send it to me. Oser had obviously been unable to understand the acronyms and abbreviations of the military gobbledygook in the folder that came from the War Department. When I obtained from Vienna a certified copy of the notice in the official register where my appointment was announced, the fiscal officer increased my pay, adding what was due to me retroactively as well as a supplement for the uniform. However, even with all the new insignia duly sewn on my uniform, I still remained to many of my "old men" from the times at Duga Resa, "Mr Cadet," and they continued to go out of their way to be as helpful and cooperative as they could.

On one occasion I went with two of my men went behind the lines to the quartermaster's supply station. We were given lodging in a dormitory dug underground which was called "Hotel Maulwurf" (Hotel Mole). There I noted graffiti on the walls, undoubtedly scribbled by some disenchanted German:

It's tough without chow
For three days in a row!
Think with what fuss
Would Hindenburg cuss
If his waiter were slow.

At another time my unit was stationed in a place all by itself. I was, accordingly, the senior officer in charge. All the local population except the Jews, who lived apart in a separate settlement, had been evacuated. A deputation of Jews called on me and invited me to a wedding. I accepted and went to the reception. Cake and strawberries were served. All the Jews spoke German as they generally considered themselves to be German nationals. The women went through the ritual of dressing up the bride in a special room and would show her only at the moment when the wedding ceremony was to be performed. There was of course, some weeping and a show of emotion. The wedding ceremony was performed in the open on a podium, the rabbi poured a glass of brandy to the bride and groom, said his prayers, and then pronounced the couple to be man and wife. After the ceremony we were invited to join the other guests for refreshments.

In another part of the same village from which the native population had been evacuated my men discovered a huge cache of food in a barn: potatoes, vegetables, and grain. The cache was deep under ground, the victuals covered with layers of straw and earth. Out of this cache we used only what was necessary for our daily needs and had the grain removed to the supply station as it would no doubt become spoiled long before its owners would be allowed to return home.

Each soldier had a subsistence allowance which he could take either in victuals or in cash. The officers' rations were 50 percent higher. I selected a Slovenian from Trieste, Brecelj by name, as my personal cook. Since I could not possibly eat all the food to which I was entitled we arranged the matters so that the cook would buy with my allowance something else, except liquor, even though it might be more expensive. The arrangement paid off especially when we entertained visitors. When we were eventually assigned to the regional command in Mikulicze where a kitchen was set up for the command's numerous staff I took my meals in the officer's mess.

Every soldier was entitled to either ten cigarettes or two cigars per day, while the officers received either 25 cigarettes or five cigars. Since I never smoked I used to give my tobacco allotment away. Oser once made me light a cigar arguing that I should smoke what was so generously provided, but I got sick after a few draughts and I threw the cigar away. I never touched tobacco since and I never missed it in the slightest. Soon I noted that notwithstanding a regular supply of tobacco the men were buying and selling cigarettes on the side to supplement their allotment. From time to time I was able to send a package of food to my mother as in the villages that were still inhabited food was available at prices which, although higher than those set by regulations, were still affordable.

Some of my "seniors" were gradually reaching the age at which they could be separated from the military service. I would send those back to the base where they were discharged. However, quite a few of the "seniors" still remained as the core of my unit. Their replacements were younger men who were not fit for combat; among those was a Czech who was an expert mason, several Slovenians and Italians from Trieste, Czechs, Sudeten Germans, and Poles. Our administration and reporting procedures were rather complicated. I instructed a young man from Dalmatia in accounting and preparation of complicated monthly reports and secured his promotion to the rank of sergeant. Unfortunately, the military command tolerated prostitutes in the evacuated territory whom I had no authority to chase away. My accountant soon contracted venereal disease and had to be sent behind the lines for treatment. His replacement was an Austrian German, a serious and capable young man who took over his predecessor's duties and whom I was shortly able to get promoted as sergeant. Since he was a technician by profession I suggested that he apply for reclassification as engineering sergeant. His request was eventually granted but at that time I would already have been transferred to Gorizia and in the process of being separated from military service and returning to the civil service in the Slovenian Littoral — a move that was approved only shortly before the war ended.

We also were sent three young Italian prisoners of war as help. One of them was an experienced carpenter, young, strong, and universally useful, another was a blacksmith whose services likewise could be put to a good use; but the third prisoner, a city man from Bologna, a manufacturer of filigree souvenirs which he used to sell to the tourists, was too weak for physical labor. Food rations for the prisoners of war were lower than normal and we could account for them only at the regulation level. Nevertheless, we were giving them the same rations as to the rest of our 200 men as it would be unfair to treat them differently.

For a time, medical ensign Osias Sigall, a German Jew from Prague and an ardent German nationalist, was attached to our group. His task was to examine men who reported sick and authorize, if necessary, their hospitalization. It was on his instigation that we bought in Lutsk,[17] a center of rest and recuperation for military

[17] Lutsk 50° 42' N 25° 15' E

personnel on furlough, out of our personal funds, a phonograph and many records of contemporary melodies. Among other things, Lutsk was the origin of the well-known ditty: "Mein liebes Fraulein Backhaus, wo ist hier...."

For a shorter time we had the services of Peter Klee, a German settler and a building contractor in Bukovina. He was an army clerk and very skillful in assisting me with engineering work, he worked on road design and helped in construction of the bathhouse and delousing shed. Unfortunately, Klee soon contracted a venereal disease, probably at one of his visits to Lutsk, and had to be sent to the hospital for treatment while we remained without a replacement.

Another ensign, a young Viennese German, attached to our group was rather ineffectual. All he could do was accompany men to the work site. He was an only child and his widowed mother was obviously very attached to him. He used to tell me how his mother would take him for walks. Once, in Vienna, they came to a small park where children in their early teens exercised on a cross-bar, two of them were especially good at, performing feats that seemed to be downright amazing for such young teenagers. His mother pointed them out and commented in a low voice: "See those two? Like monkeys." One of the two young gymnasts overheard her and, without interrupting his exercising, came back with "You're right: two swinging and two watching!" [*Zwo henga — zwo stenga* — in Viennese dialect] Although the young man, having no initiative, could hardly be of help to me, I did not mind because, he always willingly and readily carried out what was asked of him.

During one of the retreats when the front was shifting we marched in rain through a village called Piatidni.[18] There I saw a small fox-terrier like dog standing in front of a house and I mentioned that it was a nice little dog. A few hours later, as I was inspecting the wagons, I saw the little dog on one of them. To my inquiry how the dog got there I was told that since I said the dog looked nice the men picked him up and took him along. I named the dog Piati after the village where we found him. My duties often required me to cross unpopulated areas, for example in going to a forest where the men were felling trees and I often wished I had a dog to accompany me. Piati, however, could not be trained to come along. As soon as I untied the leash in an attempt to have him follow me on his own he would run back home. When I returned, the dog was already there. The ensign played with Piati a great deal with childish delight notwithstanding my wish that he refrain from it lest the dog became confused as to who was his master. Later on I would take the dog with me to France where he became even more undisciplined. Once I had to spend the night in an abandoned trench near Gouzencourt. The dog was lying next to my bunk. Above my bunk there was some kind of a shelf on which rats were running hither and thither. When there was too much commotion the dog raised his head and the rats would instantly disappear, it worked like a charm. On the

[18] Piatidni 50° 52' N 24° 13' E

ramparts above that trench the dead had been hastily buried during the battle in graves so shallow that their extremities would occasionally be sticking out. More than once, Piati had to be rebuked and chased away from gnawing at the remains of those unfortunates. Piati came home with me when the war ended and took to chasing automobiles on the street. No amount of discouragement could wean him from that annoying and dangerous pastime. Eventually he paid for his amusement with his life when a truck ran over him.

In the village of Tumin,[19] from which the local population had also been evacuated, we were joined by reserve Lieutenant First Class (*Oberleutnant*) Desider Meller, an architect from Budapest. He was the liaison between the division and the regional command from which he transmitted orders to the companies. At that time only our company was assigned to his command and he had virtually nothing to do. Of course, he took his meals with me. As there was nothing else to do he was looking for an opportunity to be recommended by the division for one of the higher decorations the so-called Military Cross of merit (*Militaer Verdienst-Kreutz*) but the division could not be swayed. Failing that he got the idea that we should both try together, recommending me for a lesser decoration "Signum Laudis with Swords." I did not like his scheming and I asked him to leave me out of it but he went ahead anyway. The result was that only my decoration was approved.

Socially Meller was friendly and good company. Of the civilians in the occupied Tumin only one woman with two daughters was allowed to remain. Meller, although married which was apparent from the letters he was frequently mailing home to Budapest, took the younger of the two daughters to his quarters where she lived as his mistress. He remained with us for at least half a year until the company was sent to a construction project elsewhere and Meller managed to stay behind. Many years later, in 1934, a colleague of mine and I were sent by the Yugoslav Ministry of Public Works to Budapest to inspect and study modern highway construction in Hungary. In Budapest we stayed in a pleasant hotel from where we were daily taken by car to the various sites in the interior of the country. One afternoon I found some time to look Meller up in the telephone directory. When I called the exchange and requested in German that I be connected with Meller residence, the operator at first declined to handle the call because, the official language being Hungarian, she was neither required nor willing to converse in a foreign language. Eventually I managed to elicit her cooperation by explaining that, being from abroad, I did not speak Hungarian. When the connection was finally established and Meller came on the line I invited him over for a chat. I recognized him immediately when he came and mentioned the wartime service in Tumin. To my surprise he would not acknowledge our acquaintance and refused to admit that

[19] Tumin 50° 51' N 24° 36' E

we were ever in the service together even though it had been only seventeen years since that time. He left as he came, a stranger. And yet I still have the nomination for decoration issued under his name.

Even before I came to Tumin, the difficulties with food supply set in. The supply of horse feed began to fail first. Depending on the size of the animal, three to five kilos of oats were allotted per animal. Instead of oats we were sent dried corn but the horses did not want that because it was too hard on their teeth. All that was left was hay. The men requested that they be allowed to cook the corn, they called the meal "*varenka*." I approved the request and they happily brewed their *varenka* all day long until it became edible. The next problem was a shortage of salt. My frequent requests to the division pleading that the supply be reestablished went unanswered for several months. Left to my own devices, I summoned Koničkov, the only Jew who still remained in Tumin, and asked him if he could get us some salt. Koničkov was about 35 years old, diminutive, of a pleasant personality, dressed according to the Russian style with high boots. To my question he responded that he had plenty of salt and could supply all I wanted. I ordered two sacks; the price, although higher than that officially set, was not unreasonable. Because we were allowed to buy food only at the pre-war maximal prices established by the government I had to show somewhat higher quantities to justify the cost. Because the rest of the food supply, too, began to dwindle, Koničkov supplied everything else that we needed. Koničkov also offered me Russian golden coins in unlimited quantities quoting a price that was not at all excessive but, being unwilling to embark on currency speculation, I turned down the offer.

In the meantime, many companies did not want to pay higher prices for food and their men were therefore dependent entirely on what was available from the commissary, in other words, they were starving. Such was not the case with my men because I was, under the exigency of circumstances, buying supplementary food at higher prices, a practice that was at the time strictly forbidden. Finally the quartermaster's central accounting office in Kovel called in all the commanders for a conference at which the regulations were reviewed and a verbal permission given for purchasing food at higher prices from civilian suppliers and accounting for the purchases by showing a correspondingly higher quantity at the government approved price. I had been doing that all along and it was a relief to hear that my practice has thus been given a cloak of legitimacy.

It was obvious from this state of affairs that the provisioning of the armed forces was no longer functioning. I missed no opportunity to make strenuous representations and complaints concerning the situation. Once, when I visited the divisional headquarters, the clerks and officials charged with the procurement and distribution of provisions came to look me over and openly expressed their surprise as to where could such wiry individual find so much tenacity and persistence in fighting for what was due to his men and the temerity to berate them, Army officials, for their inefficiency.

At that time the Germans and Austrians set up the nucleus of an autonomous Poland with Lublin[20] as its capital. The governor of that, still far from independent, Poland was the local Polish bishop. It was to Lublin that I was sending our buyers who secured there the necessary export permits for the purchased food of which there was no scarcity. In this way we could legally expend all the allowances and acquired, over and above what was necessary for survival, reserves that might come in handy should another, and perhaps even worse, crisis arise. If one considers the ever present and chronic food shortages in contemporary communist Poland with the situation during the World War I when food was plentiful, it is clear what effect the communist rule had on the state of the country's economy.

From Tumin we were ordered to the Stohod River, an affluent of Pripet River to the east of Vladimir Volinsky, where we started construction of a wooden bridge. Our camp was on the eastern bank which was more exposed to a possible Russian attack. There was no civilian population in the area. We were in the process of putting down the decking when a horseman rode up on a stallion, the horse was rearing and bouncing around and the rider had to use considerable skill to keep it under control. Hastily the rider told me: "In case you have not heard: I bring orders of a general retreat," and he galloped away. We promptly gathered our equipment, loaded our wagons and headed south. Our route led through neatly kept fir forests but on the plethora of trails it was difficult to maintain the southerly direction nor could we make out from the map which trails were the right ones for us to take. I had to stop at every junction and inspect the tree trunks to find out on which side grew moss: that was the northern side. My examination of trees attracted interest of the men who did not know the procedure I used in deciding which direction to take. In this way I estimated which way was south and we continued in that direction. According to the map, it was 20 kilometers to the nearest bridge and indeed we managed to reach it. Had we become lost in the forest we might well had been captured by the Russians. At the bridge we merged into an interminable column of retreating troops. Throughout our retreat the rain was pouring down incessantly and, as I have already mentioned, it was only with great difficulty that one could get the wagons across the soggy meadows. Later the papers brought the official bulletins about heavy fighting along the Stohod River.

Following the retreat we were detailed to the construction of a defense line under the command of Major Messerklinger. We had to get up at six in the morning and go to work in the sector to which we were assigned where we built forti-

[20] Lublin 51° 18' N 22° 31' E

fications. The place was not populated. At the edge of the forest we constructed tents woven from thin branches, our tent was sufficiently large to contain two foldable cots and a rudimentary table. Major Messerklinger wanted to take his meals in our commissary and brought with him his side-kick, Major Hoffinger who had no official connection with us. With my ensign and I there were four of us at meals. After supper the two majors used to linger on late into the night and we had to play cards with them for money. I was not interested in wasting time at a card game particularly because I was up all day as my duties required. After the two majors would leave, the ensign and I would figure out how much money we had lost.

We were building the so-called fox shelters (*Fuchsloch*). Along a fortified trench with several side corridors serving as supply routes for food and ammunition we made excavations large enough to contain twenty men. The shelters were deeper than the bottom of the trench so that the earth layer above them was sufficiently thick to protect against a direct hit by artillery shells. A shelter had two entrances one from the left and the other from the right. At each entrance we built a supporting frame, as in a mine, and continued tunneling further placing new supports as we went until the two ends of the tunnel joined. The result of our efforts was a tubular shelter where the troops took cover during artillery bombardment. The work was progressing well but it was, of course, very tiring. The supporting frames were shipped to us from a military sawmill. By coincidence a prolonged bombardment, the so-called drumfire (Trommelfeuer) erupted on the front. The Russians had good artillery and on that occasion they happened to have plenty of ammunition, too, so that the bombardment lasted for a week or perhaps even longer. Near us was tethered an Austrian balloon which, linked with the ground by telephone, served as a forward fire observers' platform. The balloon had to be at an altitude where it could not be hit by shrapnel that might come its way. Also nearby was a field hospital the commander of which, a medical doctor, was scheduled to go on leave. Before his departure the doctor went for an after lunch stroll and was hit by a random shrapnel.

At the time the army had just introduced radio broadcasts which were quite a novelty even though all we could hear over it consisted exclusively of official reports and announcements. Generally, there were no airplanes to be seen. Only once an airplane flew over us and the event represented quite a sensation.

One evening, after completion of the fox-shelters we were ordered to dig up an auxiliary jumping-off trench between the front and another line. The night was dark with no moon. We were taken to the designated place and shown the direction of the auxiliary trench to be built. In the darkness I could barely make out a soldier at the distance of one yard. The frontline before us was literally ablaze with flares but that light was of no help to us at all. Both sides were continuously exchanging fire. The fire was obviously not directed because the bullets were whizzing all around us. Accompanied by a soldier I was making rounds supervising the work and urging the men to hurry so that we could get out of the situation as soon as possible. After each round my escort and I sat down under a tree where we could hear the rustling sounds as its branches and leaves were getting hit by bullets. At our next round I heard a man in the trench exclaim: "The lieutenant is here!" I looked down just when someone lit a match and saw a bloodied face. It was Teskera, one of my "seniors," a very tall man who carelessly looked out of the trench and was hit by a stray bullet. Because it was illegal to use any of us in a combat zone, I ordered the men to stop their work, load the fatally wounded Teskera and our equipment on the wagons, and leave. It was already past midnight. Teskera was immediately taken to the hospital where he died the next morning. I wrote his widow in Croatian language expressing my sympathy. I recommended him for posthumous decoration and he was awarded a silver medal for bravery. A person thus decorated was entitled to a monthly rent of seven and a half golden crowns (equivalent to $15), I do not know, however, if Teskera's widow was entitled to any benefits.

Major Messerklinger probably had to face a court of inquiry because he had sent us into a combat zone although a company of miners, trained and available for such duty, was stationed nearby. I heard that Messerklinger became mentally ill and was transferred away from the front line duty.

Soon another general retreat was ordered and our company, too, was to depart immediately from the front. We joined an immense line of retreating soldiers and military vehicles. At the same time the Russian revolution broke out. At first Russian soldiers fraternized with the Austrians, visits between them were the order of the day. Soon Austrian prisoners of war began to come back from the Russian side. The Austrians interned the returning prisoners of war for the purpose of re-indoctrination.

Our company was somewhat elevated in its station by being assigned to the corps command (*Korps Komando*) in a place called Mikulicze, close to the army command in Vladimir Volinski. The population of Mikulicze had been evacuated and the troops were billeted in private homes. The commander and his staff were lodged in a small castle where they had their offices and the officers' mess. I was enticed to join it with my ration-and-a half allotment. The commander was General Kritek who was later succeeded by General Weber. The latter used to be the military Governor of Montenegro. He brought along his personal aide-de-camp Dr. Kumičić, an attorney from Zagreb. The front was quiet in expectation of further revolutionary developments; the army's main activity was repatriation of the returned prisoners of war.

Our transfer to the command post was a form of recognition of our good performance: we were in a more quiet zone behind the front line and were given correspondingly better assignments. I recommended one of my Dalmatian soldiers, Damjanović by name, for promotion to the rank of sergeant and put him in charge of the operations. He was a serious and intelligent farmer and I was pleased to have selected him. I also arranged to have another Dalmatian, Pilić, a merchant from Zadar, promoted as corporal and delegated to him the task of securing supplies. Although many Dalmatians had been recalled to the base they still were in the majority in my company. The transferred personnel was replaced by various other nationalities: Czechs, Poles, Slovenians from the Littoral, Italians, Sudeten and Austrian Germans — all emergency call-ups. Among them were some tradesmen and intellectuals, and all of them were easy to work with.

One indication of the more relaxed pace of our situation was a herons' nest on a barn opposite to the house in which I was billeted. The nest must have become too confining to one of the fledglings who fell out of it while still unable to fly. Men picked him up and placed him in an empty, sunny room where they fed him with frogs fished out of nearby canals; as they would open the door the bird would screech and gulp down the frog with relish. He had plenty to eat. After a few weeks we opened the door and let the bird out. He made a beeline to his nest where his siblings greeted him by screeching and his parents by rattling of their bills. Grown up herons do not screech but rattle their bills as a greeting. The errant bird was generally well received by his kin. In this bucolic setting I acquired twenty chicken and a rooster so that the company could have eggs when enough of them accumulated.

The winter was very severe and it was in winter that a fire broke out in the castle because of an overloaded stove in the commander's office. The building was constructed of wood and the stove had not been adequately insulated from the wooden frame. It was, of course, my responsibility to put out the fire and build a new stove. With my Czech mason, who was a very experienced man, we decided that the new ceramic stove would be so constructed that the hot air and smoke would be guided through a number of reentrant ducts whereby as much heat as possible would be transferred to the walls of the stove before reaching the chimney by which time they would be considerably cooler. The stove we built was well insulated from the wooden wall by solid ceramic blocks. The commander appreciated our work very much.

The Western Front in France

The idyll in Mikulicze was suddenly and unexpectedly disrupted by the news that our company was to be transferred to the western front. A special train was to take us to France. We took along large amounts of foodstuffs, fourteen horses, six wagons, a field kitchen, all the chicken and my dog Piati. The men took the news in stride because they had confidence in me. The looks the officers in the mess hall were giving me seemed to convey their concern for us while the staff executive Britto meddled into the our preparations to such extent that General Weber felt it necessary to counsel him in my presence not to create unnecessary and bothersome distractions for me. For the rest these preparations signified that the war on the Russian front was about to peter out as the separate peace to be concluded in Brest Litovsk between the communist regime on one side and Austria and Germany on the other was fast approaching.

Our transport took several days to take us through Germany and Belgium to the German occupied part of France. At the larger stations we were given substantial meals, something to which we were not accustomed at the Russian front. In Germany people would cheer as our train was passing through when they realized we were Austrians. Not so in Belgium, however. There our soldiers who had become accustomed to waving back to the population were rebuffed by Belgians who were gesturing vigorously with both hands pointing back toward Germany with an unmistakable message that they wanted us to go home. At one station in Belgium some children came up to the car asking for bread or anything but the train engineer chased them away telling them, in French, not to abase themselves before the enemy. Throughout our journey through Germany and Belgium there was a feeling in the air that something of great import was soon to happen. In the papers I was receiving from home I read that a development of historical importance was about to take place, namely, the shedding of the German yoke, self-determination and the founding of our own free and independent state.

Eventually, at the end of March 1918, the train reached Cambrai[21] in France where we finally got off. We came to the German zone of occupation with places such as Cambrai, Arras, Peronne[22] and Bapaume,[23] well known from the military bulletins. The population of Cambrai had not been evacuated. The local population refused to have anything to do with the Germans, the merchants refused to serve German personnel in their shops. The French government was able to make arrangements with the Germans for shipping provisions for the civilian population by the Somme River. The locals were distributing the received foodstuffs autonomously

[21] Cambrai 50° 10' N 03° 14' E

[22] Peronne 49° 56' N 02° 57' E 112 miles SW from Cambrai

[23] Bapaume 50° 07' N 02° 15' E 88 miles WSW from Cambrai

without interference on the part of the Germans. The latter, on the other hand, had requisitioned numerous private apartments which they were assigning as needed through an office in the City Hall (*Hotel de Ville*). Our company, too, was assigned suitable quarters and places where we could keep our wagons, tools and other equipment. I was assigned a room in a two story house in Rue de Saint Lazare. The house was wedged in between two neighboring houses and so narrow that it could have only one window only on each story. The kitchen was on the ground floor, one room on the second floor and another one on the third. Access was by a narrow staircase. There also was a cellar which was put to good use as an air raid shelter as the air raids were occurring with increasingly greater frequency.

I was given a room on the second floor; the bed was decked with clean linen of good quality. I wondered how could I be so lucky to get such nice quarters. Soon I discovered that the local gas works, a target of frequent air raids, was just across the street.

On my first day in Cambrai, as I was walking along the deserted streets to my reporting place a tank, captured by the Germans in battle, rolled along the road rumbling ominously. The crew could not be seen which made the encounter even more sinister. At that time the Germans did not yet have tanks nor did they have aircraft that could challenge the raiding Allied airplanes. A talkative German soldier who had combat experience with tanks told me that the advent of this new weapon on the scene had a shattering effect on the German morale. The soldiers dreaded most the tanks equipped with flamethrowers that could advance to the trenches and set them ablaze instantly.

The wailing of the air raid sirens was heard increasingly more often in daytime as well as at night and everyone, the military as well as civilians, ran for cover to the cellars — the only available air raid shelters. When, after those first impressions, I finally reported for duty to the local commander, air raid wardens shouting "*In der Keller*!" ("To the cellar!") kept appearing at the door. I was unaware of the significance of those warnings but the commander was anxious to break off the conversation which he eventually did by simply disappearing through the door leaving me alone in the office. When the all clear sounded I went out. Quite near the house stood an ambulance into which medics were loading the wounded and the dead, just then they carried out a corpse of a woman all covered with blood. From that time on I was keenly aware of the necessity to take cover during an air raid.

Once when I was waiting out an attack in a cellar I struck up a conversation with an 18 year old youth who told me that he was required to perform compulsory labor to Valenciennes daily. Among other things, he told me that the civilians were not allowed to grow vegetables or fruit in their gardens nor were they allowed to keep chicken or rabbits so that they were dependent exclusively on their "revitaillment" — rations doled out by the authorities. My orderly once brought in the landlady who would like to have some eggs as the canned food distributed by the authorities was not very satisfying. I agreed and asked the orderly to give the landlady what she wanted. Later the landlady presented me with two beautiful towels made in Lille. We were only slowly using up the provisions we brought with us as the German military rations were quite adequate, there was plenty of bread but the flour contained chaff and sometimes, according to news reports, even finely ground sawdust. Because of that bread I got a rather unpleasant indigestion. On one occasion my orderly showed me, without comment, an open can of paté containing a sharp white tooth. I made no comment apart from telling him not to serve that stuff. The pay received by officers and men was in German marks at parity, that is, one German mark for one Austrian crown.

We were in Cambrai only a few days when the Germans attempted a gas attack directed against the front which was to the south and southeast. However, the wind suddenly changed direction and blew the gas back toward Cambrai. The choking gas surprised me on the street as I was going home, a native coming toward me, also choking, warned me: "*Ce le gaz.*" I reached my house in a hurry where I could sit out the gas attack that backfired in the relative safety of my second floor room.

Every morning I would see another house that had been reduced to rubble by aerial bombings which were becoming more and more frequent. Destruction was all around us with houses razed to the ground and the magnificent fountains outside of the town rendered unusable by damage, poisoning or by piles of debris and refuse thrown into them. Any efforts at reconstruction had to be deferred until the end of the war.

On our first assignment we were sent to the railway station in the nearby Rumilly. The site was bombed out to such an extent that no trace was left of the depot, the rails had been ripped up from the ground and scattered in all directions. Attempts were being made to sort the things out to an extent and we were to help with this effort.

Following the work at Rumilly we moved out of Cambrai to the south and closer to the front. All the surroundings were destroyed and desolate with no living soul in sight. Next to the ruins of a fountain we found an abandoned vegetable garden with quite a few asparagus still growing. It was dangerous to pick them, however, because there were hand grenades scattered all over the garden, no one knew who had done that and why. Near the abandoned railway station at Villers Plouich, 40 miles SW from Cambrai and 50 miles from Paris, we established ourselves in an abandoned trench where we slept in makeshift bunks without roof or other shelter. The field above the trench served as an emergency graveyard, the dead had been hurriedly buried in shallow graves so that their extremities would in places be sticking out of the ground. From that place men were sent to work on similar rubble clearing projects as that in Rumilly.

Not far from us was a German command post. The Germans sent out a soldier with a request that I provide my *curriculum vitae* which they needed to recommend me for a decoration. It was a practice with the German military, as well with the Austrians, to decorate all the liaison officers when they were attached to their command. The liaison officers would sometimes reciprocate by arranging for their hosts to be decorated in turn. Because I was not a liaison officer and, moreover, coveted no such undeserved decoration, and a German one at that, which I would consider to be almost an embarrassment, I evaded complying with the request by politely declining to furnish the information saying it was not a custom with us to, so to speak, apply for decoration. The Germans, however, were not to be dissuaded and after several visits to my quarters they agreed to obtain the necessary data from the company clerk who kept the personnel records.

It was not long before I was summoned to call on Captain (*Hauptmann*) Luffert, representative of Austrian units serving on the western front. Luffert's headquarters were in Leuven, Belgium, where he installed himself in a castle. That was some kind of orientation meeting for the commanders of Austrian military units in Luffert's area of responsibility. After the conclusion of the official business we were invited to dinner attended also by the countess who owned the castle. The meetings lasted several days so that I had plenty of time for sightseeing in Liege and Bruxelles.

Back at the front one fine morning as I went, as usual, to see my company which was busy at work, I met a stranger who introduced himself as the military officer appointed to take over my company because I was being transferred back home to Gorizia. I handed over my official affairs to him and, after four months on the western front, took leave from the company and departed via Belgium, Aachen, Cologne, Munich and Vienna for my new assignment in Gorizia. With me went my orderly Brecelj, a Slovenian from Trieste. We were delayed in Cologne because our luggage had been held at the customs inspection in Aachen and it took a full day before it finally arrived all mixed up and trashed by searching so that I had to buy two new suitcases. To get on the express train for Vienna we had to change trains in Munich. The conductor tried to bar me from boarding a first class car and relented somewhat only when I threatened that I would complain to the stationmaster. He would allow me to board but not Brecelj. I insisted that Brecelj go with me and I offered to pay the fare differential if required. The altercation drew the attention from other passengers who were watching at the windows. When we finally got on board I selected a suitable compartment and gave Brecelj directions in Slovenian where to place the luggage. The passengers in the compartment were hostile and one of them uttered the derogatory word Germans used for Czechs: "Bohemian." When

I fixed him with a stare he became flustered and said: "*Ich bin ein koeniglich bayerischer Postbeamte*" ("I am an official of the Royal Bavarian Post Office"). At that moment a senior Austrian officer came along and, noting the hostile conduct of the passengers, drew me to the side suggesting that I let him handle the situation. He then went into the compartment and informed all present that they had offended an Austrian officer who had sent him to demand satisfaction in a duel from that Bavarian Royal Majesty's Postal Official in accordance with the Austrian military custom. On hearing this, the impertinence and bravado of the passengers evaporated and they began to tender their apologies of which the officer took note and declared that they were satisfactory. That being settled, I sat down but my orderly for whom I had to pay the first class fare differential, had to stay in the corridor.

Here I want to mention that I never had a tailor made uniform, the one I was issued at the officers' training school was good enough for me. It was that uniform that I wore on my only furlough in Ljubljana at the end of December 1916. At that time I made an overnight stop over in Vienna to continue my trip the following day. There I met an acquaintance from my studies in Vienna. He was an artillery officer, younger than I, and was clad in his own tailor made and neatly pressed uniform. He was embarrassed walking on the street in my company because of my none too illustrious apparel and, noting this, I made use of the first opportunity to take my leave of him.

When the express train reached Austria the German passengers left. A military chaplain came in the compartment and, being alone, we soon struck up a conversation on the topics of current interest. From his accent I gathered that he was a Czech. Throughout the war I kept my subscription to the Slovenian paper *Slovenski Narod (Slovenian Nation)* and journal *Glas Slovenaca, Hrvata i Srba* (*Voice of Slovenians, Croats and Serbs*) published in Zagreb which I was receiving regularly even while I was in France. My conversation with the chaplain had the tenor along the lines consistent with the policies advocated by those papers. On taking leave from each other, just before we reached Vienna, the chaplain confided: "I am a Czech," which, of course, was obvious to me all along.

On my arrival to Gorizia I was assigned to the military staff for the flood control of the Isonzo River and its affluents. The headquarters was in Gradiška where I had previously held my civil service appointment; the commander was engineering counselor Jakob Hassler, my former civilian chief activated as a captain in the reserve. The army command for flood control projects was formed because the military operations in the area had wrought considerable damage to the water works which we were expected to gradually repair.

I found the same lodging I used to have before the war and took my meals in the officer's mess except at the time when I was confined to my quarters by a severe influenza when Brecelj would bring the food to me. I, in turn, let Brecelj go home to his family in Trieste which was not too far away as often as he wanted. It was in Gradiška that I was promoted to lieutenant first class (*Oberleutnant*) in the engineering corps and almost at the same time I was sent through the official channels the German Iron Cross, second class. Captain Hassler, an ardent German nationalist, handed the decoration to me with undisguised envy remarking that I probably would never care to wear it, which was true.[24]

I called on the engineering high counselor Mahnič, Chief of the Building Administration at the government offices in Trieste, who told me that several personnel appointments were under consideration and that I would be separated from the military service because I was needed for work on reconstruction projects in the area.

I had been both on the Russian as well as on the western fronts, yet, I was fortunate that I did not have to take part in the destructive activities of the war as my duties consisted of directing the construction projects intended to shelter and protect human life. Thus, although my military career had peremptorily demanded four years of my life, it did not represent a total loss after all, because I gleaned valuable administrative and professional experience in its course.

During the war the Government Administration in Trieste where I was employed was paying my base civil service pay which was being sent to my mother. I also was making monthly payments on a 5,000 Krone loan to pay for my studies in Vienna that I obtained from *Kmečka Hranilnica* (*Farmer's Savings*) in v St. Vid near Ljubljana, a savings institution. The savings institution had been sending me the principal in monthly installments when I studied in Vienna; I was required to inform them of all the examination I passed. By the end of the war I had saved 700 Krone so that I was able to pay off the saving institution with good money before the currency was devaluated. I choose not to profit by the devaluation of currency which was imminent.

Throughout the war, I was contributing 20 Krone a month to the support of Štefan Baloh, a student at the Technical University in Vienna. Baloh, neither a relative nor a particularly close friend of mine, came to Vienna from Idrija a year later than I; he was doing none too well in his studies and I used to tutor him free of charge as he was very poor, neglected and virtually starving. He was not conscripted to the army and so he could remain in Vienna. I took pity on him and agreed with another Slovenian reserve officer, that we would each contribute 20 Krone a month toward his support. My colleague, however, was soon killed in military operations in Rumania and I remained the sole contributor to Baloh's sup-

[24] Note by editor. The author's family had brought to the United States all author's decorations received from Yugoslavia and also a decoration received from Bulgaria, but the German Iron Cross and two Austrian medals seem to have disappeared during World War II.

port. When I was finally settled in Ljubljana, actor Lipah who had at that time been studying dramatic arts in Vienna called on me. The objective of Lipah's visit was to secure further contributions for assistance to Baloh who had not yet completed his studies and, moreover, had a drinking problem. The university students in Vienna took the initiative in this action and they requested that all contributions for that purpose be channeled through their committee. I agreed to their request. In 1921 or, perhaps, in 1922 Baloh was finally graduated as a civil engineer after taking some eleven years to complete his studies and came to Ljubljana with a request that I help him find suitable employment. On my request, Professional Engineer Zelenko, director of personnel at the Railways Administration, found a position for Baloh. Soon, however, Zelenko notified me that the Railways Administration would have to let him go because he was unsatisfactory. My pleading with Zelenko to have pity with the man and keep him employed were only partially successful and Baloh was transferred to Novo Mesto. There, in a severe winter, he went to inspect the track with a rail motor car, got drunk and, not having adequate clothing, contracted pneumonia of which he died.

In closing my memoirs on World War I my thoughts again run to the men of the Dalmatian *National Guard* who were entrusted to my command. Initially all of them were senior men exempt from combat duty; most of them were farmers, many of them illiterate and without skills for technical and construction work required by our assignments. Nevertheless, they willingly applied themselves to their tasks and, in their later years, gained skills not ordinarily possessed by people in their class.

In thinking of those men, I still vividly remember soldier Mate Runje, a tall, healthy man and how he, being illiterate, in his free time dictated a letter to his "dearly beloved wife," a "book" in the Dalmatian parlance, to another soldier who could write. In his narration Runje was reliving events from his homestead, inquired about the cow, his arid fields on Charso, and gave advice what do about this and that in his absence, and in conclusion embraced again in his thoughts his dearly beloved wife. The writer skillfully followed the dictated narrative and faithfully set everything down while the other soldiers were quietly sitting around them, some smoking, others in reverie on their own homes. I left the scene, deeply moved and resolved to do all I could to ensure that these men who were entrusted to me, should safely return to their homes, their children and dearly beloved wives. One of the soldiers, a serious man, somewhat younger than the rest, called on me upon his return from leave and told me about the tribulations endured by his relatives at home: it was the same with mine. Runje's reveries were also mine own, my thoughts were devoted to my loved ones who had to endure the ravages of war at home and to my loyal and devoted men for whom I always remained their "Mr. Cadet."

This concludes the description of my first long journey in the uncertainty and in harm's way. My journey ended in November 1918, three and a half years after I left my homeland at a festive farewell in Metlika in the ever bright Bela Krajina. It would be three and a half years, in November 1918, before I came back from a long journey to Russia, thence across Germany and Belgium to France from whence the government of the Littoral had me recalled back to civil service as it needed personnel for the reconstruction of areas that suffered the devastation of war. In July 1918 I came from France, still as an officer, first to Gradiška where I found the director of the reclamation project of the Isonzo River and its affluents who also was still in the armed forces assigned to repair and reconstruction of the damage caused by the war.

At the end of October 1918, Hassler and I were directed to report without delay to the military district headquarters in Ljubljana to be separated from military service. On October 31, 1918 we left by an official horse drawn carriage via Gorizia, Ajdovščina, Vipava, and Postojna. By the time we reached Ljubljana, Slovenia had, on October 29, 1990, already declared its independence from Austria in the spirit of the Declaration of May.

It was then that the first Slovenian government was established and not in 1945 as it is now claimed by the communists.

Because Trieste was occupied by Italy, I could no longer return there and so I remained in Ljubljana to begin a new chapter of my life. I had to wait three months without employment before in February 1919 the new government reassigned me to the Building Administration in Ljubljana where I had always wanted to work. It took the downfall of an empire before I could get an engineering job in Ljubljana.

Liberated Homeland

Return to Ljubljana

Ljubljana was in such a festive mood as never before. People were embracing each other on the street and all houses were decorated in national colors. Huge quantities of military materiel and supplies were arriving from the front with the retreating Austro-Hungarian army. People, starved and impoverished by the war, were busy looting what they could use, primarily food. I, too, got hold of a wagon, loaded it with food and brought it home to the courtyard. However, before I could unload it, the new Slovenian government ordered that everything must be returned to the stores which the new government had secured by posting guards around them. Hundreds of starved military horses were grazing on the meadows around the city and in the parks covered by snow. Guards were posted on all the roads leading to the city. The unusual situation required a hasty establishment of a provisional Slovenian government with referees or ministers which was for the time being administering all public affairs on the territory of Slovenia. Presiding over this government was knight Pogačnik of Podnart, an old parliamentarian in the old system. Other members of the government were: Dr. Brejc, minister of the interior (police), Dr. Lovro Pogačnik, minister of war, prelate Kalan, minister of agriculture, professor Verstovšek, minister of education, Anton Kristan, minister of social services, Dr. Ivan Tavčar, minister of provisions, Dr. Pavel Pestotnik[25], minister of transportation, telephone and telegraph, Dr. Ravnihar, minister of justice, and some others. The proceedings of this provisional government were recorded in a special protocol supervised by the president's secretary. Unfortunately these proceedings have never been published and have most likely been lost although they are covering an important period of Slovenian history.[26]

The retreating Austrian army was pursued by detachments of Italian army which, advancing through Trieste and Gorizia, was moving to Charso. It was only at Logatec that the Italians were confronted by a batallion of Serbian soldiers under command of major Švabič who had just been released from captivity in Salzburg.

[25] Dr. Pavel Pestotnik (1879–1955), professor of philosophy, *Sokol* ideologue ideologue, founder of *Sokol I at Tabor*, liberal politician, member of the *Yugoslav Nationalist Party* (*Jugoslovanska nacionalna stranka — JNS)*), emigrated in 1945 and lived in Denver, Colorado.

[26] Note by editor: The official position in today's Republic of Slovenia is that the first Slovenian government was the one the communists have set up in Ajdovščina in 1945.

The major told the Italians to retreat if they did not wish to have a military incident. Frightened the Italians departed. What the domestic, Slovenian armed forces under direction of knight Pogačnik were unable to do was accomplished by a small Serbian detachment.

As soon as I arrived from the Littoral, I had to find professional employment in Ljubljana. At the end of 1918 it was not at all clear to the Slovenians in the Littoral how would the Italians act. Some even thoughts that they would be protected by the Allies if Trieste and Gorizia were not given to Yugoslavia. It turned out, however, that with the exception of some dissatisfaction over the Italian occupation of Rijeka by the adventurer Gabriele D'Annunzio, expressed by President Woodrow Wilson at the Paris peace conference, the Allies had no interest to do right by the Slovenians and Yugoslavs. For this reason many Slovenians had to leave the Littoral as the Italians simply would not allow them to be employed in Italy and if they opted for Italian citizenship they were transfered somewhere in the South of Italy while the Littoral was being populated by Italians. My future wife Ivanka Sket, too, who was a teacher in Vrtojba near Gorizia, was dismissed by the Italians in 1923 because she was not an Italian as they stated in the decree. Early in 1919 the influx of Slovenians from the Littoral was not yet as large as later when their migration caused a scarcity of employment. Still, upon my return to Ljubljana in early November of 1918 I had to wait about three months before I could get a government job.

In January of 1919 engineer V. Remec on instructions from Dušan Sernec, who was at that time in charge of public works at the Slovenian government, employed me in the service of the *National Government of Serbs, Croatians and Slovenians* (*Narodna vlada Srbov Hrvatov in Slovencev*) in Ljubljana. Thus I was finally able to secure employment in Ljubljana that was denied to me by Austria. I was at first appointed as an engineering practitioner which was my title in the Littoral when my service was interrupted by war but after one month I was promoted to the position of the *engineering adjunct.* I was assigned to the *Central Office of Government Geological Operations* in Ljubljana where I was, because of successful execution of several assignments, soon promoted to *construction commissioner.*

One of my first assignments was to take over the *Department for the Distribution of Coal* that had been administered by the Austrian construction counselor, professional engineer Jaksche. In this connection it was also necessary to review the operations of the Trbovlje coal mine which was at that time owned by foreign capital. In the process of carrying out my inspection I noticed that the German superintendent of the mining company had great respect for the Slovenian Land government and feared that the mine would be nationalized. It also turned out that the department for the distribution of coal had practically no financial resources.

Next I was sent to Velikovec to take over all operations of the local construction department and move all inventory to a place where it would be safe from the impending Austrian occupation. After that I was supervising the lead mine in Mežica. When I completed all these and other assignments which were mainly of administrative character in the transition from the Austrian domination to the inclusion in Yugoslavia I was recalled back to the Ministry of Public Works as a civil engineer. The Government Geological administration awarded me a bonus in severance pay amounting almost to twice the amount of my annual salary.

University and Enginnering School in Ljubljana

Under the Austrian rule Slovenians could not even think of a technological university as even the demands for a classical university, limited to letters, law, and philosophy, would not be granted because the Hapsburg monarchy was systematically Germanizing the country. If the Austrian Empire did not disintegrate at the end of World War I today's Slovenia would be as German as the Austrian Carinthia. Engineers living in Slovenia were by and large foreigners, Czech in government civil service and, of course, Germans in the enterprises operating with foreign capital.

Yugoslavia brought the opportunity for establishing a Slovenian university which included a faculty of technology. The principal promoters of Technological Faculty were professors Karl Hinterlechner, Milan Vidmar and Rihard Zupančič. The latter became the president of the newly founded university. About him it used to be said that he stated during a discussion of the need for laboratory facilities: "I need only chalk and a blackboard." As for engineering drawing, he said that he, as professor of advanced calculus, would solve every problem by integration. Milan Vidmar, on the other hand, was of a different mettle and he used his good connections with the German firm Siemens to acquire laboratory equipment which for many years represented the backbone of the electrical engineering instruction.

The technological faculty had its beginnings in technological courses for advanced students. The first course, corresponding to a winter semester, began in the second half of May 1919 and lasted until the middle of July. The second course, corresponding to the summer semester, started in the middle of August and lasted until the end of October. The curriculum of the first two years was modeled after the curricula of the Austrian and Czech technological universities so as to facilitate the transfer to other institutions to our students if the third and fourth year curricula could not be established in time. Approximately forty students enrolled. I was entrusted with the assistantship for descriptive geometry, situational drawing and lower division surveying. Descriptive geometry was taught by professor Jaroslav Foerster; problem session were conducted by the principal of the Real Gymnasium docent Josip Mazi whose assistants were Milan Fakin and myself. Mazi arranged for space in the building of the Real Gymnasium. The following year, my friend civil engineer Ciril Juvan took over the lectures of descriptive geometry.

The space for higher education courses was made available by Ivan Šubic, principal of the trade school on Aškerčeva cesta. In the meantime we formed the *Committee for Construction of Buildings for the Faculty of Technology.* I was treasurer of that committee and we succeeded in acquiring resources generally contributed by various industrial enterprises, institutions as well as individual donors. In the fall of 1920 the building was completed. The poplars growing in front of it were planted by my cousin Ignacij Gorenc. The Faculty of Technology was so pressed for space that it moved in even before we could organize a formal opening and dedication. Such haste annoyed some members of the committee and one day on seeing President Zupančič our committee chairman said: »You could at least told us to get lost!« As secretary I was entrusted with administering the committee's funds that were left over after the construction of the *Old Technology Building.* I kept these funds until 1945 when the communists confiscated them.

Next year when the *University of the Kingdom of Serbs, Croats and Slovenians in Ljubljana* has been established I became a lecturer for elements of construction and encyclopedia of civil engineering. Although my friend professor Ivan Vurnik was inviting me to become an associate professor at the university, which would be possible because my lecturing in the first two years would be adequate to establish myself, I did not want to leave my position at the *Directorate of Public Works* (*Gradbena direkcija*). Several professional engineers of the *Directorate of Public Works,* among them my superiors Ivan Škarja and Karel Orel as well as my colleagues Ciril Pirc and Ciril Juvan, also contributed their knowledge, expertise and time to the development of the Technology Faculty with a special permission of the Ministry of Public Works (*Gradbeno ministrstvo*) in Belgrade. After five years of successful activity, however, the ministry changed its policy and let us know that we had to resign our civil service appointments if we wanted to continue teaching at the university. We decided to remain in the civil service and abandoned our academic activities.

In cooperation with the *Engineering Chamber* I wrote *»Regulations for Compensation of Professional Engineers for Yugoslav Engineers«* which was published in 1926. As text to my lectures on the elements of building construction I published *Elements of Wood and Brick Construction* (*Lesne in opečne zveze*) which came out in 1926. After World War II, when I was in emigration, I wrote to my wife to send me copies of these lectures which might be useful in seeking employment. Because the communists had long before that confiscated all my books, my former student and assistant, professional engineer Mencinger was kind enough to give my wife his copies of the publication. Because the books were dated as printed in 1924 the communists, who carefully examined every book that people wanted to send abroad, would certainly refuse to permit the shipment. Here our friend Ivan Bonač, who had a bookbinding business, helped out by printing on the cover pages »*REPRINTED in 1946*«. That was enough to allay the suspicions of the communists and I duly received my books.

Many alumni of the higher education courses and at the Technological Faculty later assumed leading positions in the professions and academia in Yugoslavia. One of them was Venče Koželj, who after completing his studies in Ljubljana went for postgraduate work to Germany where he studied under professor Emde, known by his book on higher mathematical functions, then became professor of theoretical electrical engineering at the Technology Faculty, and dean of the faculty. I knew his father who was an established visual artist. Once he stopped me in the street and inquired how was his son doing saying that he was concerned because he never saw him studying. I was able to assure him that his son was an excellent student.

City Councilman

In 1927 the minister of the interior appointed me as member of the Ljubljana City Council. I became chief of the *Department of Public Works*. As I was also president of the Ljubljana Flying Club »*Our Wings*« (»*Naša krila*«), mayor Dinko Puc[27] delegated me as the municipality's representative on the *Committee for the Construction of the Airport* at Devica Marija v Polju near Ljubljana. The other two representatives of the City Council were Vice Mayor Evgen Jarc and chief of the treasury Ivan Tavčar.

Ljubljana Aeronautical Club (*Ljubljanski Aeroklub*) was an association to which belonged pilots and aircraft designers. Anton Kuhelj designed and built his own airplane in which he was flying over Ljubljana. Rape, also a well known pilot, had flown over the Kamnik pass (1500 meters) in the Kamnik Alps, a feat which in view of the contemporary aircraft models and unreliable engines required considerable piloting skills. People were talking a great deal about pilot Colnar who crashed with his red airplane for unknown reasons. Colnar's tomb was at the Žale cemetery. On the tombstone were a few cylinders from his engine and an inscription:

You loved the heights *Our hearts deeply wounded*
Soaring to the sky *In sorrow now cry.*

[27] Dr. Dinko Puc (1879–1945), attorney in Gorizia. Moved to Ljubljana in 1920, member of the *Yugoslav Nationalist Party* (*Jugoslovanska nacionalna stranka*) , Mayor of Ljubljana, retired from politics and the JNS in 1935. The Nazis arrested him and interned him in Dachau where he died.

In emulating the custom of World War pilots, the *Aeronautical Club* used to drop a wreath from an airplane on Colnar's tomb each year on All Souls Day, a maneuver requiring considerable precision in view of the crowds present at the cemetery on that day.

The municipality of Ljubljana and the German firm *Siemens* were joint owners of the local light rail transport. The city owned 60 percent of shares and *Siemens* 40 percent of the enterprise called *Light Rail Company* (*Maloželezniška družba*). The president of the company was Dinko Puc, the Mayor, its vice-president Milan Vidmar, university professor, who represented the firm *Siemens*. When the Light Rail Company bought a transformer, Director of the company, professional engineer Jarc called on me and told me that the acquisition of the transformer involved transactions which should be investigated. Immediately I formed in association with two other professional engineers a committee and requested documentation pertaining to the transaction to which I was entitled by virtue of my position as City Councilman. I turned out that the city paid twice as much as the transformer actually cost and that Professor Milan Vidmar, Mayor Dinko Puc, and perhaps others profited by the transaction. Professional engineer Foerster called on me and implored me not to bring the matter into the open. One of my two associates on the investigatory committee informed about the affair the clerical politician Adlešič who in a public speech hinted at possible corruption. I wanted to present the matter at the conference of the liberal party but chairman Ivan Tavčar, who also was involved, would not recognize me and he closed the session before I could make a statement.

Something similar happened with the purchase of two diesel engines for the municipal power generating station in *Kotnikova Street*. The advice of mechanical engineer Gulič that the city acquire modern turboaggregates was turned down. Instead, by mediation of a respected university professor, two already obsolescent Diesel engines were purchased in Austria. The two engines were originally intended for submarine propulsion but were never used because of the termination of hostilities.

I was City Councilman at the time of the Great Depression. At that time many people were without means to make a living and so I was besieged by persons seeking employment with the city administration. Often they would seek me out at home when I was at lunch. By interceding with the colleagues on the City Council I was often able to help people get a job. Because I made numerous recommendations of this kind the mayor once invited me for a chat at which he pointed out that he must be be very careful as each appointment represented an obligation for the city for almost one hundred years. He said that each job was connected with a pension and perhaps assistance to surviving or indigent members of the employee's family.

In view of the circumstances I simply did what I could but the final decision was a prerogative of the Mayor or the City Council as a whole.

In time, however, my financial investigations became a nuissance for the mayor. He found a way out by dusting off a law according to which I, a civil service employee of the central government which had oversight over the City, could not serve on the City Council. He persuaded the Ministry of the Interior to terminate my mandate as City Councilman. Having achieved that he sent me a letter thanking me for all the good things I did for the City and, shedding crocodile tears, expressed regrets at my departure from the Council.

Following example might give an indication of the forces that operated in determining election of deputies in the Parliament. In May 1935 the Ljubljana regional committee of the *Yugoslav Nationalist Party* (*Jugoslovanska nacionalna stranka*) unanimously nominated me as candidate at the so-called Jevtić's elections. At that time party bosses Marušič and Puc had the final say who should be the nominated as candidate and the governor's palace guard decided on Dr. Fuchs and Rajko Turk. The difference was that I would be nominated in a genuinely democratic way against the votes of Marušič and Puc, while Fuchs and Turk were actually selected only by the two party bosses who nominated them. This was the feeling of Rajko Turk, too, who conditioned his acceptance of the nomination on my agreement, as he felt that I should be the nominee.

Reserve Officers Association

Since 1925 I was the elected president of the *Ljubljana Subcommittee of the Reserve Officers' Association* (*Društvo rezervnih oficirjev*). I held this office all the time until the *Ljubljana subcommittee* and the *General Administration of the Reserve Officers' Association* parted in discord. I accepted election to the presidency solely because of the patriotic purpose of the association. Under my leadership the number of members of the *Reserve Officers' Association* in the relatively small Ljubljana region increased from two hundred to twelve hundred so that the *Ljubljana Subcommittee* was the largest in Yugoslavia.

In the Ljubljana Subcommittee the Reserve Officers' Association had a faithful membership dedicated to the motto »*For King and Country*«. Ljubljana Subcommittee demanded from its inception that the entire structure of the association be pervaded with the idea of complete equality. It is interesting to note that considerable efforts were required before we managed to delete from the Association's rules provisions which limited the rights of newly accepted, that is former Austrian, officers. Although on paper we did obtain equality discrimination unfortunately still continued in practice. The deplorable acts of the *General Administration* would not take place if the *Ljubljana Subcommittee* were represented in it as it was entitled because of its large membership. To objectively present the background and history of the dispute the *Ljubljana Subcommittee* reprinted the most important relevant documents and submitted them to the membership in a confidential memorandum.

The most significant of these was the matter of the *Committee for the Memorial to King Peter I in Ljubljana* of which I was chairman.[28] The *General Administration* was incapable of grasping the inestimable possibilities of the patriotic motto »*For King and Country*«. The membership of the *Ljubljana Subcommittee* demanded decisive action of the *Association* on matters of ideology and condemned the materialism and lack of ideals of the *General Administration.* The *General Administration* was mainly concentrating on collecting the additional contributions from the membership such as were, for example, 20 Dinars for a membership pin, 20 Dinars for a membership card holder, 200 to 500 Dinars for uniforms, 40 Dinars for a railroad discount card.[29]

Author in the uniform of a captain of the Yugoslav army

The General Administration demanded all these contributions for itself while the administrative costs had to be borne by the Subcommittee from its own resources. Dissatisfaction of the membership was growing from day to day. Under such circumstances the administration of the *Ljubljana Subcommittee* announced to its membership and the public that it is assuming the leadership of the *Committee for the Memorial to King Peter I in Ljubljana.* The Committee was joined by all representatives of the Slovenian public and cultural life regardless of party affiliation or cultural alignment. The membership unanimously approved the Administration's action and decided to make a suitable contribution from its own resources for the purpose.

The action was realistically conceived but at the beginning financial resources were needed. The *General Administration* acceded to the *Subcommittee's* request to use a sum of 100,000 Dinars which it owed to the *General Administration* as a

[28] The memorial was a stone sculpture representing King Peter I on a horse. It was erected in front of the City Hall (*Magistrat*) in Ljubljana. When the Italians, bearers of two thousand year old culture and admired throughout the world as creators and guardians of the works of art, occupied Ljubljana, they destroyed the memorial with dynamite.

[29] Monthly income of a moderately well situated civil servant used to be about 2000 Dinars.

loan for the erection of the memorial. Within the scope of this a widespread action for collecting the contributions was organized but the action was derailed when the *General Administration's* unexpectedly and prematurely, with threats and in a most offensive manner, demanded repayment of the loan from the *Subcommittee*. The planned quick action for the erection of the memorial was thus at its very beginning arrested and blocked because the *General Administration* went back on its written promise.

Shortly after the New Year of 1928 the General Administration publicly came out in the name of all reserve officers against the representatives of a political party. Members were confronted with the unpleasant fact of having to discuss in their apolitical organization matters pertaining to partisan politics. On request of its membership the administration of the *Subcommittee* sent the *General Administration* a statement that the members of the *Association* are bound solely by their motto: »*For King and Country*« and that the *Association* must remain strictly a patriotic, professional and above all a nonpolitical organization. In addition and in accordance with the bylaws the *Subcommittee* demanded that political acts of the *General Administration* be discussed at the next general assembly. Unfortunately the *General Administration* was at that time already too involved in its political activities. To the principled and in all respects appropriate statement of the *Ljubljana Subcommittee* the *General Administration* responded with personal threats, primarily against myself as well as against the other members of the *Subcommittee* and refused to recognize their rights as members of the organization.

News of the dispute between the *General Administration* and the *Ljubljana Subcommittee* soon reached the papers. The administration of the *Ljubljana Subcommittee* consisted of twenty one members and nine deputies. That the details of the dispute came out was nothing unusual. The administration of the *Subcommittee* had repeatedly stated that it is not responsible for statements which it did not release itself. This should suffice to the *General Administration* which could, using the available legal recourses, find the person responsible for informing the media. Two Ljubljana dailies had on two occasions sharply criticized the *General Administration*. That criticism included accusations which were actionable and redress could be obtained in courts if it could be shown that the accusations were untrue. But the General Administration did not take those serious allegations to courts. Instead it began to hint, in absence of any proof, that I had written those articles in the press. That was done out of sheer hatred against my person. Naturally, I neither wrote nor inspired any of those articles in the press. Overloaded with professional work and activities in many cultural associations I simply had no time to write for the papers nor had I ever told anything to the reporters. The *General Administration* knew full well that it would have the right to discover the identity of the author of those articles if it went to court in which case it would come out that its accusations were unjustified. The *Ljubljana Subcommittee* advised the *General Administration* to take recourse to courts but the latter choose not to file a suit. Instead it was unloading its attacks against the administration of the *Subcommittee* in the daily press of an ethnical and political party. Those articles had a tendency to create discord and dissension among the members of the *Association* in Ljubljana.

In the course of its dispute with the Subcommittee the General Administration resorted to political intrigue. It dug up an announcement supposedly sent by a political organization to its operatives summoning them to attempt influencing the various subcommittees to come out against the General Administration because of polemics the latter had with a representative of that political party. The Ljubljana Subcommittee was not even among the recipients of that announcement, we found out about it only from the copy sent to us by the *General Administration*. The *Ljubljana Subcommittee* consisted of reserve officers who were members of diverse political parties but not a single one came forward with objections to the protest made by the administration of the *Subcommittee* against the political activity of the *General Administration*. Reserve officers who were of political persuasion other than mine had on numerous occasions expressed their solidarity with the *Subcommittee's* administration in its efforts to preserve the non-political character of the organization. This was contrary to the claims of the *General Administration* to the effect that the *Subcommittee* acted on instigation of a political party. Because of diverse political views of the members of the *Subcommittee* it would be impossible to adopt any measures that would exhibit partisan tendencies.

Throughout the dispute the *General Administration* carefully avoided any direct contact with the *Subcommittee's* Administration. It even sent an official of the police to make inquiries on its behalf. On the other hand the *Subcommittee's* administration always endeavored to establish a dialogue with the *General Administration*. The latter reneged on all its express promises that it would meet with us and made a discussion of any kind impossible. The source of information of the *General Administration* was a former member of the *Ljubljana Subcommittee* who had been accused of an infamous act which it would not be appropriate to discuss here. Even a representative of the *General Administration* stated that he was informed by a high military authority that their informant was under suspicion. The *General Administration* was thus aware of the quality of its source and knew it was unreliable. The extent of personal hatred against me was apparent from the fact that the suspected person was informed of the confidential report about him I furnished to a representative of the *General Administration*. The suspect, however, did not go to court. For understandable reasons he preferred to miss the statutory time limit by which he could have filed a suit. All my reports were based on most reliable sources and were otherwise not available to the public. Concerning the informant in question the *Subcommittee's* administration kept trying to discuss him with the *General Administration* but all our efforts were in vain.[30]

In May of 1928, shortly before the general assembly of the *Reserve Officers' Association*, the *General Administration* issued a decision by which it removed all members of the *Ljubljana Subcommittee* from office and replaced them by its own commissioner. The purpose of this maneuver was clear: revenge because of frank words and preventing its critics to attend the meeting. The decision provoked indescribable anger of the membership of the *Ljubljana Subcommittee* as it obviously showed contempt for their democratically and consensually expressed desires. Moreover, the *General Administration* had by its decision violated the Association's bylaws as those provided for no commissioners. Because representatives of the *Ljubljana Subcommittee* were thus prevented to take part in the general assembly, the administration of the *Subcommittee* decided not to comply with the arbitrary act of the *General Administration* as such compliance would erode the validity of bylaws. Since the *Supervisory Committee*, too, came out in support of the *General Administration*, the administration of the *Ljubljana Subcommittee* called an extraordinary meeting for the end of September at which the membership unanimously approved actions of the *Subcommittee* and reelected its members.

Having received a unanimous vote of confidence, I took the initiative to personally call on the *General Administration* in Belgrade in October 1928 on the occasion of the anniversary of the Allied penetration of the front at Salonika. The purpose of my visit was to see if the dispute could be resolved by direct discussion. The reception accorded to me at the *General Administration* was personally very humiliating for me but I got over this and proposed that both administrations appoint representatives who would resolve the dispute. It was promised to me that a written proposal along those lines would be considered. In view of the humiliating reception accorded to me by the *General Administration* the *Subcommittee's* administration reluctantly approved after the fact my attempt in Belgrade. Still it sent a letter to the *General Administration* in the sense of my proposal. The response of the *General Administration* was surprising: all members of the administration of the *Ljubljana Subcommittee* were simply expelled from the *Reserve Officers' Association*. It was clear that the *General Administration* was motivated by reasons that were contrary to the lofty ideals of the *Association* and that it was acting out of personal hatred and vengefulness.

[30] This chapter is based on records in the author's archive. There is insufficient data in the archive to either identify the informer or the nature of the unsavory act of which he was suspected. It is known, however, that shortly after the end of World War I, Italy conducted extensive espionage against Yugoslavia. It is possible that the activities in question consisted of espionage which either could not be proven or the government, for reasons of foreign policy, did not want to prosecute.

Because of such arbitrariness of the *General Administration* the *Subcommittee* decided to again appeal on its membership which at the extraordinary meeting in November 1928 decided that the *Ljubljana Subcommittee* will become an independent private association. That decision followed quite logically and the membership could have made it already when the *Reserve Officers' Association* was making difficulties for the *Committee for the Memorial to King Peter I in Ljubljana*, particularly because the latter was independent of the *Reserve Officers' Association* and only wanted support of the organization which had created it. The reasons for declaring independence of the *Subcommittee* were thus not rooted in separatist tendencies but were simply a consequence of the arbitrary acts of the *General Administration* and a desire to preserve continuity for the benefit of patriotic activities.

Concomitant to the dispute were also legal complications. According to the *Law on Associations* (*Zakon o društvih*) which was on the books in Slovenia, the *Ljubljana Subcommittee* had, at the time when it was established, applied for government approval of its activities. Accordingly, in April 1925, on our application, Ljubljana's Departmental Governor (*Veliki župan*), took official notice of the existence of the *Ljubljana Subcommittee* and approved its activities under the bylaws of the *Reserve Officers' Association* as promulgated in 1922. Administration of the *Subcommittee* had, in accordance with the *Law on Associations* applied for approval of modified bylaws in the form adopted at the *Reserve Officers' Association's* general assembly in 1927. Ljubljana's Departmental Governor, however, did not take note of the new bylaws but requested instead amendments in accordance with the requirements of the *Law on Associations* then in force and at the same time forbade the *Subcommittee* to operate under the new bylaws.

The *General Administration* did not react to the corresponding proposals of the *Ljubljana Subcommittee*. Thus the *General Administration* was operating under the bylaws from the year 1927 while the *Ljubljana Subcommittee* operated under bylaws from the year 1922. Out of that accidental discrepancy arose a considerable legal problem which drew attention of the most prominent legal circles. The position of the persecuted *Ljubljana Subcommittee* was under the bylaws of 1922 particularly strong because it could not be reached by the *ukase* of the *General Administration* issued under the draconian bylaws of 1927. The blame for this accrued to the *General Administration* which paid no heed to the demands of the Governor.

Pursuant to the consensus expressed by the membership the administration of the *Ljubljana Subcommittee* applied for recognition of the independent status of the *Ljubljana Subcommittee* as an independent association on the basis of the modified bylaws of 1922. The Governor declined to approve the application because the law required consent of the *General Administration* which, of course, would not be forthcoming. This terminated the *Subcommittee's* administration's task to seek recognition as an independent association and the expelled members of the administration turned over their duties to the deputies who were still members. The *General Administration*, however, refused to recognize those deputies, too, and demanded that the *Subcommittee* be administered by a tortuously assembled commission, the four members of which did not agree with the methods of the *General Administration.*

As it was impossible to obtain recognition of the *Ljubljana Subcommittee* as an independent association, a group of three members applied for approval of the bylaws of *Reserve Officers' Association of Ljubljana.* In mid January of 1929 the Departmental Governor approved the application and the association was duly established. The area of its activity was the entire Yugoslavia. This association, too, proffered an olive branch to the *General Administration* which, seeing that all opportunities for further financial gain were gone, had not the slightest interest in further cooperation. I remained president for life of the *Committee for the Memorial to King Peter I, Liberator.*

As already mentioned, the *General Administration* was interested only in what financial gain could be squeezed out of the *Ljubljana Subcommittee.* Seeing such prospects gone, it began spreading truly fantastic calumny about the administration of the *Ljubljana Subcommittee's* self-sustaining resources. The fact, however, was that the *Subcommittee* at its meeting in 1927 decided that each member was to contribute 50 Dinars for the Memorial to King Peter I. When because of the unheard of conduct of the *General Administration* the Memorial fund drive got into difficulties, the regular general assembly of the *Subcommittee*, consensually decided that a contribution of 60,000 Dinars be made to the Memorial fund from its own operating resources. All that was done with appropriate documentation and the *General Administration* was informed of it by a financial abstract the receipt of which it acknowledged without comment. Later, however, the *General Administration* reported the *Subcommittee* to the Ministry of the Interior alleging that it did not consent to making a contribution to the Memorial from the operating funds of the *Subcommittee.* When questioned about this by the Governor, I could easily show him that everything was done according to the regulations and the Governor made a report to that effect to the Ministry of the Interior. Seeing its stratagem fail, the General Administration began spreading rumors that my report to the Governor was not truthful. Fortunately I had irrefutable evidence in my possession so that I could take the matter to courts where I was vindicated.

Members of the *Ljubljana Subcommittee* as well as members of the *Committee for the Memorial to King Peter I* were persons who held most distinguished positions in the society. To speak that irregularities of any kind could have occurred in the operation of associations or committees which they led was absurd. If even the slightest misuse of the resources were committed the social consequences for those concerned would be far reaching and unfathomable.[31]

All this was not without personal consequences for me. As it happens everywhere, the principal method of attack by corrupt politicians is to go after the livelihood of the person who opposes their dishonest activities. Thus, soon after the New Year of 1929, I suddenly received a decree ordering that I be transferred »because of official need« to Novi Sad in Serbia. For me that was equivalent to banishment to Siberia. I immediately sensed that it was a political maneuver particularly because the Directorate of Construction in Ljubljana asked the Ministry of Public Works that I be allowed to stay until March in order to introduce my replacement to the duties of my position. The ministry hysterically insisted that I report at my new job by the end of January. There was nothing left for me but go to Novi Sad.

The General Administration was thus successful in having me transferred away from Ljubljana where I had removed a good milking cow from its clutches. Later I learned from parliamentary deputy Albert Kramer[32] how such transfers were orchestrated. In my case, the General Administration gained access to a certain Antić, Minister of the Court, that is, some kind of Court's factotum. Antić simply presented his calling card with my name written on it to the personnel manager at the Ministry and my fate was sealed.

In Novi Sad I was well received because they were in need of engineering expertise in the construction of canals and levees on the Danube. I rented a room from a Hungarian family where I also took my meals. I was very busy with inspections in the field for which I was receiving additional compensation. The Serbs were very hospitable and they always treated us to lunch or dinner when we had to carry out an inspection.

Sometimes we would go boating on the Danube. Once we found ourselves in front of a larger steamer which was coming right at us. It is possible that the captain either did not notice us or did not care. Under international rules of navigation boats of lesser power have right of way in such circumstances but on the Danube, at that time, probably no one ever heard of this. The boatman was rowing franti-

[31] After the memorial to King Peter was completed *Committee for the Memorial to King Peter I* in Ljubljana had some funds left over. These funds were used for scholarships to students up to 1941. When the Italians occupied Ljubljana there was still 50,000 Dinars left. I called a meeting of the Committee which authorized me to distribute the entire residue to active military officers and their families who were stranded in Slovenia by the occupation.
The *Committee for the Memorial of King Alexander I*, whose treasurer I was, had a large sum left over. The Committee entrusted the funds to my safekeeping and I deposited them in a savings bank. When the communists confiscated all my belongings they seized that deposit, too.

[32] Albert Kramer (1882–1943), lawyer and philosopher, founder of the daily *Jutro*, minister and leading member of the *Yugoslav Nationalist Party* (*Jugoslovenska nacionalna stranka*) in Slovenia, in 1941 he declined to leave the country with the Yugoslav government. He was trying to reconcile the various liberal parties.

cally but we did not seem to move. I remember that I thought that the shoes would make it difficult for me to swim and that it would be a good idea to take them off. In thinking on how to swim, the steamer came within a few meters of our boat and at that instant our boat pulled away as if it were released from a chain. Perhaps we were pushed away by the wake flowing from the bow.

Naturally, I was frequently traveling home to Ljubljana which became quite a nuissance and living in a split household — my mother in Ljubljana and I in Novi Sad-was quite a financial burden. I thus called on my friends for assistance in arranging for my return to Ljubljana, either by a transfer back to civil service or by finding employment with a suitable private enterprise. Member of Parliament Albert Kramer opened many doors for me at the ministries in Belgrade. In Ljubljana an opening for the manager of operations at the municipal *Light Rail Company* presented itself. My friend Anton Ditrih was very active in this search which was soon rewarded with success: soon he was able to let me know, that the mayor told him that my appointment was assured because. all the members of the governing board of the *Light Rail Company* were agreeable and the firm *Siemens Shuckert* in Vienna, a major shareholder with the *Light Rail Company* likewise was in agreement. Director Jarc told me to be available by the beginning of September. Thus I was in a position to tender my resignation from the government service. The *Directorate of Public Works* in Novi Sad and the *Ministry of Public Works* in Belgrade were not happy about that because they wanted to keep me on their staff. Novaković, the principal of the general department at the ministry in Belgrade and the minister's right hand, who was very favorably disposed toward me, inquired if I would consider staying if I were transferred back to Ljubljana. I replied that in view of my experiences transfer alone would not be enough in absence of prospects for promotion. Novaković acceded to that and after a leave of two weeks I was already sitting in a new position in Ljubljana. I overheard some former colleagues in the office next to mine commenting: »Incredibe, terrible, what politics...!« but this did not upset me.

Time brought realization that the good of the country is more important than the personal ambitions of the one time leadership of the *General Administration* of the *Reserve Officers' Assocoation*. The *Association* underwent a thorough reorganization and from 1935 until the German occupation in 1941 I was President of the *Reserve Officers' Association in L:jubljana* and Vice President of the *Reserve Officers' Association of Yugoslavia in Belgrade.*

Memorial to King Peter I

In December of 1926 the administration of the *Ljubljana Subcommittee* of the *Reserve Officers' Association* issued a proclamation that it decided to erect in Ljubljana a memorial to King Peter I. The proclamation was signed by leading political, religious and cultural personalities and all cultural and professional associations. Activities to collect the funds necessary for construction of the memorial began immediately. Two years later the administration of the *Ljubljana Subcommittee* of the *Reserve Officers' Association* changed its name to the *Committe for the Memorial to King Peter I. In Ljubljana*. When the funds collected through contributions reached a satisfactory amount the *Committee for the Memorial to King Peter I* under pressure of bilious criticism of unsuccessful artists announced a contest for the conceptual design of the memorial and at the same time for selection of the most suitable place. A considerable number of artists responded to the invitation but none of the proposed designs was selected. The reviewing board evaluated the submittals and awarded prizes to the three best proposals but the committee selected none of those as it turned out that the contestants were not capable to create something beautiful.

In view of the situation the Committee authorized me to call on sculptor Mestrović in Zagreb and persuade him to accept the commission for creating the memorial. Mestrović listened to my offer with kind understanding, showed me his studio where he was just finishing a large than life sized sculpture of a native American brave on horseback commissioned by the City of Chicago, and convinced me that because of that work he could not at that time assume the responsibility for creating a memorial to King Peter I in Ljubljana. He suggested that we approach sculptor Lojze Dolinar in whom he had full confidence to carry out our commission professionally and to the highest standards of the arts. On Mestrović's advice the Committee was indeed able to retain Dolinar who then presented several conceptual drawings out of which the Commission would select the one deemed most appropriate.

Next came the problem of selecting the most suitable place for the memorial on which the Ljubljana artists again could not come to agreement. At that time the dominating artistic school in Ljubljana was that of architect Plečnik, obsolescent as his creations might have been considered by his more progressive colleagues. As authorized by the Committee I called on Plečnik and requested his advice. Plečnik made a sketch of the memorial — King Peter I. on horseback according to Dolinar's idea–standing in front of the City Hall between one of the three arcades. With that sketch I then secured the City Council's approval of the location. There was a very contentious debate before the City Council. The artists, among whom was the impressionist Jakopič, opposed the proposal vigorously. I had to summon up all my resources using my prestige as presiding elder of *Sokol I*, president of the *Professional Engineers' Association*, president of the *Reserve Officers' Association* and cash in on many political IOU's to win approval of my proposal prepared on the advice of architect Plečnik. Since the Committee already had sufficient financial resources it commissioned sculptor Dolinar to proceed with the construction of the memorial and have it placed in front of the City Hall. Dolinar was still working out the details of the various versions of the memorial. The Committee thus requested Plečnik to assist the sculptor in selecting the most suitable version. For a while Dolinar was willing to humor the architect but after a while he sensed that

Members of the Committee for the Memorial to King Peter I, the Liberator (1931); first row: first row second from the left is Committee's secretary Štefanija Prebil, the author is sixth from the left, sculptor Lojze Dolinar seventh, attorney Janko Žirovnik eighth, in the second row Etbin Bežek is second from the left and in the third row author's brother Albin Bevc is first from the right

Plečnik's suggestions degenerated into nitpicking and was no longer willing to act on them whereupon Plečnik, offended, withdrew from the process. The Committee then selected the most suitable variant from among those proposed by Dolinar.

The memorial represented King Peter on horseback in natural size. It was sculpted from dark grey domestic marble matching the historical architecture of the ancient City Hall. Dolinar presented me with a bas-relief sculpture of King's head, an exact replica of that on the memorial. That work of art was lost when the communists seized the power and confiscated all my property.

Dolinar also created a memorial pin with the King's head. All members of the Committee received one of those. King Alexander received a gold plated version of the memorial pin.

The formal unveiling of the memorial took place on September 6, 1931 in presence of an enormous crowd of onlookers who came from all parts of Slovenia. It was estimated that approximately one hundred thousand people came to see it. The motion picture archives in Ljubljana still has the film of the event but does not

Memorial to King Peter I, the Liberator, in front of the City Hall in Ljubljana

allow the public to veiw it. The *Committee* arranged for a 75 percent discount on all fares to Ljubljana for which special coupons were available at a nominal cost. From the purchase of these coupons accrued so much money that the *Committee* could use the excess funds for scholarships for needy students.

The unveiling was attended by representatives of the king, both Slovenian ministers of the government, the divisional general, all the signers of the original proclamation, the Governor, mayor and so many people that the square in front of the City Hall could not accommodate them. The overflow crowds were thus gathered on the Marijin Square and Vodnik Square where they could follow the proceedings over the loudspeakers installed for the purpose. The proceedings started with my opening address and the unveiling of the curtain covering the memorial. Next spoke the mayor who in the name of the municipality accepted the memorial in its possession and care. Following was a parade of a military unit, the honorary participants and the crowds gathered at the two nearby squares. Numerous festivities were taking place throughout the entire city.

The next day Queen Marija came to see the memorial. She was greeted by the *Committee* who waited in front of the memorial to receive her. After the queen inspected the memorial she requested that individual members of the *Committee* and sculptor Dolinar be presented to her.

In April of 1941 the Italians occupied Ljubljana where many relatives of various military personalities lived at the time. The *Committee for the Memorial of King Peter I* held a secret session in which it decided to distribute the residual construction funds among the first victims of the Italian occupation to mitigate to the extent possible their plight.

Marriage

I met my future bride Ivanka at the *Sokol* rally in Belgrade in 1930. We were introduced by our mutual friend Dr. Franta Mis,[33] an excellent physician and head of the city's public health department in Ljubljana, who had studied public health at Harvard University.

Ivanka was daughter of Dr. Ivan Sket and Katarina, née Kramberger. When we met she was teaching physical education at the intermediate school in Split, Dalmatia. The Skets had seven children: Franček, Ivanka, Malči, Janko, Boris, Stanko, and Milena. Malči died of diphteria as a child, Franček died in World War I of typhoid fever in Galicia. Boris died in 1943 as a partisan near Višnja Gora,[34] Stanko was interned in Dachau by the Germans where he died in 1944.[35]

---

[33] Franta Mis (1900–1970) Studied medicine in Zagreb, graduated as medical doctor in Prague and did postgraduate work in public health at Harvard University, in Detroit, and Toronto. In 1927 he returned to Ljubljana where he was active in preventive medicine and public health. He was instrumental in establishing the polyclinic for the schools and service for combating tuberculosis. Because of his protest against German withholding of drugs sent to Ljubljana by Red Cross he was interned in Dachau. After World War II he was chief of the sanitary and epidemiological service in Ljubljana. He was an excellent lecturer and organizer of public health programs.

[34] Boris Sket, bank employee in Ljubljana, intended at the time of Italian surrender to join the partisans in Bela Krajina, on a territory controlled by the communists. His acquaintances — it would not be right to call them friends — promised that he would be safe with the leaders of the so-called *Liberation Front*. Boris waited for a courier who would bring him appropriate papers but the courier did not show up. Afraid that the Germans would arrest him as an active member of the communist underground he left to join the partisans on his own. The partisans stuck

Dr. Sket, from Šmarje pri Jelšah, studied law at the University in Graz, Austria, where he met his wife Katarina from Rogoznica at Ptuj. On completion of his studies he was first employed as a law clerk in Celje and Ljubljana, but soon opened his own law offices in Gorizia. As an excellent attorney he soon acquired a large practice. His offices in the street *Tre Re*, directly across the street from the courthouse, in which he employed six people, had the latest technical equipment such as dictaphone, copiers, and the like. In court, where at that time, no verbatim records were being made he would make stenographic notes, a skill not generally possessed by attorneys. When Dr. Sket appeared in court the judges were put on notice that everything had to be done exactly by the book. The Skets lived in a beautiful house in town, they had an automobile, too, which at that time was still a rarity.

Ivanka Sket, authors bride (1930)

Sket's property included a foundry, military barracks *Reiner Kaserne* housing one thousand troops, and two villas on the *Corso* with splendid gardens all of which Dr. Sket took over from a bankrupt businessman Arturo Polli during the involved process of disentangling his financial affairs.

Ivanka attended the intermediate school in Gorizia where she was one of the three girls in a class of forty boys. One of her schoolfellows was Andrew Kobal, who later became an author, psychologist and high ranking officer of the Central Intelligence Agency (CIA) in Washington.[36] Ivanka loved to ride her bicycle, swim in the Isonzo river, and exercise at the Sokol Hall.

When World War I erupted, Gorizia , on the Austrian-Italian border soon found itself near the front. Artillery shrapnel was falling on Gorizia and also on the Sket's garden. Some professional informer told the authorities that at night light signals can

··· him in a detachment which was soon attacked by the *National Guard* and the Germans. Still shocked by the communist mass murders at Turjak, Kočevje and Jelendol the *National Guard* and the Germans killed everybody, including Boris. A day after Boris left for the partisans, the courier he expected brought orders directing that Boris was to be sent straight to Bela Krajina as "radiotelegraph operator" where he would join Zoran Polič who knew him from St. Lenart in Slovenske Gorice.

[35] Stanko Sket received his Juris Doctor degree at the University in Ljubljana during World War II. As an accomplished linguist who took postgraduate studies at the University in Rome he was employed at the Regional (*banovinski)* Department of Public Works as interpreter for the Italian language. Soon after the Italian surrender the local police came to arrest him because they in all likelihood took him for his brother Boris. Stanko attempted to escape when they brought him out of the house but unfortunately he was not successful. In prison he was savagely beaten. How author's wife attempted to save him by interceding with the chief of police Hacin and how he died in Dachau is described elsewhere in these memoirs.

[36] Andrej Kobal: *Svetovni popotnik pripoveduje* (*Memories of a World Traveler*), Mohorjeva Družba, Gorizia, 1975, p. 17.

be seen out of Sket's windows. The Austrian authorities searched the Sket house and found in the attic a paper picture of the illustrious emperor Franz Joseph on which someone punched out the eyes — this was probably done by the children. Dr. Sket was summarily arrested because of such terrible affront to the image of the tyrant and shipped off to a concentration camp. Because of this, and also because of the war his family, left without resources, had to leave Gorizia. Katarina took the youngest two children with her to her relatives at Kicar near Ptuj, and sent Ivanka and Janko to Šmarje to the home of Dr. Sket, called the Halar. After some time, when Austria was already losing badly at the front, Dr. Sket was released from the concentration camp and conscripted into the Austro-Hungarian Army as a private. The slight advantage of this was that he had to receive some kind of sustenance pay for his family and thus Katarina could send Janko and Ivanka to school in Celje where they were rooming and boarding with the family of the German protestant minister Lohman.

When the war was over, the Sket family returned to Gorizia, already occupied by the Italians. At first the question who will be awarded Gorizia and the Julian March seemed to be still somewhat open because of Woodrow Wilson's declaration on national self determination which ran contrary to the allied promises to Italy. The latter went to war at the side of the Allies because it was promised the Slovenian Littoral and the Dalmatia.

Ivanka hung out the Slovenian flag on the Sket house and wore a hat bearing a golden braid with inscription "*WILSON*." The Italians ripped off the flag and slapped the young lady. Thus they once again demonstrated their gallantry toward women.

Dr. Sket found his properties in a terrible disorder. The barracks were immediately occupied by the Italian army, all his other possessions, except the villa, were under some kind of government sequestration. Naturally, he immediately instituted the necessary legal proceedings which, however, dragged on for more than ten years after he left Gorizia . He had to leave because after Gorizia was awarded to Italy he, being a citizen of Yugoslavia, could not practice as attorney there.

In Yugoslavia Dr. Sket established himself in St. Lenart in Slovenske Gorice,[37] then in Šmarje pri Jelšah, and finally in Celje. After incredible difficulties[38] he succeeded in litigating out something over one million Italian Lire as compensation for his property. Out of this he loaned Ivanka and me 30,000 Dinars so that we could buy a small house in Šentjernej na Dolenjskem which the communists confiscated in 1947.

Ivanka was graduated from the teacher's college in Tolmin, qualified as a teacher and began teaching in Vrtojba, a small place near Gorizia to where she could commute on her bicycle. However, because she was not a subject of Italy, the

[37] Today only "Lenart."

[38] Only a person who has tried to secure restitution of or compensation for property confiscated by the communists in Slovenia knows how difficult such proceedings can be.

Italians dismissed her after three years of employment. In Yugoslavia she started teaching first at Sladka Gora near Šmarje, then in St. Lenart and finally in Beltinci. She was member of the *Maribor Sokol District* (*župa*) and completed several advanced courses in physical education. The most important of those was the physical education course of the *Czechoslovak Sokol Federation* in Prague in 1927. Thus she could be coach at the *Sokol Assocaiation*. At one of the *Sokol* rallies she met the presiding elder of the *Sokol Association* at Split who pointed out to her that with her qualifications and experience she could obtain a much better teaching job at a higher level school in a bigger city than was her current job in the sticks of Prekmurje. Indeed she was successful in securing a teaching appointment at the intermediate school in Split where she was teaching when we met.

Ivanka in Sokol dress uniform (1930)

We met several times after our first meeting in Belgrade and once I went to visit her in Split on a short vacation. Then I went to St. Lenart and formally requested her parents to give her to me in marriage. Dr. Sket consented and we decided the

Trainers course at Tyrš Hall in Prague (1927); Ivanka Sket is sitting in the first row second from the right

wedding would take place on May 9 at *Sveta Trojica (Holy Trinity) v Slovenskih Goricah*. Our honeymoon voyage was to Split, by steamship along the picturesque Dalmatian coast. Ivanka remained in Split for a few more weeks to wrap up her affairs. When she came to Ljubljana she immediately became member of *Sokol I. at Tabor* where she taught children's gymnastics and later became coach.

Our apartment at Emonska cesta 4, in the house owned by industrialist Jelačin had to be somewhat enlarged. We added three rooms and a new bathroom. This required considerable work and was quite expensive. My brother Albin moved to a new room far to the other end of the apartment which was spread along a long balcony. The view from that house inspired the painting "Križanke" by impressionist Rihard Jakopič. After my brother's premature death we decided to move to a more modern apartment in a two tenant villa at Erjavčeva cesta 26 (now 28).

Memorial to King Alexander I

Shortly after the assasination of King Alexander I in Marseille on October 9, 1934, leading national, cultural, educational and professional organizations elected a *Committee for the Memorial to King Alexander I.* Ljubljana's Mayor Ivan Hribar was chairman of the committee; I was elected treasurer because, as president of the *Reserve Officers' Association*, I had at my disposal an excellent organization with membership willing to promote the project and solicit contributions for the construction fund. We soon collected the first million Dinars and were well on the way to reach the two million figure.

One day I was visited by a young man who introduced himself as Marjan Tepina, said that he was a student of architect Josip Plečnik and had been retained by Mayor Hribar to prepare plans for the memorial. He demanded that I pay him 2,000 Dinars out of the Memorial fund as his monthly emolument. Expressing regrets I declined as the Committee never discussed such an appointment, much less authorized the payments. I told him that such matters are decided upon by the entire Committee and that he should wait until the Committee directs me to pay him. I assume that Tepina informed Mayor Hribar of that and he soon convened a session of the Committee. Tepina's matter came up at that session. Hribar was embarassed and apologized for having acted on his own but requested that we approve his action retroactively so that he might save his face. He said that he went to see Plečnik to prepare the plans for the memorial who declined to do so and recommended that Tepina, his former student, be retained for the task. Plečnik offered to advise Tepina in the process of his work. Tepina, on his side, also engaged help of all his friends and relatives who put pressure on various members of the Committee so that eventually the Committee agreed to retain him. Hribar recommended that he be paid a monthly stipend of 2,000 Dinars but proposed no specifics as to the form of the memorial which the Committee would give to Tepina in the form of instructions. Having done that the Committee had nothing else to do but wait until Tepina would come forward with his plans. In the meantime I was busy with collecting additional funds for the memorial.

Eventually we were invited to a session convened to acquaint us with Plečnik-Tepina plans. We found out that Tepina was proposing to construct some sort of an arcade in the garden of café Zvezda, commonly known as Kazina. The arcade was conceived as the first stage of restructuring and modernizing the area of the city bounded by *Congress Plaza (Kongresni Trg), Šelemburg Street, Prešeren* (now *Čop) Street* and *Wolf Street*, a periphery already built. According to the new plan a square accessible to the public would be built in the middle of that periphery. The work was to be paid for by the Committee as a present to the municipality. In addition to the construction of the arcade the Committee was expected to buy the garden of the restaurant Zvezda. This would play into the hands of political foes of Zvezda whose operating space would be reduced considerably. In the midst of the arcade Plečnik and Tepina would place a small bust of the deceased monarch, about one foot tall which any passer by could single handedly remove. This was the first time that the Committee heard of an urban renewal plan as part of the memorial. A closer examination of the plans showed an inscription over the arcade: "*De mortuis nil nisi bene*" (Say only good things about the dead). The Committee could not bring itself to approve those plans and it was decided that they be submitted to the broader circle of the organizations which had established the Committee.

Following the meeting members of the Committee began to object to the manner in which the plans were ordered and to the impertinent and offensive Latin inscription. They demanded that they be given the plans for closer examination and further discussion. The plans which Tepina had the temerity to submit, however, disappeared without trace and could never be found. Even the secretary of the Committee did not have a copy nor could he tell us where the plans were. The majority of members of the Committee was preparing organized opposition to the arbitrary ways of Chairman Hribar. At the meeting of the entire Committee an overwhelming majority decided to reject the Hribar-Plečnik-Tepina plan for the following reasons: execution of the entire plan would require decades and a change of urbanistic plans of the entire city, acquisition of space for the proposed enclosed square, and reconstruction of all existing buildings on the perimeter with respect to the interior orientation of the arcades. All this would be exclusively a project of urban renewal if the city decided to undertake it. The arcade would not represent a memorial to the deceased monarch and, consequently, the Committee could not use the funds collected for the purpose of building the memorial. Tepina's appointment as Committee's architect was accordingly terminated. [39]

---

[39] Tepina later made his career as a communist. He was a mediocre architect but during World War II he joined the communist partisans where he soon become a political commissar. In that capacity he reportedly shot the son of Professor Lovše and Mrs Pavla Lovše who was his subordinate simply because the lad disobeyed him. After WWII Tepina become Mayor of Ljubljana until 1967. The United States government fawningly invited him for a tour of architectural and engineering objects of interest. In 1967 he was sent as Yugoslav consul to Trieste.

Being thus rebuffed, Ivan Hribar called a session of the inner committee, informed us that he was resigning as Chairman and demanded that the rest of us follow his example. I did not think that I was obliged to resign as I had been opposed to the arcades and the plan for building them had been rejected. To snub me Hribar bestowed upon those members of the Committee who, succumbing to his pressure hastily resigned, the privilege of calling him by first name, pointedly omitting me from the chosen.[40]

Hribar then went to Governor Natlačen and told him that he resigned as president of the Committee. He made representations that the entire Committee had resigned and requested the governor to intervene in the crisis. Indeed I was the following day visited by Dr. Franc Stele[41] who informed me that the governor appointed him as commissioner to take over the affairs of the *Committee for the Memorial to King Alexander I* and demanded that I hand over the Committee's treasury. I declined because I had not resigned and had, in fact, decided to call on the governor and inform him of the Committee's majority position how did the crisis arise and also because I felt that the governor's decision would have to be contested in courts. A meeting of the majority members of the entire Committee was called for a meeting in Café Zvezda but the authorities forbade it by placing a policeman at the door who was turning away the arriving members. The majority then met in the offices of lawyer Janko Žirovnik. At that meeting I was instructed to deliver the Committee's funds to the court where we would contest the governor's *ukase* for safekeeping.

I then went to the governor who received me and listened to my clarifications of which he took note by stating "I see that I was incorrectly informed." Following that, Stele no longer insisted that I hand over the funds of the Committee. As the police would continue interfering with our meetings, attorney Žirovnik advised and then carried out the election of a new inner committee by mail. Chairman of the new committee was Josip Pipenbaher, vice-chairman Janko Žirovnik, secretary Joško Pogačnik while I remained as treasurer. The new *Committee* then got in touch with the governor and reached an agreement with him that Stele would be governor's representative on the *Committee*. Stele did not interfere with the activities of the *Committee* at all.

The memorial was created by sculptor Lojze Dolinar. The statue was of bronze, larger than life in size. At the sides of the granite pedestal were bronze basreliefs illustrating Prešern's "Baptism at Savica" and Gregorčič's "Save our boat from the

[40] When the Italians occupied Ljubljana in 1941, despairing Ivan Hribar, melodramatic to the end, said goodbye to members of his family gathered at lunch, went out and drowned himself in the Gruber Kanal of the Ljubljanica River not far from his house.

[41] Elsewhere in the author's archive the name is given as Dr. Skoberne.

storm." Memorial pins were issued with a basrelief replica of the memorial on them. The formal unveiling was a solemn affair. In attendance was the heir apparent accompanied by Prince Regent Paul and his wife Olga. The mechanics were arranged so that the young prince would pull a cord which would signal to hidden personnel who would actually pull away the flags covering the statue. Young Prince Peter was not sure that he would be able to pull away the large and heavy flags because he did not know of the arrangement. I encouraged him: "Your majesty, just give it a tug." He did and in an instant the memorial was unveiled.

Attending the unveiling of the memorial were various associations including the *Sokols* (*Falcons* — liberals) and *Orli (Eagles* — Catholic conservatives*)* who arrived in their uniforms. At the end of the proceedings these associations were sup-

Prince Peter II unveils the memorial to his father King Alexander I, Founder of the Union, at the Congress Plaza in Ljubljana in 1940; the author is behind the right shoulder of the general

posed to march away from *Congress Plaza* in accordance with a plan prescribed by the police. The *Sokols* were supposed to march out via *Vega Street* to the *Napoleon Square* and disband. But the *Sokols* did not want to disappear unnoticed and decided that they would march through the center of the city. When the police Commissioner noticed their intention he came over to coach Tone Gorjanc and reminded him that the *Sokols* were to leave by *Vega Street*. Gorjanc retorted: "Commissioner, sir, when everyone leaves by Vega, we will leave by Vega, too." He placed himself at the head of his column and the *Sokols* marched first down *Šelemburg Sreet*, past the main post office, then by *Prešern Street* and *Miklošič Street* to their home at Tabor. People were cheering them and *Sokols* in turn were cheering Prince Peter, heir to the throne.

The Italians, known the world over as creators and guardians of works of art, who occupied Ljubljana in 1941, could not stand the sight of memorials to Yugoslav kings who were symbols of Slovenia's liberation from the Hapsburgs and defense against the Italian covetous pretensions toward the Slovenian land. In August 1941, the Italians blew up with dynamite the memorial to King Peter I. Witnesses reported that they saw the head of the demolished memorial in the Isonzo River where the Italians dumped the debris.

Concerning the memorial of King Alexander I, the Italians first demanded of Mayor Adlešič to dismantle the memorial and store it in the museum. Adlešič declined to do so at municipal expense and had the impertinence of demanding that the *Committee for the Memorial to King Alexander I* pay for the removal. The Committee, however, had already distributed all residual funds to the Yugoslav officers who were stranded in Slovenia by the Axis occupation. The Italians then attacked the memorial themselves. At night they first attached a steel cable to the sculpture, melted the feet of the horse with a welding torch, and pulled the statue down with a tractor so that Dolinar's work of art crashed down on the plaza. For a while people were saying that the remnants of the memorial were stored in the museum but I never found out what has really happened to them.

Not the slightest disapproval of this barbarous act of the Italians has has ever been expressed outside of Yugoslavia although there always were widespread lamentations whenever a bomb or shrapnel would touch some decrepit ancient house in Italy. The international public opinion remained indifferent to the destruction of these two works of art. No one thought of prosecuting the criminals, and no one demanded compensation for reconstruction of these two creations of Lojze Dolinar.

Inspecting the Autobahn

In September 1936 a contractor and professional engineer, Josip Dědek orgainzed a field trip to Germany for a small group of engineers so that we could familiarize ourselves with the German *Autobahn* construction which at that time represented an engineering novelty.

In addition to Dědek the group included professional engineers Petrov, Jančar, Porenta, Meier and F. T.[42] On September 8 we departed from Ljubljana in Dědek's car and arrived in Villach the same evening. The next day we reached Heiligenblut

[42] The name cannot be recognized in the manuscript.

from where we took the mountain road by the *Pasterza* glacier to Zell am See and through Salzburg to Munich. On the way we inspected the new German *Autobahn*, some segments of which were still under construction, and we could watch how the construction personnel used specialized machinery called *vibrators* in paving the concrete sections of the road. As such machinery was only known to us from the professional literature it was very interesting to see it in action and it was a good preparation for our future project in Yugoslavia: the road from Ljubljana to Kranj.

In Munich, a metropolis which at that time prided itself of being the capital of Nazism (*Hauptstadt der Bewegung*), we spent some time on visiting the museums and other historical sights. I also bought some toys for my children. In touring the historical sights we came across a building on Königsplatz with a crypt of those sixteen Nazis shot at Hitler's abortive putsch in 1923 when they attempted to seize power. Each of the tombstones arranged in a row bore the inscription: "*Hier*" (*Here*). This symbolized the ritual at fascist gatherings where at the roll call the leader would include the names of those who died in some of their numerous brawls and other violent undertakings; when the leader would call the name of the deceased comrade all the present would shout:"*Presente!*" (*Here*).[43] In visiting that place we were unpleasantly surprised because everybody who entered that Nazi shrine was required to render a Nazi salute. Had we refused a very nasty scene might have resulted. The consular representatives of European countries were at that time so fascinated by Hitler that they did not think of protesting to the Nazi authorities and demanding that they do not impose this practice on foreign visitors. Typical for that period was the attitude of Winston Churchill who as late as 1938 (!) when it was clear that Hitler was well on the way of carrying out his well known program, wrote "*Were England to suffer a national disaster I should pray to God to send a man of the strength of mind and will of Adolf Hitler.*" As an adventurer Churchill always was attracted to personalities such as Hitler, Stalin, and Tito but in this case he had, perhaps unwittingly, revealed the true thinking of the British.

From Munich we went to the Bavarian Alps and the Eibsee lake under the massive Zugspitze which we ascended by aerial tramway, an interesting construction of mechanical engineering. The following day we drove through Garmisch and Mittenwald to Innsbruck, thence over the Brenner Pass and Vipitena to Italy. In Vipitena we stayed overnight in Hotel Posta Vecchia and the next morning headed toward Bruneck on the Dolomiti road past Corvara, Baddia and Arabba and over passes Pordoi and Sella back to the valley of Adige River. I never thought that nine

[43] Even after the Americans occupied Milan and rounded up and imprisoned the fascists the latter continued with the tradition. When they learned that the Italian partisans had shot Mussolini at San Giuliano di Mezegra on the banks of Lake Como so that he would not embarrass Italy at some kind of Nueremberg trial, one of them called out: "Benito Mussolini!" whereupon the entire group hollered "Presente!"

years later I would meet the Americans in Bruneck and be saved from the British and their communist allies.

It was night when we descended into the valley of Adige to continue our way toward Lago di Garda. We were all somewhat sleepy and so was probably Dědek who was driving all that time. That could have been fatal. Suddenly we heard a loud noise and sparks began to fly around the car. I looked out and saw that the rear wheels of the car were hanging over the embankment of the Adige River. Dědek fought hard to extricate the car but the vehicle would not move. Suddenly the car jerked forward and pulled onto the roadway; it was still running so that we could reach the nearby Desenzano di Garda where we stayed overnight.

Next morning we found that the differential was damaged. The repairs could not be immediately made and Dědek had the car shipped home from Verona by rail. We continued on our way by the same train. We stopped in Venice to take in the sights and returned to Ljubljana by way of Trieste. At home my wife told me that she could not sleep that night when we had our encounter with the Adige River. That and a few other similar incidents make me think that thoughts are by some hitherto unknown process transmitted over a distance under certain circumstances.

Sokol

The Sokol Idea

At the end of the eighteenth century the French Revolution brought forth new ideas. The political momentum of the French Revolution by itself was not as important as was the revolution in human thinking. The revolution, although initially confined to France, soon made its impact felt throughout the world. The revolution destroyed absolutism and people shed the fetters of dogmatism in the world view and everyday life. Influenced by the French Revolution Europe was swept by the ideas of humanism and return to nature which gave direction to future progress. It is true that the principles of the Revolution were after a while somewhat reduced and transformed by nationalist ideas, primarily under the influence of romanticism in history and literature but its principal ideas still dominate in all fields of human endeavor.

In the wake of the French revolution and its ideas the Czech nation began to awaken from the deep slumber in which it sank after the political debacle at White Mountain (*Bíla Hora*) in 1620 where Czech protestant armies were defeated by the combined forces of the German empire and the catholic league. Germans then dominated the Czechs for centuries afterwards. The renaissance came to its full expression in the first half of the nineteenth century, particularly after the demise of Bach's[44] absolutism.

The history of the Sokol movement is intimately connected with the life and work of its founder Miroslav Tyrš. As a young man Tyrš took part in the uprising against Austrian absolutism and fought on the barricades in Prague in 1848. Tyrš studied philosophy and became professor of aesthetics and founder of the Sokol movement. The name *Sokol* means Falcon, a proud soaring bird. In the twenty two years that were left to him after the founding of the movement Tyrš, with indefatigable work and help of idealistic trainers as well as with the moral encouragement and financial support of Jindrich Fügner, succeeded in transforming his initially local gymnast association into a nationwide movement with fully developed professional and ideological basis and well formulated goal which is still universally

[44] Alexander Bach (1813–1893), Austrian politician, minister of justice and of the interior. Instituted Bach system of bureaucratic control of the Hapsburg lands including stringent measures of secret police. The Bach system met with opposition especially in Bohemia and Hungary, and after the Austrian defeat in the Italian war of 1859 Bach was dismissed and a new system introduced.

desirable today regardless of one's nationality. The goal is: enabling everyone to achieve the dignity of a free man through improvement of his physical and mental fitness and to preserve the freedom of a nation or regain it if it had been lost.

From its inception in 1862 the Sokol movement was the most important and most active agent of Czech national renaissance. The great contribution of Tyrš, and other Czech national leaders such as Masaryk and Beneš, was that they established an organization for their people which was educating the youth in the spirit of resistance against authoritarianism and dedication to the attainment of the everlasting goal of a free nation.

All the problems and upheavals of the twentieth century associated with the various and sundry dictatorships of the extreme left and extreme right could not undermine the Sokol movement. Its extraordinary vital force has its roots in the idea that inspired it in the past and still inspires it today. To understand the Sokol movement it is necessary to recall the history of the time when it was established.

Tyrš Hall in Prague

Miroslav Tyrš studied the history of ancient Greece and formulated his conclusions in his first dissertation on the *Fundamentals of Gymnastics* which became the foundation of the Sokol educational system. In his next paper, *Our Task, Direction and Goal* he formulated the foundations of the Sokol idea which crystallized itself in a hundred years of practical application. These can be set forth in the following basic principles:

- Not only individuals but all nations must attain freedom from oppression of any kind.
- To achieve this freedom individuals as well as nations must constantly strive for improvement.

- The future is made through learning and improving oneself by one's own efforts.
- Force and power in itself cannot achieve a permanent change, durable results are achieved through the evolution of the mind. The effort of the individual determines progress that is made.
- Life is a constant struggle in which the winners are those who improve themselves and advance while those who neglect themselves and hesitate succumb. This holds for the individuals as well as for the nations.

Tyrš concluded: "Our first and universal task is to maintain our people in a generally good condition which prevents a nation from becoming extinct, that we maintain that physical, mental and moral health which allows no corruption, no hesitation and no backwardness — that most heinous crime that can be committed against a nation." He set forth the following directives as a plan for action.

- To achieve and preserve freedom against oppression individuals as well as nations must be physically and morally strong.
- It is therefore necessary in the spirit of these principles to develop in the individuals everything that leads to the improvement of life (industry, health, strength, progressive thinking, charity) while everything that is contrary to this must be removed (mendacity, selfishness, debauchery);
- It is necessary to develop in the individuals the spirit of brotherly love and unselfish help to others;
- It is necessary to develop in the individuals the sense of duty to society and subordination of lesser interests to the general good (in accordance with the ranking of values: individual–society–nation–humanity);
- Everyone has access to the Sokol education which is intended for everyone regardless of race, social standing or religion.

Moreover, Tyrš taught the following:

- The Sokol movement, which is for all social strata and classes, represents moral and physical education of an entire nation, its development for strength, courage, charity and resilience — it is for this reason that we must strive to include the entire nation.
- The Sokol movement wants to achieve unity of the entire nation. The Czechoslovak Sokol has always worked to this end as has the *Yugoslav Sokol* which has been and will be the protagonist of the idea of national unity, brotherhood and love.
- *Sokol* develops the feeling for all-Slav cooperation, brotherhood, equality and democracy.

- *Sokol* is not a political organization, its mission is progressive education and development of the nation.

Building the Sokol Hall at Tabor

The 1903 *All-Sokol* rally in Ljubljana demonstrated that although the Sokols were relatively few in numbers their will was strong and so were their muscles. Between 400 and 500 gymnasts performed calisthenics on the track field below the Tivoli Park, roughly at the location of today's fairgrounds. The calisthenics were relatively simple in keeping with the level of development. Ivan Tavčar[45] greeted the attending public in a rather long speech. The public vigorously cheered the *Sokols* in the hot afternoon as the temperature was climbing to 86°F. The entire organization was carried out with great simplicity and on a minimal budget. The first *Sokol* public performance in larger numbers successfully achieved its purpose of gaining the desired publicity. The calisthenics were directed by Viktor Murnik, the well known coach of the *Ljubljana Sokol*. Murnik used to write and interact with the *Sokols* a great deal but he did not seem to be particularly committed to progressive ideas. That was the cause of friction at the annual general meetings of the *Ljubljana Sokol* and the root cause of contests for stronger leadership in the association.

Although the presiding elder of the Ljubljana Sokol, Kajzelj, was very diligent and caring he did not succeed in channeling the energy of Sokol membership to productive work and right direction. Coach Murnik undoubtedly deserved credit for the development of the *Ljubljana Sokol Association* but he was too much a technician who remained rather aloof with respect to the true brotherhood and democracy. Too conceited with his importance Murnik wanted to be first everywhere and recognized and admired as such. He was secretary of the Chamber of Commerce (*Trgovska zbornica*) and was graciously devoting all his free time to the *Sokol Association*, primarily through the monthly journal *Slovenski Sokol* (*Slovenian Sokol*) in which he was publishing Tyrš' ideas about the Sokol movement in addition to articles of technical nature describing the "Sokol system." This was by no means an easy task particularly because Murnik did not seem to have full mastery of the Czech language. His rather simple translations stimulated a great deal of thought and activity in all larger *Slovenian Sokol* associations and undoubtedly brought good results.

[45] Ivan Tavčar (1851–1923), writer, attorney and liberal politician, member of the first Slovenian National Government in 1918, mayor of Ljubljana.

Under Murnik's leadership selected and well trained crews of Slovenian *Sokols* took part in international competitions held in various places in Western Europe and were always very successful. This was good for the promotion of Slovenia's name but, of course, flaunting it was not appropriate. Murnik knew how to select some eight or ten of the best gymnasts and train them to perfection. There is merit in this, to be sure, but the practice tended to push new, younger *Sokols* away from the true purpose of *Sokol* development which emphasizes the development of all. Working with the entire membership is more important than training an elite group to perfection.

While exercising and aerobics is conducive to discipline it does not represent all of the *Sokol* activity. That includes also development of patriotic and democratic social attitudes which must be part of every *Sokol's* life. Achievements in competitive sports are of temporary nature, social development, on the other hand, is the foundation on which a healthy Sokol organization and a morally sound nation must be built.

The initiative for establishing the *Sokol I* association came from Pavle Pestotnik who, after arriving from Prague where he received his doctorate in philosophy, began to seek a way for younger *Sokols* to break out of the inertia prevailing at the *Ljubljana Sokol*. Members at that association were practically without technical and ideological leadership that would provide an impetus to the youth and channel their energy to planned and well defined task within the association and outside of it. Pestotnik soon became coach at the *Ljubljana Sokol*. The young were gathering around him and let him know that they expected him to lead them out of the doldrums. It was not, however, advisable for the youth to attempt taking over the leadership at a general meeting because that could be damaging to the *Sokol* organization. Therefore Pestotnik, Bricelj, Počivalnik, Mulaček and Fran Medič decided to establish a new Sokol association as Ljubljana was big enough to accommodate two. Franjo Medič was among the most active proponents of this idea.

Dr. Pavel Pestotnik

Toward the end of August of 1907 the secession from the *Ljubljana Sokol* was carried out. Some twenty gymnasts have already been exercising since June in the small gymnasium at the elementary school at Ledina in very modest circumstances. Elected as presiding elder was attorney and well known alpinist Tominšek. Although the latter was not well versed in the association's affairs, much less familiar with the *Sokol idea*, he was a good man and known in public to be of impeccable character. Pestotnik was elected deputy presiding elder.

In the second half of September 1907 *Sokol I* held a general meeting in the garden of *Bierhalle*, a restaurant located off St. Peter's Street almost back to back with the elementary school at Ledina. About seventy people attended. Coach of the *Ljubljana Sokol*, too, came to greet the new association. He conveyed his wishes for success and somewhat awkwardly offered his assistance which did not go over too well. At first *Sokol I* made every effort to conduct itself loyally to the senior association and even inquired whether the *Ljubljana Sokol* wanted to assume the name *Sokol I* which would then make the new association *Sokol II*. But the *Ljubljana Sokol* made it clear that it would not change its traditional name. Among themselves, members of *Sokol I* quipped that in view of this *Sokol 0* — zero hinting on lack of activity–might be a more appropriate name for the senior association.

The Catholic Congress of 1902 revived the catholic organizations both along the religious as well as political lines. A few years later the future *Orel* (*Eagle*) was founded as the Catholic alternative to *Sokol*. The principal promoter of *Orel* was Štefe who operated out of the editorial offices of Slovenec, daily newspaper of the *Slovenian People's Party* (*Slovenska ljudska stranka*). Athletics and cultivation of the body had for a long time been contrary to the precepts of ascetic Christianity and Catholicism but modern times modified the religious view to the extent that the strength and beauty of the body were no longer considered pagan but rather a duty to maintain one's health which is considered a gift of god. Enemies of *Sokols* began to intensify their propaganda which could be heard even in the church where instead of preaching the God's word hatred against the *Sokols* was being incited.[46] The oversight and invisible leadership of the *Orel* was in the hands of the highly placed clerics headed by their militant and uncompromising bishop Anton Bonaventura Jeglič. The bishop achieved considerable success in the development of spiritual and party indoctrination within his party. *Orel* became politicized while *Sokol* was supposed to be apolitical. Certainly the original *Sokol* idea was to build an organization for national defense against the German expansion similarly as the German Turnverein had been created with the intention to defend the Germans against Napoleon's conquests. Catholicism, however, which is "for all — for the entire world" and consequently cannot be nationalistic, could not approve the

[46] Not everybody succumbed to the hate campaign, however. Thus Rudolf Smersu, president of the *Slovenian People's Party* and president of the *Slovenian National Committee* on August 28, 1980 wrote to the author as follows: "Let me use this opportunity to tell you that I used to live at Tabor in Ljubljana and that my sister Ema was an enthusiastic athlete with the *Sokol* association there while I was participating in the athletic activities at the *Orel*. Our family was one where diverse views were respected. I was brought up so that I refused to take part in any name calling and that I was striving for cooperation in all walks of public life. To this end I often saw my schoolfellow Dr. Marjan Zajec who was member of the secretariat of the *Yugoslav Nationalist Party* while I was in the secretariat of the *Slovenian People's Party* with Dr. Köstl and others. I am very pleased that we are now working together in politics."

Sokol's nationalist orientation. The pejorative term for *Orels* used to be "*Owls*" and both sides used to sing derisive and taunting songs. The more the hatred against the *Sokols* was fanned up, the more intensive became the work in their associations. The hate campaign did not harm the *Sokol* movement, on the contrary, many were only stimulated to more active and intensive participation. The *Orel* organization did not accomplish much although its name was often mentioned in connection with the *Orel* activities in Moravia.

It was in such atmosphere that *Sokol I* had to earn with hard and persistent work its right to exist and be recognized. As early as in 1908 *Sokol I* could show the results it was achieving in gymnastics with a public performance. The public performance included a complete gymnastics program conducted in the small space between the elementary school at Ledina and Bonač's factory. The beginning with forty gymnasts was modest but it succeeded in breaking the ice. By that time the association was already well organized, its various sections had their own procedures and pursued well defined goals. Pestotnik was present everywhere advising and directing activities in the gymnasium, in the leadership of the training staff, at meetings of the governing board and the various sections. He was particularly interested in the educational section and acted as an ever present liaison between all groups and individual members.

Pestotnik presented to the association a plan for future activities and development. It was expected that the organization would be complete within three to four years, that is, there would be male and female gymnast members and when an adequate number of trainers came aboard, youths and children would be included in junior membership. The program was carried out as planned and by 1911 *Sokol I* was a full fledged *Sokol* association. The membership was about one thousand. At first the operating funds consisted only of the individual members' contributions. Management had to be frugal because the association's coffers were at times completely depleted and the eastern part of Ljubljana had few, if any, wealthy supporters as most of the so-called better citizens were members of the *Ljubljana Sokol*. Although lean, *Sokol I* was truly democratic relying only on its own work and prospect for the future.

The first coach was Albin Kandare, an excellent gymnast particularly in the more advanced gymnastics. He, however, remained at *Sokol I* for only one year probably because he did not quite grasp the *Sokol* idea and had succumbed to the indoctrination of his Jesuit trained catechist Pečjak. Janko Dimic, a typesetter at the *Teacher's Press* (*Učiteljska tiskarna*) by profession, took over the duties of principal coach after Kandare. He was a politically moderate social democrat and universally popular among the membership. His health was seriously impaired in the Russian captivity during World War I and he died a few years after his return home. Although he was by no means rich his wife gave *Sokol I* a respectable sum from his modest estate. The *Association* was deeply grateful for her generous and noble gift. Other highly effective trainers were Slapničar and brothers Rupnik. The detachment led by Slapničar, in particular, was always perfectly polished at every public performance.

In 1909, again on the initiative of Pestotnik, *Sokol I* was instrumental in establishing the *Ljubljana Sokol District (Ljubljanska Sokolska župa).* In this group of *Sokol* associations were included, along with *Sokol I, Sokol II* and the *Sokol Associations* of Moste, Štepanja vas, Ribnica, Litija, Šmartno, Šentvid and Kamnik.

Sokol I was prolific in expending its energy and skills everywhere. Three to four week training courses were conducted in all Sokol associations. These courses were the key to advancement. At *Sokol I* the training courses were more comprehensive; they consisted of forty-three hours of instruction. Thirty eight gymnasts and trainers completed that course. Drenik and Murnik of the *Ljubljana Sokol*, too, provided a few hours of instruction each. The entire region was bustling with activity.

Ljubljana Sokol, however, was not quite happy with the establishment of the *Ljubljana Sokol District.* Smrtnik, presiding elder of the *Sokol Association in Celje*, who was a friend of Murnik's, reproached Pestotnik accusing him that he was fragmentizing the *Sokol* organization. Pestotnik had to explain to him that what was good and working for the Czechs should also be acceptable in Slovenia. The important thing was training enough good *Sokols* to become involved in *Sokol* activities. Stronger *Sokol* associations should make available their surplus personnel to weaker, smaller associations thereby stirring them up from inactivity. This, of course, required decentralization.

Pestotnik held ideological lectures on the objectives of the organization and explained the *Sokol* gymnast system in all associations of the district. For him there was no Sunday, no rest. He covered the travel expenses of his lecturing out of his own pocket.[47]

The results were soon apparent, new young members appeared in the gymnasium and at public productions. Successful district rallies and competitions were conducted under Pestotnik's leadership at *Sokol I*, in Ribnica and in Kamnik.

Administrative management and gymnastic activities are closely related. If one is failing the other cannot thrive. Very effective managers were: Fran Vrečar, nicknamed "Rožica," Ljubo Drmel, association's first secretary and principal librarian for many years, Lozej a conscientious and competent accountant. Those *Sokols* had a track record of more than thirty years experience in the organization's affairs.

---

[47] Pestotnik's way of doing things represents an interesting contrast to that of the, at that time still concealed, communist Franjo Lubej about whom many years later the disciplinary court of honor of *Sokol I at Tabor* established that he used to charge his dinners to the account of *Sokol I* while he still held the position of principal coach and occasionally worked evenings in the association's office.

In 1908 *Sokol I* set as its goal the building of its own hall. In Bohemia even small *Sokol* associations had their own halls for gymnastics, meetings, and management of their affairs. It should be likewise in Slovenia and in Ljubljana in particular. The same year Pestotnik prepared the bylaws of the *Association for the construction and maintenance of a hall for Sokol I.* A year later the bylaws were submitted to the authorities and approved. The first president was Kostevec, a merchant on the St. Peter Road. The governing board consisted mainly of older, well established businessmen. The difficult work was progressing slowly. To keep the idea of a hall of our own alive we involved the youth in the committee's activities. *Sokol I* was carrying out fund raising activities but although these fund raisers were quite successful it was still necessary to find large contributors. The first substantial gift was received from Anton Bončar, restaurant owner on St. Peter Road.

As part of its fund raising activity, *Sokol I* staged a "buy-a-flower" campaign throughout the city. Every *Sokol* sympathizer helped, particularly young women. There were so many collecting stands and booths that everyone had a chance to pick up flower and make donation. That action brought over 200,000 *Krone.* The building of the *Sokol* hall had to wait until the end of World War I, however.

Plans for the *Sokol Hall at Tabor* were prepared by my friend architect and university professor Ivan Vurnik. He had traveled throughout Yugoslavia in search of an idea for his design. Eventually he found a building in Niš that provided inspiration for the concept and style of his design. With the *Sokol Hall at Tabor Vurnik* consolidated his renown as an established architect whose reputation was known beyond the national borders. I suppose that plans for Tabor may still be found in Vurnik's archive and possibly one could get them there and have copies made. Of course we had our own set of plans at *Sokol I.* I made arrangements with the City Council for a grant of a parcel of land on which the *Sokol Hall* could be built.[48]

At that time the future site of our hall was crossed diagonally by the Vidovdan Road. Because the municipality included the road into the grant, traffic had to be rerouted around the land, past a pavilion, the *Srca Jezusa* (*Heart of Jesus*) church and the military barracks. The Association readily fenced in the acquired land. I did all the surveying necessary for the exercising field. In the following seven years I spent all my free time directing and supervising the construction of the hall. I worked closely with professor Vurnik on all matters concerning the structure. The most important decisions were: insulation of walls from humidity, especially in the basement of the building, steel and concrete straps all around above the walls to resist earthquakes, wooden roof designed by structural engineer Stanko Dimnik,

[48] Note by editor: Publication *Sokol I at Trg Tabor*, issued at the occasion of formal opening of the gymnasium on August 14 and 15, 1921, states that the *Municipality of Ljubljana* sold the land to *Sokol I.* Quoted are particular terms from the contract of sale between *Sokol I* and the municipality concluded on February 21, 1920.

Arhitect Ivan Vurnik

and copper roof covering which was my idea. Considerable resistance had to be overcome concerning the latter because of the cost; today that roof still remains one of the more remarkable sights in Ljubljana. In front of the hall Vurnik had envisioned a pedestal for a memorial to Tone Malej.[49]

Once I was showing the still unfinished building to industrialist Franjo Medič and mentioned that plaster of Paris would be particularly desirable as material for finishing the inside walls because it is durable and need not be painted. Medič asked how much plaster of Paris would be needed. My preliminary estimate was five railway freight cars and Medič said that he would supply the plaster as his contribution. All I had to do is let him know when we were ready to apply it. The outside was plastered with material prepared according to a formula thought up by Vurnik.

Considerable concern was expressed when we discovered the problem with the terrain which required that the lower part of the building be sunk rather deeply below the street level. This was necessary because the building site used to be a depression called *Kravja dolina* (*Cow valley*) which was over the years packed with fill. We discovered that only after we began our exploratory soundings prior to excavation. It was necessary to remove a volume of fill equal to that of today's basement. The entire fill had to be removed before we reached firm terrain. That job cost 300,000 Dinars.

The municipality deeded the building site to *Sokol I* under condition that it is built up by a specified time. The deadline was approaching but, because of the above problems, we had barely begun pouring the concrete foundation. In the meantime the municipal administration changed and came under control of clerical politicians who were hostile to the *Sokol* organization. Those politicians pressured the government commissioner who was at that time in charge of the municipality to appoint a commission which would establish whether the site had been built up by the specified time. If it were found that the site had not been built up it would revert to the municipality. It was in late winter, shortly before Christmas, when the low temperature made it inadvisable to pour concrete. We requested our builder Tavčar-Svetina to expedite the work so as to complete the ceiling above the basement, a few feet higher than the adjoining track field. The concrete ceiling of the basement required wooden scaffolding. Unfortunately construction superintendent Srakar

[49] Tone Malej, Olympic athlete, died in Luxemburg in 1930 when he fell from the swinging rings during a practice exercise.

turned out to be an obstreperous fellow who would not listen to advice. When the evaluating commission arrived — with university professor Jaroslav Foerster as expert — the scaffolding gave way and the still fresh concrete collapsed with a tremendous rumble right then and there in front of the commission. We were flabbergasted. However, when the commission found out what caused the accident, it issued an opinion to the effect that the site had been built up. Later I had to request the construction firm that superintendent Srakar be replaced because his obstinate and arbitrary conduct interfered with the construction. The firm Tavčar-Svetina became bankrupt soon after that and the construction was taken over by builder Mudrovčič whose professional associate was university professor Hrovat.

Construction and ownership of the Sokol Hall belonged to *Association for Construction of Sokol Hall at Tabor.* It is that association which is the legal owner of the *Sokol Hall.* Members of the association were stockholders. I was one of them because I invested a substantial sum. Among the last surviving stockholders were Rajko Turk and I. It would seem that I am now the last surviving legal owner of the *Sokol Hall at Tabor.*[50]

All legal affairs were conducted by notary Andrej Kuhar. Here it is appropriate to mention that at the beginning of the Italian occupation in 1941 Mrs Stare, owner of an apartment house bequeathed her property to Sokol I. The notary advised me that Sokol I inherited a three story apartment building and that he would delay the execution of her will until the end of the war otherwise the Italian Dopolavoro, which appropriated the *Sokol Hall at Tabor,* would seize that apartment house along with the rest of *Sokol I* property.[51]

Construction of the *Sokol Hall* cost about seven and a half million Dinars. The greater part of this sum was accumulated through gifts and the proceeds of various presentations. However, the construction in progress required that we find additional funds. When we regained influence in the municipal government we could arrange for the City Savings (*Mestna hranilnica*) to write off a large part of our debt which was of the order of one million Dinars — I do not recall the exact amount.

A large part of the debt was refinanced with a loan of the *Czech Sokol Community* (*Česka Obec Sokolska*) on which we were making regular payments. That loan was obtained through the good offices of brother Vincenc Stepanek. The larger

[50] Note by editor: When the present day Republic of Slovenia came into being and people began to claim restitution of property confiscated by the communists, it turned out that Stane Flegar, too, was a surviving original owner of the Sokol Hall at Tabor, who, jointly with heirs of other founding stockholders demanded that the authorities return the *Sokol Hall* to its builders and owners. Naturally, the Slovenian government refused to return the property which it is using for commercial purposes.

[51] Note by editor: This was a convenient excuse for the predatory government of the Republic of Slovenia which afterwards seized the property claiming that it belonged to an organization of the Italian government.

firms that participated in the building also wrote off part of their charges. In brief: the management of the *Sokol Hall* became solvent. When I was still member of the City Council I arranged for the municipality to grant *Sokol I* the land adjacent to the building and the land outside the fence of the track field so that *Sokol I* owned all the land up to the surrounding streets. In the entrance hall, a memorial plaque honored Ivan Tavčar who as mayor was instrumental to the initial acquisition of the building lot. On the track field, viewed from the hall a linden tree grows on the right which in assuming ownership we designated as the memorial linden tree.[52]

Sokolski dom na Taboru Pogled z letnega telovadišča

Sokol Hall at Tabor, view from the athletic field

Youth member Stane Pirc actively participated in the dedication. In mentioning "larger firms" it should be mentioned that the excavations were made by construction firm Accetto. As already stated the principal builder was Tavčar-Svetina who was later replaced by Mudrovčič and Hrovat. The copper roof was installed by five cooperating sheet metal contractors among which were Pičman and Žitnik. Copper roof was a novelty for the sheet metal contractors because at that time such installations were rare, I can say that they were made only on historically important buildings. At that time I got used to move around on the roof freely and without any hand holds. Installation of gutters and draining tubes surprised us. The contractors initially did not use copper for these so that after the first rain all joints with the copper roof were corroded by electrolysis. We thus got our education to the effect that copper and other sheet metal material cannot be joined permanently. Plumbing and ventilation was installed by the firm *Obnova*, a German firm which had no competitors. Tavčar, partner in the firm Tavčar-Svetina was a refugee from the Slovenian littoral that had been so generously awarded to Italy for its contribution to the *Entente* in World War I. Later, during World War II, I met him again in Trieste when our group dined in his restaurant.

[52] Linden is the Slav national tree as oak is the German tree.

Once we were caught unawares with a demand to pay the Catholic church tax. I appealed the assessment on grounds that the Association had members of diverse religious persuasion and that we were not informed of the proposed assessment. The regional administration *(banska uprava)* denied the appeal stating that the

Leadership of Sokol I association at Tabor in 1937. First row: first from the left Ivan Vrečar, third Ivanka Vrhovec, fifth Dr. Pipenbaher, in the middle author as presiding elder, fourth from the right Anuška Jug, fifth Etbin Bežek; second row: fifth from the left coach Tone Gorjanc; third row: first from the left Franko Krajger, fourth from the right Boris Sket

rector at Saint Peter's church made the announcement from the pulpit and that we did not protest at that time. I appealed further to the Council of the State (*Državni svet*) but, unfortunately, the Council upheld the regional administration.

In the fall of 1988 my son Vlado was in Ljubljana and he wrote the following report on his fleeting visit to Tabor in the *Sokol Herald* (*Sokolski Vestnik*).

"Although I had visited Slovenia already in 1986 I did not have time to take a look at the inside of the Sokol Hall at Tabor. This time, however, I took the time. Tabor is still an imposing structure with its copper roof which is the largest in Ljubljana. Nowadays it seems somewhat cramped between the neighboring buildings among which is the new retirement home built opposite to the former home for the elderly. Cars are parked along the entire southern part of the fence. The door below on the left through which my father and I used to come in when he had something to do in the office

were unlocked and I went in. I did not cross that threshold for 47 years, that is, since the Italians occupied the building. Nor did I ever go in there during the communist occupation. Once the Sokol Hall seemed to me very bright and spacious, but now it seemed somewhat dark and no longer so clean as I remembered it. In front of the main hall are posted larger than life photographs of Janez Porenta, Lojze Lubej, Dušan Podgornik and two other communists who, I suppose, were expelled from the Sokol I association and later, unrelated to that expulsion, shot by the Italians as hostages. The entrance to the main hall was locked. Just then a man appeared and I asked him if he would show me the halls. He gladly agreed to it and unlocked the main hall. When, in compliance with the posted instructions I started to remove my shoes, he kindly told me that it was not necessary. The main hall is now used for basketball games and the man invited me to a match that would take place in the afternoon. The stage that used to be in the hall has been removed. The ten feet high wooden paneling that used to have high gloss black finish is showing signs of wear; it had been painted over with white paint which is showing through a worn brown finish making for a rather shabby appearance. The swinging rings in the side hall next to the exercising grounds and, if I remember correctly, in the main hall are still there but there as are a few other pieces of athletic equipment. The polished hardwood floor still seemed in good condition, a few anchors for the equipment were still in place. The man then took me to the upper halls. The marionette theatre is no longer there but my guide knew that it once was there. The smaller hall in which the Slovenian National Committee proclaimed United Slovenia in 1945 and the upper hall opposite to the hall where the marionette theater used to be are now converted into a fitness center equipped similarly as such fitness centers in the United States. Even the name "fitness center" is in English; apparently it does not bother those who once used to object to the usage of the word "Savez"[53] in the title of the Sokol organization.

South elevation of the Sokol Hall at Tabor

On the staircase there are no longer pictures with irredentist motives depicting Nanos and the Slovenian Littoral. Those were no doubt removed already by the Italians but the partisans, who lost Trieste for Slovenia probably do not want to be reminded of that either. The murals have been replaced with posters showing a young man and

[53] *Savez* — Serbo-Croatian word for "federation" appearing in the name of the Sokol organization: "*Savez Sokola Kraljevine Jugoslavije*" (*Sokol Federation of Yugoslavia*). The communists were fond to gripe against use of the Serbo-Croatian word "*Savez*" and insisted that the Slovenian word "Zveza" be used. All the other words in the title are identical to the Slovenian words.

a young woman in swim suits with well developed muscles and, in case of the woman, well rounded body parts. I thanked my kind host who showed me around and left the old Sokol Hall quite happy that I would soon return to my free and sunny California."

Harbingers of Communism

The echoes of the upheaval caused by the communist revolution in Russia reverberated throughout the world long after the end of World War I. Many prisoners of war who returned home from Russia, where they had had a direct exposure to communist ideology, became converts and advocates of communism. The war had lasted too long not to have left an imprint on those who experienced its horrors either at home or at the front, and the conditions for the spread of communist propaganda in its aftermath were quite favorable. At that time the communist organization in Slovenia and its leadership was controlled by a carefully selected group; the parlor pinks had not yet made it to the forefront. Yet as time marched on, changes were taking place in the local communist party. Soon after the end of World War I new developments caused the repute of old-time communists like the Žorgas to fall into oblivion. Even before that, the defiant internationalist Milan Lemež, another prominent local communist, succumbed to the clarion call to arms and he could be seen taking part in the war effort as an employee of the Austrian Imperial Military District in Ljubljana.

According to an apocryphal story, four young men from Poland came to study at the newly founded university in Ljubljana immediately after the Armistice. The youths took up quarters with a landlady, Mrs. Kraigher, who rented rooms to students and had several sons of her own. After the "Poles" accomplished their mission, the police discovered that they were Bolshevik propagandists from Russia and expelled them. Left behind was a communist cell that eventually evolved through the phase of parlor communism becoming the nucleus of the communist party leaders in Slovenia which has been operating between the two wars and during World War II to this day. During the communist revolution, carried on under the guise of a national liberation movement, sons of that very same students' landlady rose to the highest positions in the communist party leadership and, following the establishment of the communist dictatorship, ascended to the top positions in the communist government.

Milan Lemež and Dragutin Gustinčič, the two lone communist members of the City Council of Ljubljana, were at first fawningly cooperative toward the national government. The idle Gustinčič soon disappeared to Russia, from where he returned at the end of World War II, welcomed at first with all the honors and attention reserved for a Soviet hero, but later, in 1948, when Tito quarreled with Stalin and was expelled from the Cominform, arrested and thrown into prison from which he was released only after he contritely signed a confession admitting the error of his ways and thanking the understanding members of the political police who assisted him in the process.[54] In brief, after the expulsion of the "Poles" the leadership of the Slovenian communist party was gradually being transferred to the cell or a system of cells the Soviet emissaries had set up. In those cells there was scant tolerance for people without higher education: the dominant majority consisted of university graduates and those who aspired to higher education, although some of them never managed to attain it.

The old guard communists were losing their influence. In 1926 they could still muster a bombing threat directed against the *Committee for the Memorial of King Peter I* but that was their swan song. The new tactics adopted by the parlor pinks was infiltration into the public and cultural activities where they expected to operate under the cover of respectability. The main operating principle of the communists was: A lie repeated a hundred times will eventually be accepted as truth. I already narrated how one of those communists, Marjan Tepina by name, wormed himself into a paid position on the staff of the Committee for the Memorial of King Alexander.

I was at the forefront of the defense against the communist infiltration of the Sokol organization between the years 1937 and 1941. In the following I describe how we defended the *Sokol* organization against that alien ideology and outline the profiles of the defenders as well as of those who wanted to subvert our organization.

Comunists Recruit Two Sokols

The *Sokol* movement spread throughout the Slavic world where it was very popular and had thousands of members. In World War I *Sokols* lead the movement for the liberation of Slovenians, Czechs Slovaks and Croats from the oppression of Austro-Hungarian empire on the ruins of which they formed their own nation states.

[54] Note by editor: According to Gustinčič's first wife Anica Lokar, Gustinčič had written a treatise on the *Development of the Slovenian National Question* which he dutifully gave to Edward Kardelj for review and approval when they were both in Moscow. Soon a book with that title was published in Ljubljana under the name of *Sperans* which was an alias Kardelj used. Gustinčič's former wife claimed that Kardelj had appropriated Gustinčič's entire manuscript, that Gustinčič had objected to Kardelj's appropriating his manuscript when they were both back in Yugoslavia after the end of World War II, and that this dispute rather than Gustinčič's support of the Cominform in 1948 led to his arrest and imprisonment. In the opinion of the author, however, Gustinčič did not have the intellectual ability to produce a treatise such as Kardelj's book.

At the end of 1929 legislation on the *Yugoslav Sokol Association* was promulgated by the *Yugoslav Royal Government.* The law dissolved all other gymnast organizations and entrusted all physical education to the *Sokol* organization. The titular Presiding Elder (*starosta*[55]) of the *Sokol* organization was the crown prince who later became king, the first Deputy Presiding Elder *(podstarosta)* who was the acting head of the organization was appointed by the government on recommendation of the General Assembly of the *Sokol Federation* (*Savezna Skupština*). The *Sokol* organization retained its ideology as developed by Tyrš and its system of physical education. The local *Sokol* associations were autonomous, their governing bodies were elected at the general meetings and subject to the approval of the *District's* (*Župa*[56]) governing body. The *Districts* enjoyed a similar autonomy; the *District* governing boards were elected at the *District's* general assembly but were subject to the formal approval of the *Sokol Federation.* The General Assembly of the Federation, consisting of the delegates from the regions, elected the Federation's governing board and the presiding elders (*starešinstvo*). Deputy presiding elder who was in actual charge of the *Sokol Federation* was Engelbert Gangl.[57] The government as a rule did not interfere with the administrative, ideological or physical education matters. The only exception to this practice occurred in 1939 when Prince Regent Paul demanded the removal of the first Deputy Presiding Elder Engelbert Gangl who had, in a speech reported by the press, stressed the importance of alliance and solidarity among all Slavs in defense against German aggression which caused a protest on the part of the German ambassador in Belgrade. Gangl was repeatedly and unanimously re-elected Deputy Presiding Elder by the General Assembly of the *Sokol Federation.* Although the balloting was accompanied by intensive demonstrations in favor of Gangl, the Prince nevertheless insisted that he must go.

Engelbert Gangl

Engelbert Gangl ran the *Sokol Federation's* affairs from Ljubljana where he also was editing the *Sokol* associations' publications. The governing board of the *Sokol*

[55] *Starosta,* is the old Slav word for the chief of a Slavic tribe in ancient history.

[56] *Župa* is the old Slavic word for a group of communities, called *Zadruga*, in ancient Slav history, it usually possessed a fort which the associated communities used for defense against raiding Huns or Tartars. The *Sokol* organization adopted the term for a district to which a number of local associations belonged.

[57] Engelbert Gangl (1873-1950) Slovenian poet and writer, Deputy Presiding Elder in charge of the *Sokol Federation of the Kingdom of Yugoslavia.*

Federation arranged for the government to provide a paid personal secretary who would assist Gangl. A candidate for this post was soon found in the person of Franjo Lubej, a teacher in Ljutomer. Lubej's wife Milka also was a teacher, therefore a problem how to transfer both of them to Ljubljana because of differences in their seniorities had to be solved. We found a solution whereby Lubej gave up the teaching and served as full time secretary to Gangl paid according to the teachers' pay scale with a generous year end bonus while his wife was transferred to a teaching post in Ljubljana. It was very difficult to carry out such moves within the tightly controlled civil service structure but the feat was eventually accomplished and Lubej came and became a member of the *Sokol I Association* at Tabor. He brought with him the reputation of a successful gymnastics trainer and coach. Following his election as the principal coach of *Sokol I* at Tabor he was soon elected the principal coach of the region. Although political activities outside the *Sokol* organization were not relevant to his appointment as secretary to the vice-president, Lubej was fond of boasting that he was secretary of the *JNS* (*Yugoslav Nationalist Party)* in Ljutomer and that he, as president of the election commission in the birthplace of Anton Korošec during the Jevtić regime, had been in no small measure responsible for Korošec's defeat at the polls. Korošec, of course, obtained enough votes elsewhere so that his seat in the Parliament was never endangered.

At that time district judge Jože Rus[58] appeared on the scene in Ljubljana and became a member of the *Sokol III Association* at Bežigrad. He brought with him an ominous reputation of a querulant. At Bled, in his birthplace, all his male relatives had been expelled from the *Sokol Association* because of their disruptive conduct. In Višnja Gora he caused such quarrels that the *Sokol Association* could not recover from them for years. In Gornja Radgona he likewise created similar troubles. In Ljubljana Rus became closely attached to Lubej who got him on board as a regional deputy coach. He also became a close associate of France Veber, a professor of philosophy at the university, who became his mentor in matters of philosophy. At the meetings and conferences of the *Sokol Association* Rus attempted to blend his perception of Veber's private lessons on philosophy into his meandering and interminable speeches. People conversant with philosophy characterized Rus' philosophical recitals as confused while lawyers considered his legal expertise to be below par. Veber began to associate with Lubej and Rus with no special purpose in mind but the duo Lubej-Rus became inextricably involved with the destiny

[58] Jože Rus, district judge in Ljubljana; before the war tried to infiltrate the *Sokol* organization with the communists and later, during World War II, at the first session of the Communist "Parliament" in Jajce proposed that Tito be voted the title of *Marshal of Yugoslavia*, he became the President of the Yugoslav rubber stamp Assembly after the war and eventually declared a "hero." Milovan Djilas, in his book Wartime, Harcourt, Brace, Jovanovich, New York, 1977, p. 235, characterized Rus as a man of rather limited political potential but useful for having brought with him into the communist fold several *Sokols* who had turned communists.

of the *Sokol* organization, an involvement that had far reaching and devastating consequences.

The Stojadinović regime abolished the position of secretary to the deputy presiding elder of the *Sokol* organization and Lubej had to go back to teaching, this time at the elementary school at Graben in Ljubljana. In 1937, *Sokol I* at Tabor celebrated its thirtieth anniversary. At that time the current local government, which looked with disfavor on the *Sokol* organization, decided to have a representative of the police present at the organization's meetings and conferences. *Sokol's* legal experts maintained that such intrusion was illegal because *Sokol* was established by separate legislation and as such was not subject to the provisions of the general law on organizations. Resisting the government, however, proved to be futile. Accordingly, the anniversary general meeting was attended by a police commissioner who at the very beginning created a scene by insisting on sitting next to me rather than at the place designated for guests. At the conclusion of the meeting, the chair recognized Lubej who immediately began attacking the regime in his remarks which caused the police commissioner to intervene and order the general meeting terminated. I saved the unpleasant situation by observing that the agenda has been exhausted and that I was concluding the general meeting. The police immediately initiated proceedings against Lubej. To defend Lubej, I went to see the Chief of Police. It was a matter of establishing whether Lubej's statements were incriminating or not. There were two versions of what had been said: mine, based on my notes, and that reported by the commissioner. I managed to avert reprisals that could damage Lubej's career as a teacher.[59]

A number of *Sokol* members were Russian émigrés living in Yugoslavia. We welcomed them as Slavic brothers. King Alexander, whose fiancé Princess Anastasia was murdered by the Bolsheviks, granted all Russian émigrés Yugoslav citizenship which, at that time, could not be easily obtained. In this way the Russians were able to continue working in their professions. We became good friends with some of them. As time went by I was approached on a few occasions by *Sokol* members Golubev and Tkachov who were concerned that Lubej was talking to the membership as a communist. In a routine survey of the *Sokol*'s hall structure I discovered some graffiti, small and not very noticeable: hammer and sickle which a communist sympathizer must have made in the remote recesses of the hall. We did not quite know what to do about any of that at the time but soon we would see trouble looming on the horizon.

[59] Lubej wrote [*Odločitve* (*Decisions*), Partizanska knjiga, Ljubljana, 1979, p. 98] that he first learned that the author protected him from the police when he read his booklet *On the Ramparts in Defense of the Sokol Idea* (*Na braniku za sokolske ideale*), Slovenska Pravda, Enfield, England, 1965. If the author had in fact defended him, wrote Lubej, that was proof that he had influence with the police as a prominent member of the *Yugoslav Nationalist Party.* Lubej admits that he never suffered any consequences because of his excesses at the meeting except some pecuniary fine.

The 1938 Sokol Rally in Prague

In keeping with the *Sokol* tradition the 1938 quadrennial *All-Sokol Rally* was to take place in Prague, Czechoslovakia. Those rallies were attended by the *Sokol* organizations from all over the world. The *Sokols* from the Ljubljana Region chartered a special train for their trip to Prague. The train was routed through Hungary rather than directly through Austria, already annexed to Germany. In Hungary no one was allowed to leave the train at any stop because all passengers were listed on a common travel document visaed by the Hungarian authorities. As the regional principal coach, Lubej was designated leader of the group. He had the responsibility to ensure that everything ran smoothly. Although at variance with the usual practice on such occasions, Lubej and Rus decreed that all members on the chartered train had to wear the *Sokol* uniform. At the conference of the District's governing board the request of deputy presiding elder Gangl's wife for permission to travel in civilian clothes came up. Lubej threatened to resign as the group's leader if the request were granted and the governing board regretfully denied the request so that Gangl's wife, a member of the *Sokol* organization, had to travel separately by a regularly scheduled train.

District's Presiding Elder Josip Pipenbaher did not go to Prague. It thus became my responsibility to represent the *Sokol Association of the Ljubljana District* at the

Members of Sokol I in Prague in 1938; author is fourth from the left in the front row

rally in the capacity of his deputy. At the same time the catholic organization *Slovenski Fantje*, formerly called *Orel (Eagle)*, decided to hold its rally in Ljubljana. The law on the *Sokol* organization as the sole gymnast society in Yugoslavia was still in effect but the current regime nevertheless indicated its support for re-establishment of the *Orel* organization with the attendance of Prince Regent Paul and Princess Olga at the *Orel* rally. As was to be expected, clashes sporadically broke out between members of the two groups. On the day the *Sokols* were scheduled to depart for Prague, a police commissioner appeared at my home, told me that I was under arrest, and took me to Hotel Union where the state secretary for physical education Miletić was staying. The hotel, owned by clerical interests, served as some kind of headquarters for the *Orel* rally and my involuntary presence there attracted the attention of bystanders. Visiting state secretary Miletić wanted to hold me responsible for any incidents that might occur between the *Sokols* and members of the *Orel* organization. I refused to accept any such responsibility. According to the oath of office he had taken, Miletić was required to uphold the law on the *Sokol Federation of the Kingdom of Yugoslavia* which was in his area of responsibility but he turned out to be a rather spineless character. Some rumors would have it that Miletić became state secretary for physical education by mistake. Several persons by that name had been in the government circles and when the wrong Miletić showed up at the swearing in ceremony it was too late, or perhaps too much bother, to start all over and so he was allowed to remain in the cabinet as another yes man.

Apart from a few sporadic provocations, the expected clashes did not materialize. The *Sokols*, as already mentioned, were to depart for Prague in the afternoon of the day of my encounter with Miletić. We assembled on the track field in front of the *Sokol Hall at Tabor* and marched in orderly formations to the railway station where we were pleasantly surprised to see a vast multitude of cheering people who had come to see us off. The people entered the station and surrounded the train, those who could not come in remained outside in a densely packed crowd. That spontaneous demonstration was the best response to the state secretary who did not know what his responsibilities were.

Prague was tense as the foreboding of the approaching Munich tragedy seemed to be in the air. Included in the immense parade of *Sokols* from all over the world were also the *Czech Sokols* from Vienna, Austria, which was by that time already incorporated in the German Reich. The Germans allowed the Czech from Vienna to attend the rally on condition that they carry the German flag with swastika along with their *Sokol* banner. The ominous scene signaled the onset of the Czech national catastrophe.

In Prague Rus and Lubej behaved as if the "new order" was already in place. When we were lining up for the parade I was standing with the District's elders in the front row of our group. Lubej came and rudely ordered me to the last row. That was too much even for Trček, principal coach of *Sokol II at Moste,* which as we shall see later was all too willing to go along with Rus and Lubej. Trček came to me and took me back to the front row to the traditional place of the District's representative.

Instead of spending their time among the membership Rus and Lubej spent a great deal of time elsewhere. On our train to Prague the *Sokols* collected two thousand Czechoslovak Krone intended for the Czechoslovak Defense Fund the motto of which was "*Ne dame se!*" (*We will resist*). As deputy presiding elder of the *Ljubljana Sokol District* I took the money to the secretary of the *Czechoslovak Sokol* Köppl. Rus, although not invited, thought he had to come along. During my brief presentation to Köppl Rus attempted to start one of his logorrheas but Köppl discouraged him with a disapproving look and let me finish my address. I still recall Köppl's expression and I conjecture that Czechoslovak *Sokols* could not have been indifferent to the communist infiltration of their ranks. It is possible that Köppl learned that Rus and Lubej were seeking out the local communists.

On return to Yugoslavia our special train was detained by the police commissioner at the border crossing station in Kotoriba. The police were looking for a certain Jan Prohaska who was on the list of the passengers but whom Lubej and his assistants could not produce although as organizers of the trip they were expected to account for all the persons on the train. As I was the representative of the *Ljubljana Sokol Region* the ultimate responsibility for the accounting was mine. To break the impasse I suggested that there might have been a slipup in the organization of the trip. When the train, after a prolonged and unsuccessful search for the missing person, was finally allowed to continue on its way to Ljubljana, Lubej sent one of his cohorts to let me in on the mystery of the missing passenger. At the departure of the special train from Ljubljana, a communist whom the police intended to expel to a neighboring country was with Lubej's help allowed to board the train under the name of Jan Prohaska who signed up for the trip but did not go. It was that communist who was traveling on the special train in the *Sokol* uniform with papers borrowed from Prohaska whose name was entered on the list of passengers but who did not actually go to Prague. Lubej and Rus safely delivered that communist to his friends in Prague and, as they later hinted, used the occasion to establish a contact with the international communist organization in addition to their already established liaison with the communists at home in Yugoslavia. Through means of its own the police discovered that one of the prospective travelers did not leave for Prague and substantiated its suspicions by verifying that a person was missing from the train on its return. Lubej and Rus had thus helped one of the leading communists to travel to Prague camouflaged as a *Sokol.*

Many years later, when I was already in the United States, I got a letter from my friend Etbin Bežek who was allowed to visit his daughter in Forli, Italy, and there read part of the Slovenian edition of these memoirs. In his letter from Italy Bežek described how he had by accident met the person whom Lubej had smuggled to Prague and thence to the Soviet Union. Bežek wrote that in November 1949, when he was held in a packed prison cell, he and his cellmates used to pass the time by telling stories from their lives that others might find interesting. When his turn came he told the story of the smuggled passenger. When he finished, a man who had just recently been pushed in the cell came up to him and asked him if he had ever met that Prohaska. Bežek told him that on the last day in Prague on coming to the dormitory to pick up his luggage he found in the room Lubej, Rus, and the man masquerading as Jan Prohaska. Lubej introduced Prohaska's impersonator to Bežek and told him that he was taking his farewells as he was on his way to Russia. To this the cellmate replied: "Well, I am that Jan Prohaska." Bežek was, of course, very surprised to have met him again and under such circumstances as the man must have either fallen out with the communists or had taken the side of Cominform which at that time had an ideological dispute with Tito's communists. In prison Bežek got to know the man somewhat better, he learned that his real name was Franc Kolar from Idria and that he was married to the daughter of Lascko, a farmer and a partisan hero from Ptuj. Kolar even invited Bežek to come and see him once they both got out of the prison but the opportunity for that never presented itself.

In that letter Bežek also wrote about his encounter in prison with another *Sokol*, Poharc, and secretary of Deputy Presiding Elder of the *Sokol Federation* Gangl. It was in July or August of 1945. A certain Drago Šumak, former *Sokol* from Ljutomer, interrogated Bežek about the disciplinary proceedings at Tabor. Bežek refused to sign the *process verbal* of the interrogation because it contained an untrue statement and Šumak had him thrown into an underground solitary cell. Being without food and water he knocked on the door at night and asked the guard for water. The guard opened the door and at the same time, in the doorway of the cell No. 1 opposite to his, appeared Poharc. He looked drawn and did not react when Bežek waved to him. The encounter lasted less than a minute. Bežek did not get water. The same night, the communists said, Poharc had hanged himself.[60]

Bežek wrote further that he had no regrets about his two dozen years with the *Sokol;* that had meant a lot to him. When he frequently encountered Rus and Lubej on the street he would hold his head high thus letting them know that he felt better than they. Naturally, he had no personal contacts with them.

---

[60] Statements of this nature, when made by the authorities, are always suspect as to the real cause of the prisoner's death. It is entirely possible that Poharc had succumbed to torture.

Communists Attack

The inseparable duo Lubej-Rus had thus entered into a conspiracy for bringing the *Sokol* organization under the control of the communists. To this end they laid the groundwork in the regional technical committee and in among the coaches. Instances in which the two tried to meddle in the affairs of the governing boards of the associations and the region became increasingly more frequent. A typical example was an attempt of the coach of the youth at *Sokol I* to cover up the loss of the youth banner which the standard bearer left behind on the train when he was changing cars at the border checkpoint in Kotoriba. The association found this out only when the banner turned up in Zagreb and was by mistake sent to Maribor. The governing board reprimanded the standard-bearer and suspended him from his duties for the period of one year. The principal coach, however, hysterically objected to the board's action claiming that it was the coaches' responsibility to mete out any penalties that might be in order. Nevertheless, the governing board persevered and took appropriate action.

Events that unfolded in the *Ljubljana Sokol District* and in *Sokol I at Tabor* may be better understood in the light of the following background. Riko Fuks, a Member of Parliament, came to warn me that Governor Marko Natlačen was determined to dissolve the *Sokol* organization of the entire district or any of its associations, if only an opportunity under which he could invoke the law on organizations should present itself. The local authorities had, as already mentioned in connection with the closing of the general meeting of *Sokol I at Tabor*, maintained that the *Sokol* organization, too, was subject to the law on organizations regardless of the law on the *Sokol Federation of the Kingdom of Yugoslavia* which exempted the *Sokol* organization from the provisions of the law on organizations. I communicated the deputy's warning to the appropriate persons in the Association and in the District.

The calm following the *Sokol* rally in Prague did not last long. Soon after the Prague rally a national committee for the celebration of the 20th anniversary of liberation was formed in Maribor. The *Maribor Sokol District* assumed the responsibility of organizing participation of *Sokols* from all parts of Slovenia. For this purpose representatives from Maribor came to a meeting of the *Ljubljana Sokol District* and explained that a nonpartisan national meeting was being planned. The speakers would include Anton Korošec, leader of the clerical party, a representative of the *Sokol* organization, and so on. The program would include a parade and a mass in the open. The *Maribor Sokol District* had already approved the program in principle and promised that the Sokols would attend. Accordingly, it was inviting the *Ljubljana Sokol District* and trying to enlist its cooperation in securing as large participation as possible. This was an opportunity for Lubej and Rus to attack the regime. Their counter proposal was that the *Ljubljana Sokol District* agree to attend only if the following conditions were met: Korošec should not be allowed to speak, the keynote speaker should be a member of *Sokol*, the *Sokol* organization will march at the front of the parade, there was to be no mass or, in the alternative, members of *Sokol* would not attend it. The counter proposal was not acceptable to Maribor but Lubej in Rus had brought to the meeting several members whom they had indoctrinated in advance and who, consequently, supported them. In the situation that was thus created it was impossible to make a rational decision. Presiding elder Pipenbaher, accordingly, adjourned the meeting to a later date.

When the meeting reconvened, Lubej and Rus brought along deputies of the members of the regional governing board and the entire trainer's committee to secure majority in support of their position. President Pipenbaher excused himself and left me with the unpleasant responsibility to conduct the meeting.

Lubej and Rus defended and recommended the conditions for participation already mentioned above. These conditions represented an affront directed against Korošec, who was at the time State Secretary of Interior, and whose responsibilities included the direction of the police and gendarmerie.There were other aspects, too, which would make an unpleasant impression if the *Sokol* should make them public. I analyzed these points at the meeting and repeated the warning of Riko Fuks concerning the danger that the *Sokol* organization in Ljubljana could be dissolved. A repressive act of this kind would, of course, suit Lubej and Rus quite well as it would allow them to continue their agitation. This was a well known communist mode of operation discussed at some length in the works of Lenin. The communists would be happy with the activities of *Sokol* being curtailed by the authorities as their mode of operation is to foment strife wherever they can and then offer themselves to lead people out of the crisis situation without, of course, the slightest intention of doing so. All my efforts to forestall such course of action were of no avail. The majority, achieved by packing in the deputies, voted to accept the conditions proposed by Lubej and Rus and to carry them out. As chairman of the meeting I observed that the motion of Lubej and Rus was adopted. In view of this, I tendered my resignation as District presiding elder, asked my deputy to take over, left the meeting room and went to my office. My deputy, however, also resigned and so did all the rest down the line until they were all gathered in my office. The resignations of the regular leading members of the governing board were quite spontaneous, without any prior arrangements. The meeting thus could not continue. The following day president Pipenbaher, too, resigned, and so there was no one who could implement the decision adopted at the meeting. The governing board of the *Federation*, however, declined to accept our resignations and instructed the District governing board to remain and continue. The planned celebration festivities of the twentieth anniversary of national liberation in Maribor was eventually cancelled for other reasons.

New Ringleaders of Subversion

On examining the books of the District's accounts, presiding elder Pipenbaher discovered certain irregularities, namely charges made by Lubej for meals and sustenance on days when he worked in the District's office. Those irregularities as well as Lubej's arrogant and rude conduct were cited in a disciplinary complaint of Pipenbaher against Lubej which he addressed to the governing board of *Sokol I at Tabor* where both Pipenbaher and Lubej were members. The governing board did not promptly institute a disciplinary proceedings because the association's bylaws provided that the entire governing board should act as the court of honor which required a considerable amount of preparation and time on the part of the board members to take the matter up. The affair caused considerable disturbance and to compound it Lubej continued his rude and unbecoming attacks on the District's presiding elder. When the latter resigned, I as his first deputy and at the same time presiding elder of *Sokol I at Tabor*, automatically became the target of Lubej's attacks. Eventually the governing board decided to institute disciplinary proceedings against Lubej. The immediate result of this decision was Lubej's suspension from his position as District coach and other offices he held in the *Sokol* organization. Lubej defied the governing board. For the forthcoming general assembly and elections of the regional *Sokol* organization at which Lubej, being under suspension, should not be allowed to run for office, he and Rus canvassed the entire membership and prepared the terrain for elections at which their slate of candidates expected to prevail. The two won over all the trainers and the technical committee of the District. So widespread was their agitation that even the *Federation* sent one of its non-Slovenian deputies of the *Federation's* presiding elder to act as observer at our general meeting. Because the Association's governing board was being confronted with the difficult task of repelling the attempted takeover by the communists, I reluctantly agreed to run for the office of the District's presiding elder.

Before describing the outcome of the general assembly, let me digress by introducing two other personalities who played out their roles as fellow-travelers and communists. Lubej and Rus formed their own list of candidates for the governing board under which they would, of course, be able to hang on to their offices of principal coach and his first deputy. Their candidate for the regional president was Stane Vidmar. As a young man, Vidmar had impressive achievements in the *Sokol* organization to his credit. During World War I Vidmar was with Ljudevit

Pivko in Italy fighting against Austria as a Yugoslav legionnaire.[61] At the apex of his fame he was a member of *Sokol I at Tabor.* Following an election at which Vidmar headed the opposition slate that lost out, he transferred his membership to *Ljubljana Sokol* but later on came back again to *Sokol I at Tabor.* Several years prior to the general assembly, however, he deliberately allowed his membership to lapse by failing to pay his dues. The dues in the *Sokol* organization always were very nominal and certainly well within the means of Stane Vidmar, owner of a factory. The *Association's* leadership tried to persuade him to remain a member but being unsuccessful in its efforts it had to drop Vidmar's name from its rolls. When Lubej and Rus ran him as a candidate for the office of District presiding elder, Vidmar had not been member of the *Sokol* organization for several years.

The other personality who played a key role at the general assembly was Zoran Polič. Polič was son of well to do parents in St. Lenart in Slovenske Gorice. He studied law at the University in Ljubljana where he was graduated in 1938. At the university he was for a short time president of a student chapter of the Ljotić organization *Edinost.* The Ljotić organization was considered by some to have totalitarian leanings. The principal trainer represented to the governing board that there was a need for Polič to come back to Ljubljana after his graduation. The governing board went along and found a job for Polič as editor of the *Sokol* column at the Ljubljana daily paper "*Jutro.*" The job was a modest one but it was enough to tide Polič over until he could secure a job as lawyer in the Department of Finance. His family apparently did not support him while he was studying at the university since he depended on a scholarship consisting of a free room in the university student dormitory and free meals in one of the better restaurants which were arranged for by Etbin Bežek, who was kindly disposed toward Polič. Bežek was repaid for his kindness by imprisonment when Polič, as Secretary of the Interior of the Slovenian People's Republic, was in charge of the political police. Not only that: after Bežek had served the term to which he was sentenced, he was soon arrested again and sent to the concentration camp in Strnišče from whence he was released only after President Truman told visiting Edward Kardelj that Yugoslavia should disband concentration camps if it wanted to receive aid from the United States.[62]

[61] Ljudevit Pivko wrote in *Proti Avstriji (Against Austria), Rame ob ramenu,* Prva knjiga, Klub dobrovoljcev v Mariboru, Maribor, 1925, p. 98), that Stane Vidmar deserted from the Austrian army almost the first day when he came to the front. The Italian office of military intelligence reported his desertion to Pivko who immediately requested that Vidmar be directed to his detachment of Legionnaires. In the Legion Vidmar organized gymnastics according to the Sokol model and was very effective in training the men.

[62] So far as it is possible to determine Truman was the only American President who demanded that Tito show at least a modicum of respect for human rights. President Eisenhower's fawning letter to Tito designed to induce the dictator to acquiesce to the return of Trieste to Italy is a matter of record, President Nixon felt it necessary to personally travel to Belgrade to honor the communist dictator. President Carter sent Senator Joseph Biden, Chairman of the Senate Committee on Foreign Relations to pay respects at Tito's funeral; surprisingly he did not go himself.

When Polič returned to Ljubljana, he became the principal coach at *Sokol I at Tabor.* Although no one expected Polič to be subservient to any particular person or group on account of the benefits he enjoyed while he was a student, it was nevertheless surprising to see that he allowed the communist conspiracy to involve him in a confrontation with the governing board. Perhaps the communists viewed Polič as a somewhat less reliable communist sympathizer than we at the time thought he had been. As is narrated in greater detail below, Milan Apih, an old time communist, wrote in the Ljubljana weekly journal *Mladina* about Polič on April 1, 1988: *"We, communists, who were operating clandestinely and in the strictest possible secrecy, were carefully avoiding Polič. Polič was considered to be among those whose "patriotism" made them favorably disposed toward the regime and its police which would welcome any hint concerning our clandestine activities."* In other words, the communists considered Polič to be quite capable of being an informer.

District General Assembly

Two slates of candidates appeared for the forthcoming elections to be held at the Regional General Assembly: one headed by me that had no candidate for the principal coach, the other headed by Stane Vidmar, which included Lubej running for principal coach. The General Assembly took place on a Sunday morning; time was short because delegates from out of town had to catch their various local trains that were leaving around noontime. This circumstance suited the opposition very well and Jože Rus made full use of it by mounting a filibustering speech in which he obstructed the expeditious conduct of the general assembly by an interminable, indigestible, and vacuous logorrhea interspersed with lies and distortions. Zoran Polič, a member of the delegation of *Sokol I,* defected by sponsoring Vidmar's list, rationalizing his treachery by hypocritically claiming that I was needed as presiding elder of *Sokol I.* Rus's obstructionist tactics succeeded in delaying the meeting until the out-of-town delegates were leaving, at which time the vote could finally be taken and Vidmar's list won. A member then posed a question from the floor as to whether Vidmar was a member of the Sokol organization; on consulting the roster, I responded that he was not. The following day the District administration notified the *Sokol Federation* of the outcome of the elections, but glossed over the fact that Vidmar was not a member of the *Sokol* organization. However, the *Federation,* having been fully informed by the representative of the federal president who attended the assembly as an observer, voided the elections on grounds that Vidmar had allowed his membership in the *Sokol* organization to lapse. At the same time the federal governing board reappointed members of the former regional governing board to the offices they had held. Because Pipenbaher resigned I had to shoulder the greater part of the responsibility.

Lubej and Rus were making political hay out of my response to the inquiry as to whether Vidmar was a member of the *Sokol* organization. The facts, however, were the following: Shortly before the regional General Assembly was to take place Rus appeared in the office of *Sokol I* and informed the clerk on duty that Vidmar desired to be reinstated as member. In accordance with the established procedure, the clerk set Vidmar's application, along with all others that had been submitted, for consideration at the next regular monthly meeting of the *Association's* governing board which was to take place a few weeks after the regional General Assembly. At its regular monthly meetings the governing board would act on applications for admission of new members who, if accepted, were to be on probation for one year. Lists of new members were forwarded to the regional administration which maintained a roster of all members. The question from the floor was posed by the District recorder who knew that Vidmar was not a member of the *Sokol* organization. Thus, on the day when the general assembly was held I did not know that Rus had applied on Vidmar's behalf for the latter's admission nor could have the Association's governing board acted on the application because the first regular session on which it could have done so was to be held several weeks after the general assembly. My response to the interpellation from the floor after the announcement of election results simply stated a fact, the District's former governing board took no action to disqualify Vidmar's slate. The federal governing board acted on its own initiative. Stane Vidmar himself was not much interested in the whole affair. On the day when the elections were held he was traveling abroad and after he returned he took no steps to ensure the approval of his slate. He, as well as Zoran Polič, had been enlisted into the communist conspiracy by the aggressive tactics of Lubej and Rus.

It should be pointed out that the communist conspirators frequently attempted to provoke us to characterize them as "communists" in which case they would immediately institute a civil suit for defamation of character.[63] Curiously, this happens to be one party the membership of which when imputed to someone without proof constitutes the offense of defamation. It was clear to everyone that we were dealing with a conspiracy dominated by communist ideology. In the course of protracted proceedings, the governing board of *Sokol I at Tabor*, acting as a disciplinary court, established that a conspiracy existed, imposed disciplinary sanction on its ringleaders, and expelled the infiltrated communists from the association without ever using the characterization "communist." That characterization had already tainted the atmosphere of the association. It was interesting listening to

[63] Note by editor: After fifty years of communist rule, alleging that someone is a communist can again be cause of action in libel suits in contemporary Slovenia.

Janez Porenta, member of the Lubej-Rus conspiracy, whom I had helped during the time of the Great Depression securing a fairly decent job, when he came to seek another favor from me. While a member of *Sokol I* he had been renting an apartment in the annex of the *Sokol Hall at Tabor* for a very nominal cost, something which was not easy to come by. When, under the influence of Lubej and Rus, he decided to drop out of the *Sokol I Association*, the administrator asked him to vacate his quarters so that they could be made available to somebody else. Porenta thus came and requested that I intercede on his behalf. I pointed out that I would be in a difficult position if I tried to help somebody who had repudiated the association's leadership at a time when it needed support and suggested that I might nevertheless try to see if he could be accommodated provided that he disassociate himself from the conspirators. Porenta apologetically declined, saying that he could no longer go back because he was already too deeply involved with the "*collective*." He was a simple man whose vocabulary would ordinarily not include such expressions; obviously he learned them in the indoctrination sessions carried out by the two communist agents. "*Collective*" is in the vernacular communists use as is pointed out in Arthur Koestler's *The God that Failed*."[64] Another such member, appearing as witness in disciplinary proceedings, launched into a speech maintaining that, in contrast with the governing board, the conspirators were seeking "*social justice*," another expression customarily found in the communist jargon.

Disciplinary Proceedings

It came as no surprise that Lubej and Rus worked on the emotions of the association's trainers to such extent that because of the instigation of Zoran Polič, the entire group defied the association's leadership by demanding Lubej's reinstatement. The ringleaders, five of them altogether, were, along with Lubej, summoned before the disciplinary court and, in accordance with the bylaws, suspended from their functions. Because the accused refused to submit to the discipline, the governing board temporarily closed the gymnasium while it sought replacements for the suspended trainers. When new coaches were appointed and began their work, the communists tried to persuade members to stay away from gymnastics. They were quite successful in their efforts because the communist infiltration had already made considerable inroads. The association's bylaws required that every member under 26 years of age exercise regularly. In accordance with the bylaws the governing board requested all members who had not been taking part

[64] A. Gide, R. Wright, I. Silone, S. Spender, A. Koestler, L. Fischer: *The God that Failed*, Bantam Books, New York, 1952.

in gymnastics to resume their regular attendance if they wanted to remain on the rolls of the association. When the time by which the absentee members should comply was up the governing board summarily expelled all who did not regularly appear at the gymnastics. Many of those who were expelled got into the situation because of the pressure brought upon them by the communists or because they were part of the "*collective*" to borrow a word from Rus's vernacular. Some, who were more interested in gymnastics than in the "*collective*" opted for transfer to other *Sokol* associations to avoid the strong arm tactics of the "*collective*." The only groups that were not hurt were boys and girls whom the "*collective*" did not seem to influence. Nevertheless, many members remained loyal and were able, under the leadership of the new coaches, to resume their regular gymnastics.

Under such circumstances the governing board opened the disciplinary hearings which lasted six months, ending in the fall of 1939. The weekly hearings would drag on from 8 o'clock in the evening until the small hours in the morning. All six defendants had their own defenders; one of them was a professional attorney and a well-known communist. At the hearings that attorney could not get used to the *Sokol* custom whereby members would address each other as "*brother*" and kept referring to his clients as "*comrades*" (*tovariš*) which could be amusing if the matter were not so serious. Stane Vidmar, too, appeared in the role of a defender. It is now known that in 1945, at the end of World War II, Vidmar, anticipating the forthcoming nationalization of all industrial plants, presented his factory in Savlje to the "people," that is, to the communist regime, a generosity in marked contrast with his earlier parsimony out of which he was refusing to pay even the very nominal dues of the *Sokol* organization preferring to let his membership lapse. After the so-called liberation Vidmar temporarily emigrated to Switzerland. Throughout the proceedings the "collective" was resorting to rowdy conduct, threats, and verbal abuse very much in the style for which the communists became notorious during the time of the Axis occupation (1941 to 1945) and even more so after the so called liberation. The defendants and their cohorts used to assemble in one of the halls adjoining the hearing room where they would sing funeral dirges and shout execrations at the governing board. On one occasion the Association's building manager, having heard some suspicious noises in an adjoining room that was kept locked, went to investigate what was going on but, after unlocking the door, could not open it because somebody was holding it shut from the other side. With the help of a friend he succeeded in pushing the door open and discovered behind it a young member of the "*collective*," Klemenčič by name, who had climbed by rope over the balcony in an adjoining hall and, picking the locks as he went, was trying to get into a room housing the main electric switchboard. The door behind which he was apprehended was the last obstacle for him to overcome before he would reach the switchboard. His intention was to turn off all the lights in the building whereupon the "*collective*," its courage worked up by the funeral dirges, would fall upon the members of the governing board and beat them up. Klemenčič was summarily expelled from the *Sokol I Association*. By a strange turn of fate electricity was Klemenčič's nemesis. Having risen to the rank of a colonel of *OZNA*, the communist political police, Klemenčič, installed in a luxury apartment confiscated from a bourgeois, was one day taking a bath in a bathroom heated by portable electric heater with exposed heating elements. Suddenly he began to feel electric shocks and he cried out for help. His wife rushed into the bathroom and when Klemenčič indicated that there was a problem with the electric heater she threw it into the bathtub thus instantly electrocuting her husband.

On another occasion the "*collective*" formed a phalanx across the street intercepting members of the board as they were leaving the *Sokol Hall* and tried to provoke an incident by heaping verbal abuse upon them. In his defense statements Zoran Polič was resorting to such lies that I called him a *morally depraved person*. On hearing me characterize his associate in this way, Lubej grabbed a pitcher to hurl it at me but cooler heads in his group managed to restrain him at the last moment.

Polič's mode of operation was revealed in 1941, when during the Italian occupation he was snooping around the house where I was attending a meeting of the *Reserve Officers Association*. Outside of the building he intercepted an acquaintance and, not knowing that he happened to be my brother-in-law, told him that I was a millionaire, that I owned a large estate in Dolenjska, and urged him to avoid me because I was a dangerous person, a calumny that could only be spread by a morally depraved person. Later Zoran Polič became a permanent member of the communist government, both in the hills with the partisans during the occupation as well as in Belgrade after the war. Apparently my characterization was not too far off from that of Milan Apih, one of his former communist comrades, made fifty years later in the weekly journal *Mladina* where he wrote that other communists considered Polič capable of being a police informer.

It should be mentioned that Jože Rus could not be called before the disciplinary court of *Sokol I at Tabor* because he was not a member of that association. Thus he was able to agitate freely among the members, exert a decisive influence on the strategy of the opposition, and subordinate all their activities to the ultimate goal which was approaching with the fatal events that would unfold during the World War II. When in 1939 Germany occupied Czechoslovakia, Rus, although not a member, appeared at the meeting of the *District governing board* where he had no business to be. When I inquired what was the purpose of his appearance he said that in view of the gravity of the impending world events the governing board should consider taking action in which he desired to be of assistance. I politely responded that the governing board did not need his help and asked him to leave. Rus immediately went to the meeting of the Association's trainers that happened to be held concurrently in another room and told them that I had the gross impertinence of kicking him out of the governing board's meeting.

The tactics of the defendants were inspired by Rus and their defenders. Instead of responding to the charges they were attacking the governing board as a whole and its individual members in particular, all their speeches consisted of a string of arguments, lies, innuendo and outright insults. Not one of the defendants ever attempted to deny the charges. Witnesses called by the defense emulated the defendants' conduct. Unruly crowds of uninvited spectators worked into frenzy by the subversives were present at every hearing and threatened the board with violence. Even the thirty member disciplinary court was not immune from subversion. Thus, for instance, one of the members of the disciplinary court, who had access to the Association's roster, gave the list of members to the rebels. When his disloyal act was discovered he was summarily suspended. Six members of the disciplinary court defected and were unmasked in the process. Twenty four loyal members of the disciplinary court remained steadfast under incredible pressure to the end of the disciplinary process and voted unanimously to impose the disciplinary action.

The disciplinary action the board took was as follows: Lubej was expelled from the entire *Sokol* organization,[65] Zoran Polič and three others were expelled from *Sokol I at Tabor*, Viktor Repič was expelled for the period of one year. During the

[65] In the matter of Franjo Lubej the disciplinary court held that Lubej:
— received from the *Sokol District* without proper justification, contributions for lodging for the time from January 1 through September 30, 1937 at 300 Dinars per month totaling 2,700 Dinars which sum, notwithstanding the outstanding request, he did not repay to the *Sokol District*, whereby he committed a selfish act unbecoming a *Sokol* to the detriment of the District's treasury thereby causing moral and material damage to the *Sokol* organization;
— sold to the District trainer badges ordered from the firm Palčič prior to the all-*Sokol* rally in Prague, collected money for the badges from the trainers but did not pay the firm for the goods it delivered; the District's administration ordered him to pay the amount of 649 Dinars directly to the supplier, which at the date of his expulsion had not been fully paid;
— provided no accounting for the sum of 800 Dinars which he received prior to the departure to the all-*Sokol* rally at Prague subject to accounting;
— declared all members of the *Sokol* organization who were willing to take part of the jubilee gathering at Maribor in August 1938 to be traitors of the *Sokol* movement;
— verbally abused the *Yugoslav Sokol Federation* and *Sokol* member Dr. Gavrančič, at the time third deputy of the presiding elder of the *Sokol Federation of the Kingdom of Yugoslavia* and Dr. Arsenijević, at the time Secretary of the Federation, who signed the decision dismissing the District administrative board to wit: "Until we kick out these two hooligans who signed that I will not return to the District administration's office."
— violated the privacy of mail by opening in the District's office a letter addressed to the District's presiding elder;
— incurred an expense to the District by boarding at Celje, where he had to report to respond before the administrative court, the wrong train which went to Maribor instead of boarding the train for Ljubljana;

Axis occupation Victor Repič joined the communist partisans and was active in destroying the local railroads in Dolenjska. After the war he entered the diplomatic service of the Yugoslav communist regime and served for some time as Tito's ambassador in Vienna, Austria.

The expelled members first appealed the disciplinary action to the District governing board which upheld the disciplinary court's decision. A further appeal was made to the governing board of the *Federation* which, however, did not act on it. Accordingly, I presented a documented report at the *General Assembly of the Sokol Federation* in which I explained the essence of the subversion and recommended that the disciplinary action be upheld. However, at the very moment when the vote was to be taken Ivan Kovač, principal coach of the *Sokol Federation* and personal tutor of the Prince and Heir Apparent, sprung up and threatened to resign should the *General Assembly* affirm the imposed disciplinary measures. Because the *Federation* always had a hard time to find qualified trainers the *General Assembly of the Sokol Federation* decided to avoid another crisis and tabled the vote on my motion.

Ivan Kovač was friendly with some of the trainers in Ljubljana who influenced his action on that occasion. A short time later Kovač chose to administer a petty snub to the District's governing board. He sent a wire requesting that the board meet him for a conference at the railroad terminal in Ljubljana on his arrival. However, when his train pulled up, he "forgot" about the waiting board members and left the train by a somewhat unorthodox route through the station's maintenance shop. Later in the day, the District secretary found Kovač at his home and brought him to the regional office where it turned out that he really had nothing to confer about and was only mouthing generalities and platitudes. During the war Kovač was supporting the Mihajlović resistance movement. After the war, Kovač's communist protégés apparently did not accord him recognition to which he felt entitled and he emigrated to the United States where he worked for the *Voice of America*. He must have felt at home there.

Mitja Ribičič, a youth of unpleasant and mordant personality, also was a member of *Sokol I at Tabor*. Later he received training in Moscow, became colonel of *OZNA* and vice-president of Tito's government. He was totally under the influence of Lubej and was causing great problems at the marionette theater directed by Ivo

...

— in his communication of August 26, 1939, addressed to *Sokol I*, accused the governing board of *Sokol I* of deliberate dilatory action concerning his disciplinary complaint against [former] presiding elder Pipenbacher;

— accused the governing board of evading disciplinary regulations;

— accused the governing board of unheard of action in the recent disciplinary proceedings, that it deliberately refrains from maintaining order, that it is using the institution for supporting ill conceived ambitions of certain persons who are pursuing their personal material objectives, that it derives particular pleasure out of such proceedings and, that he declares that he would not respond to the proceedings as he considers the affair to have become moot because of the passage of time.

Erbežnik. Because of that he was called before a disciplinary board where at the hearing he called Erbežnik a jackass. In Maribor paper *Večernik* Ribičič published a derogatory cartoon alluding to the disciplinary proceedings at Tabor. Ribičič, too, was expelled from *Sokol I* as was his mother who was initially a member of the disciplinary court.

As the regular general assembly of *Sokol I at Tabor*, scheduled for January 1940, was drawing near Lubej and Rus decided on another plan of attack. The subversives had the list of members and were diligently agitating. The governing board countered this by organizing the general assembly so that only members in good standing were allowed to vote. Officers of the association were stationed at the entrance to the meeting hall in which the general assembly was taking place admitting only those who qualified. When, following the report of the governing board, the opposition spokesmen were recognized they, in keeping with their style, launched into an attack on the governing board. The members approved the governing board's report and reelected the board with 400 members voting for and 15 against.

The Mahovič Affair

An interesting incident related to the foregoing events came to light in 1940. In the evening after the conclusion of the general assembly, Vladimir Mahovič, member of the opposition, came to the Association's office and threw his membership card on the floor in front of the association's clerk saying that he no longer wanted to be member of "an association like this." At the meeting of the governing board following the general assembly, Mahovič's statement was noted and he was duly dropped from the rolls of the *Sokol* organization.

In the fall of 1940 I was summoned to the office of Governor Marko Natlačen. The governor handed me a police report to read in his presence saying that he strongly suggested that the *Sokol* organization pay attention to the conduct of its members. The police report began with a statement: "The undersigned Vladimir Mahovič, member of *Sokol I at Tabor*, etc..." At that time Herman Rauschning's book "*Conversations with Hitler*," (*Gesprache mit Hitler*)[66] written by former governor of Gdansk, created a considerable sensation. In it Rauschning reported the German plans for the conquest of Central and Eastern Europe as revealed to him by Hitler personally which had already been implemented with the annexation of the none too reluctant Austria, the occupation of Czechoslovakia and as they were later carried out during World War II. The book was translated into Slovenian by Boris Ziherl, a communist, who entrusted its distribution and sales to Mahovič.

[66] Hermann Rauschning: *Gesprache mit Hitler*, Europa Verlag, New York, 1940

Under the pressure brought by the German government the Yugoslav government banned the book. Mahovič was very successful in selling the contraband publication and sold a large number of copies to the officer corps of the Drava Division of the Yugoslav Army. The sales to the officers were arranged so that each officer placed his order by putting his name on a list which went to Mahovič. Mahovič then sold the list of officers to the German consul general for a sum of 500 Dinars whereupon the German consul lodged a protest with the governor. The governor told me that following the consul's demarche he directed the police commissioner to call on Ziherl and ostensibly seize a few copies of the book so that the consul would "shut up." It turned out that the police commissioner was a schoolfellow of Ziherl's and he explained in a friendly way to him the true purpose of his mission to put him at ease. Ziherl, in turn, confided all this to Mahovič who lost no time in informing the German consul about it, again for an emolument of 500 Dinars. Enraged, the German consul made a big fuss with the governor threatening that he would demand disciplinary action against the officers who had purchased the book. The governor then ordered that Mahovič be questioned by the police and had me read the report of the *process verbal* because Mahovič maintained that he was a member of *Sokol I* at Tabor. I explained that since January 1940 Mahovič was no longer a member and assured the governor that I would secure a public statement to that effect from the *Sokol Federation* to which the association would report the incident. Within a very short time the federal administration published a statement in all major papers to the effect that Vladimir Mahovič had not been a member of the *Sokol* organization since January 1940.

Rus Rebuffed in Court

After the resounding defeat of the subversives at the association's general assembly the expelled former members filed a suit in the District Court in Ljubljana against all members of the association's governing board on grounds that the disciplinary action against them was defamatory. In accordance with the established procedures of the District Court such suits were assigned to two judges, one was given the suits filed against defendants whose last names began with initials from the first half of the alphabet, the other suits against persons with last names beginning with letters from the second half of the alphabet. The suit, which was no doubt instigated by Rus, did not name me, the president of the disciplinary court, as the first of the 24 defendants but Pavel Pestotnik. This was done so that District Court Judge Jože Rus would be assigned the suit as all those filed against defendants whose names began with a letter from the second half of the alphabet.

The president of the District Court thus duly assigned the suit to Rus. If I had been named at the head of the list of the defendants, as I should have been, the suit would have been assigned to the other District Court Judge. Thus schemed the unethical judge who, naturally, took no action to recuse himself as an interested party. Our attorney, Josip Cepuder, filed a formal protest to disqualify Judge Rus for cause but the presiding judge vacillated and would not decide the matter. On the day scheduled for the hearing, Jože Rus actually came to the courtroom and went through the motions of starting the proceedings but soon after the opening formalities he got cold feet and adjourned the proceedings to a later date. The case was then assigned to the other judge who dismissed the suit on grounds that the association's disciplinary court had sole jurisdiction in the matter and that there was no basis in the law for the District Court to reopen the proceedings. The subversives appealed the judgment of the District Court to the Regional Court of Appeals which affirmed the judgment of the District Court with a scathing censure of the expelled subversives' conduct.

Following is the opinion and ruling of the District Court in Ljubljana issued on May 20, 1940:

The first level court correctly found that Defendants are members of the governing board of Sokol I and as such ex officio members of the disciplinary court of honor. It was, accordingly, not up to the Defendants to decide whether they want to act in the capacity of a member of the disciplinary court of honor; that was their duty which they assumed when they accepted their election to the governing board.

As authorized members of the disciplinary court in the course of disciplinary proceedings and gathering of the evidence established that a person who by virtue of joining the Sokol I Association submitted himself to the Sokol disciplinary court of honor committed this or that act which is for the Sokol organization dishonorable, constitutes a gross offense and defiance of the District and Association's governing boards, destruction of authority, etc., they could be made respondents to a libel suit only if they, in rendering their judgment based on proven evidence and findings added unrelated and separate insults either in the form of insulting words or in the form of libel.

The subject decision [of the disciplinary court] passages of which constitute the essence of this complaint, however, shows no evidence of such conduct. It is not incumbent on a regular criminal court to examine evidence gathered in the course of disciplinary proceedings before the Sokol disciplinary court of honor and the pronounced verdict of that court as to whether that verdict is based on evidence and, in case it were not, to impose penal sanction on members of the disciplinary court for defamation of [Plaintiff's] character. Plaintiff has an opportunity to appeal to the Sokol District and, eventually, to the Sokol Federation which are the proper arbiters for deciding if the judgment is correct.

Because Plaintiff, a private party, directs his complaint only to dispassionate determinations from the judgment of the disciplinary court of honor and because the first level court correctly found that it is impossible to conclude from the manner in which the decision is formulated that any offense was intended and, moreover, because the allega-

tions of the background leading to the disciplinary proceedings and conduct of individual members of the disciplinary court of honor with respect to the Plaintiff lack credibility, this appeal had to be denied as palpably unfounded for reasons stated in the judgment under appeal here.

Communist Attacks in the Press

The subversives wasted no opportunity to publicize their one-sided version of the events at Tabor. *Slovenec*, the main paper of the clerical party, even, was publishing the rebels' stories with a malicious glee. Particularly offensive was the journalistic viper *Slovenska beseda* (*Slovenian Word*), published by former governor Dinko Puc, which was pouring out its bile against me and the *Sokol I Association's* governing board. Unfortunately that tabloid could not be reached under the existing statutes because it was published sporadically and the dates of its issues were adjusted so that it could not be compelled, under the existing laws, to bring corrections or retractions or sued for malicious libel. By way of retaliation let me narrate here an anecdote from the time of the Axis occupation. At the time when Boris Kidrič, the architect of communist terror in Slovenia and subsequently federal secretary of national economy, was recruiting members of other parties for his "*Liberation Front*," he once held a meeting that later became known as "meeting of the cabinet members" in *The Inn at Rožnik* (*Gostilna na Rožniku*). The meeting was attended, among others, by two former governors one of whom was Puc. When it was time for Kidrič to leave the two governors were falling over each other in helping him, a former fugitive from justice, put on his fur coat — but I do not know who won the contest.

Governor Marko Natlačen

Governor Natlačen, who in 1939 was still very much antagonistic toward the *Sokol* organization, had, in my opinion, eventually mollified his position. I have already reported on my audience with him in the Mahovič affair. In December 1940 Anton Korošec, a prominent Slovenian politician and State Secretary of Internal Affairs in the Yugoslav government, died. On Sunday following his death I was out of town and when I came home, in the evening, an invitation from the governor was waiting for me. The governor was requesting that I come and see him as soon as possible. He told me that it was his wish that the *Sokol* organization, too, take part in the commemoration which would thus assume a universal national character. I promised that I would get in touch with the appropriate representatives of the *Sokol* organization and advise him of any decision that would be taken. The *Sokol* organization decided that a delegation of *Sokols* would place a wreath at the bier of the deceased leader in the gubernatorial palace and visit the governor to express its condolences. In addition, a detachment of *Sokols* in civilian clothes would join the funeral procession. We were under the impression that the governor was trying to be conciliatory in expectation of events that would require national unity. In January 1941, the governor attended a patriotic meeting of the *Reserve Officers' Association* in the national theater and delivered a formal address. Generally, all of us felt that the forthcoming events would soon require a supreme effort for the preservation of our national integrity.

Sokol I Standing Alone

The other four *Sokol* associations in Ljubljana had not shown the slightest interest nor made any effort to comprehend the meaning of the events at *Sokol I at Tabor.* The expelled former members of *Sokol I* were admitted to other *Sokol* associations as if nothing had happened and as if there had been no danger for the entire *Sokol* movement and its ideas. *Sokol I* fought out the battle against takeover by communist infiltrators without help or support from any quarter in the face of public indifference where, as usually, circumspection and cowardice prevailed. Olympic champion judge Leon Štukelj, many years later when he was visiting in Los Angeles told my son that even he, bright jurist as he was, had no idea of what was going on at Tabor. He found out that only from my booklet "*On the Ramparts*" published in England in 1965[67] which he somehow managed to obtain.

The well organized group of former members and their supporters was persistently spreading unfavorable rumors against the interests of *Sokol I.* To counter those rumors the leadership of *Sokol I* found it necessary to publish a formal report on its operations in general and on the athletic activities in particular. The rebels were irresponsibly spreading rumors to the effect that everything at Tabor was in disarray that after the rebels left all activities came to a standstill and the like. These were the last straws grasped by those who were expelled. They tried to persuade the trainer corps at the other *Sokol* associations to persuade their leadership to publicly censure *Sokol I* so that they could use this in support of their appeal to the leadership of the *Sokol Federation of the Kingdom of Yugoslavia.*

The trainers at one of Ljubljana's *Sokol* associations carried out an extended discussion how a special deputation could be put together which would persuade the

[67] Lado Bevc: *Na Braniku za Sokolske Ideale*, Slovenska Pravda, Enfield, England, 1965, *supra.*

Federation to interfere in the affairs of *Sokol I*. Elder members resolutely declined to have anything to do with that. Only the leadership of *Sokol II* at Moste where the rebels were particularly welcome succumbed to the pressure and notified *Sokol I* that it was dissatisfied by the way in which *Sokol I* handled the resolution of its internal affairs. Moreover, the leadership of *Sokol II* sent copies of its censure directed at *Sokol I* to other *Sokol* associations in Ljubljana and elsewhere. The intent to induce other associations to take similar action was transparent. Fortunately, *Sokol II* was unsuccessful in its attempt.

Leadership of *Sokol I* responded by declaring that it repudiates any kind of interference in its affairs. If anyone were in position to lecture others with an official act that would be only *Sokol I* because it was *Sokol II* where expelled members and their abettors were admitted to membership without prior consultation with *Sokol I*. *Sokol II* took in male members of *Sokol I* youths who were brought there by subterfuge by the expelled group. The leadership of *Sokol I* viewed such conduct as disloyal and certainly not brotherly but as an internal affair of *Sokol II*. Accordingly *Sokol I* characterized the circulated propaganda — censure of *Sokol I* — as an inimical act.

Concerning activities at *Sokol I*, it can be stated that the celebrations of state holidays on September 6th and December 1st were carried out normally and with a good attendance. The same goes for the memorial services of deceased *Sokols* on November 1st. Numerous children received gifts from Santa Claus and we gave a tea party for the youth. Thanks to the loyalty of the membership and new trainers the New Year's evening was very successful, it even brought more income than usually.

The situation in the gym was as follows: Young men had their exercising interrupted by only two weeks for the known reasons. All other departments were exercising regularly without interruption. The two groups of boys had an average attendance of 92 out of 120 enrolled, girls an average of 68 out of 84 enrolled, young women 40 out of 50 enrolled, women averaged 22 out of 35 enrolled; men under 26 were frequently requested to exercise and those who did not were dropped from the rolls of the association in accordance with the bylaws which required regular exercise.

The new training corps consisted of known and professionally trained personnel: twelve men trainers and nine women trainers in addition to the members in training. Let me list those that I remember: Tone Gorjanc, principal coach for boys, member of the training corps for twenty years, Stane Pirc, formerly assistant coach, then coach of *Sokol I* became the coach of the *Sokol District*; Rudolf Ryška, trainer of young men, trainer of the *Czechoslovak Sokol* and following the liberation in 1918, trainer of young men at *Sokol I*, two years coach of the *Ljubljana Sokol District*; Pavel Pestotnik, member of the trainer corps of *Sokol I* from 1907 to 1928, trainer of *Czechoslovak Sokol*, presiding elder of the *Sokol I* association and *Sokol District*; Dr. Josip Kunc, first deputy presiding elder of *Sokol I*, Janko Sket, formerly coach of *Sokol I*, trainer with many years' experience; Anuška Jug, coach, trainer of *Czechoslovak Soko*l, *Federation, District*, and *Association's* coach of many years, undoubtedly first in professional qualifications; Marta Podpac, trainer of young women, member of many years of the trainer corps of *Sokol I*, winner at international athletic contests; Milena Sket, deputy coach at present, well-known participant at international competitions with excellent results in competition and leadership of athletic groups; Ivanka Bevc, graduate of the trainer's course of the *Czechoslovak Sokol*, principal trainer of many years at the *Sokol Association* in Beltinci, District coach in Split.

Sokol I also consolidated its athletic activities. In the cleared atmosphere all elder and experienced members gladly and with pleasure assumed the various assignments and exercising departments of *Sokol I* which showed a substantial influx of members from day to day.[68]

Leon Štukelj

In conclusion it is appropriate to review the era we lived through and identify and verify the established facts. *Sokol I at Tabor* had a large hall which was among the largest and most modern edifices of its kind. The *Sokol* organization provided all the facilities for the physical education of members, youth and children, it was exemplarily prosperous. Fate brought into the association the impulsive Lubej, a man who needed someone who would broaden his horizons and channel his energy to activities in accord with the *Sokol* principles. Instead of embracing these he found himself in an ever closer association with a man with whom they both began to travel on the path of international communism. *Sokol* as a democratic organization never mourned apostates; it could not, however remain indifferent after discovering a conspiracy of communist infiltrators who intended to take it over. Communism, which in a hundred years of its long and grisly history never achieved anything, could not expect to attract free men; it needed a disguise of a highly respected organization under the guise of which it could carry out its criminal program. Leadership of *Sokol I* recognized this and took action to defend the association against an ideological suicide. In its battle the leadership had the loyal support of the large

[68] See: *Naša zaključna beseda* (*Our conclusion*), Uprava Sokola I v. Ljubljani, Ljubljana 1940.

majority of membership which made possible the eventual victory notwithstanding the grave danger the extent of which is described in the chapter dealing with the time of the Axis occupation. The leadership has for half a year stood fast and in exhausting nightly sessions prosecuted the disciplinary action in its court of honor under incredible pressure of a mob incited by two apostates acting in ways that would become known to all later during the communist revolution and its reign of terror.

The leadership and I incurred mortal enmity of the communists and *Sokol* apostates. The conscience of the public witnessing this deplorable spectacle remained frozen solid under an impervious layer of indifference. *Sokol I* stood isolated and alone among the other *Sokol* associations and nearly ignored by the national leadership; those who dared express approval during or after the struggle with the communists preferred to do so quietly and in secret lest they be set upon by the vindictive mob. After the battle at *Sokol I* was won, the cautious constituting the majority of the public, carefully avoided any visible expression of approval of our stand.

I was a member of many organizations where I frequently held the leading offices. At the elections in 1926 I was elected to the City Council of Ljubljana of which I was a member until 1934, with a break in 1929 and 1930. I was a member of the Chamber of Commerce, chairman of the City's engineering and construction committees, and member of various other committees. During the Great Depression and afterwards I was besieged by people who needed help in finding employment and indeed I was able to help many of them to secure a livelihood. However, when the subversives at *Sokol* brought their matter before the public, I noted that some of the people I had helped considered me a liability because I was in the forefront in combating the communist infiltration into the *Sokol* organization. People who used to say that I might have been in some respects a liability to an organization still remained, cautiously and circumspectly for the time being, my friends. Such cautious souls are always on the lookout as to what the future might bring and thus are always ready to provide a silent majority to the one whose ascendancy to power they may fear but to whom they compliantly give their support when it is demanded and when the war came they would invariably appear in the ranks of the so called "*Liberation Front*". Things have always been and always will remain like this. Those who have seen Gary Cooper's motion picture *High Noon* will understand what I am saying here.

At the changing of the guard, as it happened in 1945 when the communists seized power, the cautious were regaled by a few leftovers from the well furnished communist table. However when it became apparent that there were no more crumbs left, the cautious who would not recognize us in our darkest hour put on their friendly faces again and went to great lengths in telling us how disappointed they were with the communists. At important events the cautious cut such a miserable figure that no one can count on them because they represent a big fat zero on everybody's balance sheet.

Since the General Assembly in January 1940 and up to the time of Axis occupation the situation within *Sokol I* at Tabor became normalized and the association continued to prosper and grow. That time was truly a quiet before the storm.

Polič Takes Off His Mask

Thirty years after the war writings of Zoran Polič, now a well established communist minister,[69] suddenly appeared in the Ljubljana papers. The titles of his articles were: *The path of Slovenian Sokols to the Liberation Front* and *Forty Years of the Left Wing of Sokol in Slovenia.*[70]

Polič and his co-defendants before the *Sokol* court of honor had been vigorously and vehemently denying to have anything to do with the proponents of alien ideology and had the gall to take the *Sokol Association* to court on grounds that they were libeled. They had lied then as has now been affirmed by Polič from his secure haven of the communist *nomenclature*. He even had the nerve to say that there are "events we must not forget, so that what cost us so many lives and caused inestimable material and moral damage would not happen again — we must tell those persons abroad who always talk about national reconciliation and concord as if they had never done anything against the Slovenian people why we cannot accede to that which had led them on the path of treason and shame."

That insolent allusion to the *Yugoslav Sokol* was yet another outburst of a pathological liar. We Yugoslav *Sokols* have never said anything about reconciliation with the "leftist Sokols" nor about forgetting. Faithful to our Yugoslav homeland we endured all the misfortunes and shunned no sacrifice, including that of our lives. Our homeland and communism are not one and the same. The homeland will remain

69 Zoran Polič (1912–1997) During World War II member of the *City Council of the Liberation Front* in Ljubljana, member of the executive council of the *Liberation Front of Slovenia* and member of the general staff of the so-called *National Liberation Army of Slovenia* — the partisans. Delegate at the session of *AVNOJ* (*Antifašističko vijeće narodnog oslobodjenja Jugoslavije — Antifascist Assembly for the National Liberation of Yugoslavia* — Tito's rubber stamp parliament) and the *Constitution Drafting Assembly* of Tito's *Democratic Federative Yugoslavia.* Member of the first communist government of the *People's Republic of Slovenia*, first as minister of the internal affairs (police), and then state secretary of finance. In 1947 he became president of the *Council of the Republics and Regions of the Federal Socialist Republic of Yugoslavia.* Member of the *Central Committee of the communist party of Slovenia*, vice-president of the *Republic Conference of Socialist Alliance of Working People*, vice-president of the *Association of Combatants* of the so-called war of national liberation, president of the *Olympics Committee* and recipient of numerous communist decorations.

70 Ljubljana *Delo* (formerly *Slovenski poročevalec*) January 17, 1978.

forever long after communism is no more. To our homeland we dedicate our love and devotion, but with communism we cannot have any reconciliation. Against the "leftist *Sokols*"[71] we have already spoken in our final judgment, justly meted out in 1939 by the *Sokol* court of honor. From the safety of his police state[72] the brazen prevaricator tried to brand loyal Yugoslav *Sokols* as traitors but it is he and his cohorts who remain tarnished with the infamy of treason.

Polič finally admitted that the ground for the leftist Slovenian *Sokol* has been laid as early as in 1935. That happened at Čebine where Edward Kardelj, an agent of the Comintern, orchestrated the establishment of a communist congress. When in 1937 the then principal trainer at *Sokol I at Tabor* staged an imaginary crisis he succeeded that the *Sokol a*ssociation's leadership tried to resolve the crisis by inviting Polič back to Ljubljana and in this way, the "leftist *Sokols*" brought to the *Sokol* ranks a true Trojan horse which, however, did not succeed in converting the *Sokol* association to a Čebine-like communist front.

In addition, Polič asserted that the Stojadinović regime tried to impress the *Sokol* organization into the role of a militant formation supporting the *Yugoslav Radical Party* (*JRZ*). That is a fantastic fabrication. It has never been even remotely suggested, much less demanded, to *Sokol I* to assume such role. Neither the representative of the regime nor the *JRZ* party ever came up with anything remotely akin to that. It was the "Čebines" who during the course of disciplinary process at Tabor were publishing their vitriolic attacks against *Sokol I* in the Ljubljana daily *Slovenec*, at that time the mouthpiece of the *JRZ* regime. Similar attacks were published, as already mentioned, in the sporadic publication "*Slovenska beseda*" published by Dinko Puc in his newly found identity of an "old radical."

So much about organizational affairs. Even more interesting is Polič's disclosure about the ideological orientation of the "Čebines." He stated: "Concerning the ideology we have to observe that the group was at first convinced that the ideas published by the *Sokols*' principal ideologue Miroslav Tyrš in his philosophical writings would suffice particularly if supplemented by the political views of Thomas G. Masaryk."

Then comes the revelation: "The more the leftist *Sokol* group of Čebines was impelled by the situation to align itself with a definite world view and, in our circumstances, with domestic political ideologies, the more it became clear that the

[71] To speak about communists as "leftist *Sokols*" is nonsense. A *Sokol* cannot be a communist and a communist cannot be a *Sokol*. It is only possible for a communist to masquerade as a *Sokol* so that he might in accordance with Stalinist precepts corrupt the progressive and democratic Sokol organization and use it as a respectable cover to advance the criminal goals of communism.

[72] Zoran Polič lived in the so-called "*Little Kremlin*" (*Ferantov vrt*), an apartment complex near the communist parliament and the police headquarters where he was, along with other communists such as Zdenka Kidrič, securely guarded at all times.

Sokol idea outside of the educational domain cannot satisfy the class defined political factors and, specifically, that the Masaryk's views were in fact only a mask of the bourgeoisie in its fight against the ever stronger presence of revolutionary representatives of the working class."

What workers and what representatives of the working class? Among the communists expelled in 1939 from *Sokol I* was not a single laborer, the expelled ringleaders were of the following professions: a teacher, bank officer, government employee and a student. As for their ideological mentor: he was the "Čebine philosopher" judge of the district court. Those men arrogated to themselves the role of leaders of the working class. On the other hand there were quite a few laborers and people of modest means among the membership at Tabor who remained loyal to the *Sokol* ideals, a fact borne out by the outcome of the general election where the communist attempt to overthrow the leadership failed and the leadership was reelected by 400 votes against 15. From the Yugoslav official statistics it is clear that in the membership of the Yugoslav communist party only 19 percent are workers.

Polič concluded:"*It was natural for the democratic wing of the Sokols to collaborate with the communist party and eventually become an integral part of the progressive movement within the Liberation Front.*"

In joining the Dolomite declaration the *Čebines* emphasized:"*The Slovenian Sokols have from the time when the Liberation Front was created recognized the communist party as the only political entity which could by virtue of its program, organization and discipline — during the war as well as in the future — be the leader in the political life of the Slovenian nation. On this basis the Slovenian Sokols within the Liberation Front established and preserved its relationship with the communist party and adhered to their original decision not to organize political actions of their own.*"

Consistent with this the "leftist *Sokols — Čebines*" on July 8, 1945 gathered at a general assembly, adopted a resolution to merge their organization into an entirely novel gymnast association, says Polič. In other words, the *Čebines* dissolved the *Sokol* organization and submerged it into a regime controlled gymnast association called *The Partisan*.

Polič could not help but tell another lie concerning the fate of the *Sokol Herald* (*Sokolski vestnik*).[73] The *Herald* was a periodical published and edited by the District's trainer staff intended primarily for the gymnasts. The district administration had nothing to do with that periodical nor was it responsible for its contents. I must admit that I have only now learned from Polič's writings who was publishing in that journal and about what did the authors write. Neither I nor Pipenbacher ever thought about assuming the control of that publication. It is only now that Polič is boasting that the publication used to spread the ideas of the "leftist group." It may

[73] Not to be confused with *The Sokol Herald* (*Sokolski Vestnik*) published from 1965 to1993 in Milwaukee by Dušan Svetlič under the auspices of the *Yugoslav Sokol in the Free World*

well be that in 1938 that publication ceased spreading the ideas of the "leftist group" because of Lubej's suspension when the disciplinary proceedings were initiated and Pipenbacher ensured that during the pendency of his suspension Lubej could not edit the journal. I am however not familiar with the contents of the publication and Polič's insinuation that I had taken control of it simply is not true.

Polič naturally also writes a lot about "comrade" Kardelj, of his assessment and directing of the activities of the *Čebines*. In true Soviet style Polič writes how comrade Kardelj anticipated in his highly intelligent analysis the future development of events at *Sokol I at Tabor* except, of course, their outcome and the defeat and expulsion of the communists.

In his writings Polič inconsistently uses terms such as Slovenian *Sokols*, left wing of the *Sokol* in Slovenia, democratic wing of the *Sokol*, *Sokol organization* and the like. Why is he still concealing that this was simply a group of confirmed communists infiltrated into the *Sokol* organization with the purpose of taking it over. In sum: everything that represents itself in the Čebine declarations as "*Sokol*" is a brazen thievery. Those people never belonged to *Sokol* and we, *Sokols*, deny them the right to refer to themselves as "*Sokols*." Being a *Sokol* is incompatible with the communist views, a *Sokol* cannot be a communist. Another brazen deception is reference to democracy which Polič uses in referring to his "wing." Here he was faithfully emulating his role model *Stalin* who dictated a constitution to the Russian people in which abound words like democracy, freedom and justice in cynical derision to millions of victims deprived of life and liberty because of Stalin's interpretation of those terms. In his old age Polič admitted what he used to deny since 1935. After forty years members of *Sokol I at Tabor* finally saw from his admission that they were justified in passing judgment on the communists who had infiltrated their association so that they could destroy it and use it for the purposes of an alien ideology. If they came out openly as communists hardly anyone would follow them. They needed a cloak of respectability and they did their best to steal it from the *Sokol* organization. Even the communist party wanted them to "bring" in *Sokols* so that it could parade them as respectable people who supported the communist party.

Those of us who have succeeded in saving ourselves and our families from the destructive rage of unmasked communists — *Čebines* and who are in the Free World trying to resurrect and renew the Yugoslav *Sokol* can tell Polič once and for all: that we will not talk about reconciliation with him and his cohorts nor will we forget all the evil that his group under the misleading disguise inflicted not only upon the *Sokols* but upon the entire Slovenian nation. Polič has managed to climb to the top of the communist hierarchy, in any case high enough to share the guilt and the responsibility for crimes committed against the nation which his party obstinately denies.

Slovenian Press Repudiates Polič

Sokols in the free world are not the only ones who can criticize the turgid writings of Zoran Polič. In the last few years articles began to appear in the Yugoslav press which most certainly were not in accord with the communist ideology and practice of the totalitarian regime. People are increasingly more aware that communism has nothing to offer except misery, misfortune and chaos. Obdurate Stalinists like Polič are naturally concerned over that as realization of truly democratic principles and free market economy would shake and eventually overthrow their privileged positions.

Ljubljana weekly magazine *Mladina* (*Youth*) has in recent times become noted for writings of which the communists disapproved to such extent that several issues of the magazine had been confiscated. In its April 1988 issue *Mladina* brought no less than three articles[74] critical of Zoran Polič, one of the three leading communists — the other two were Franjo Lubej and Jože Rus — who helped dig a grave for the *Sokol* organization and liberty in Yugoslavia. It is interesting to read how Polič's current activity is assessed in the old country by people who got to know him as one of the principal promoters of the communist dictatorship.

In a column "*New in the dustbin of history*" which reports under the symbol of hammer and sickle on the latest moves of the communist party and its official, Vlado Miheljak wrote that Zoran Polič, member of the *Council of Federation,* considered by *Mladina* to be a club of useless and burned out politicians, attacked in an interview for *Politika Ekspress*, the Slovenian political leadership for being too tolerant of initiatives put forth by the civil society. Polič, an obdurate communist, finds every breeze and every breath of liberty dangerous and feels that it must be suppressed.

Somewhat more extensive is article by Milan Apih: "*What does Zoran Polič Want?*" From that article one can discern what Polič discussed in his interview with *Politika Express.* It is also noteworthy that Apih characterized Polič as his peer and as his — one time — *Sokol* brother. Apih may well have been member of the *Sokol* organization but since he has been, as is evident from his writing, a communist already at that time, it is clear that he could not have been *Sokol* in his heart. This does not mean, however, that opinion of someone who once was a communist should be disregarded out of hand. These days it is becoming clear that all too many Slovenians who once were communists are coming to realize that communism has nothing in common with freedom, democracy and social justice.

What, then, did Apih write about Polič? First he stated what we already know, namely that as a student at the University in Ljubljana Polič was chairman of the

[74] Vlado Miheljak: *Novo na smetišču zgodovine*; Mile Šetinc: *Med Poličem in demokracijo*; Milan Apih: *Kaj hoče Zoran Polič?*; *Mladina*, No. 13, Ljubljana, April 1, 1988.

Ljotič's student association *Edinost* (*Unity*) considered by some to be totalitarian. And he told us something else, to wit:"We, communists, who have at that time operated in secrecy, carefully avoided Polič. We considered him to be among those whose "patriotic" feelings dictated that they be friendly toward the regime and its police who would appreciate getting any hints of our clandestine activities." Thus at the time when Polič was expelled from the *Sokol I Association at Tabor* the communists considered him quite capable of being an informer. Apih felt it was wrong for Polič to have, by virtue of some obsolescent bylaws, as a one time member of the *Executive Council* of the so-called *Liberation Front*, the right to be a life member of the *Republic's Council* of the *Socialist Alliance of the Working People* and as such presume to speak with the authority of his position to which he could never be elected.

In his interview with Aleksandra Pavlevski, known for her vitriolic advocacy of repression, Polič expressed "desire for authoritarianism and an iron hand." According to Polič writers should be somehow induced to represent the realities of the regime in a more favorable light. Polič reinforced that statement with threats directed at cultural workers and the youth with specific references to repression and "sanctions."

Apih did not overlook that Polič had earlier in attacking Spomenka Hribar, when she suggested that Slovenians might start thinking about a national reconciliation, resorted to lying in absence of any grounds for his attack. That communist "lawyer" attacked even the *Amnesty International*, an organization promoting respect for human rights, because that organization advocates that people should not be treated with such savagery as they were in none to distant past treated by the Slovenian communists, and that people should not be tortured and imprisoned because of their political convictions.

Polič was enraged because Slovenians were openly talking about an "alliance of all dissatisfied" and of the reluctance of the communists to deal resolutely with such ideas. He said that it was no longer possible to tell who in Slovenia is for and who is against the "principles of Tito's era."

"What then does Zoran Polič want?" asked Apih at the end. Did he want with his universal combativeness and attacks on every thin reed of democracy which can barely poke out of our soil before it is trampled all too hard by the hoofs of backwardness to achieve a herostratic glory perhaps as leader of a new junta, at least in Slovenia, where he is so obstinately trying to arrest any progressive movement? How well would all that be in accord with the Stalinist methods which he learned to practice when the going was good.

Mile Šetinc devoted to Polič an entire page with article "Between Polič and Democracy." Already in the introduction the reader finds that people in Slovenia are fed up with Polič and therefore his writings are rarely published. *Mladina* had apparently turned him down — although I am not sure that this is true. *Delo*, however, still publishes Polič's writings from time to time. His articles, says Šetinc, never require less than half a page. In any case, it seems that Polič did make at least some progress since the time when as editor of the *Sokol* column at *Jutro* he was scribbling his meager and infrequent reports on *Sokol* affairs. Polič probably learned his meandering circumlocution from his mentor Jože Rus who was such a bore that even his party comrade Milovan Djilas[75] dubbed him as politically too awkward for accomplishing anything of value. Polič is all too garrulous, wrote Šetinc and since, as it appears, he does not have much following to help him enforce his views, Polič is looking elsewhere for forces that would support him. Possibly in the same way as the Hungarian and Czech communists who had to call in the Soviets. The *Central Committee of the Communist Party*, too, is according to Polič, "courting the reaction."

Polič and his comrades would probably like to see again some arrests, prohibitions, purges and dismissals in the editorial offices. In the interest of the situation in Yugoslavia Polič would like to act in a way which has fallen in disuse lately but, he says, he is not making threats, he only makes appeals to reason.

Šetinc then wrote about "Spring," various forms of Lenin's *New Economic Policy*, Gorbachev's *Glasnost*, and the blooming of a hundred flowers all of which in the so-called people's democracies bring excesses, that is to say, that people try to take them seriously which is of course always unacceptable to the ruling communist party and the cycle of repressions repeats itself. These cycles, according to Šetinc, are as characteristic for the so-called people's democracies as are the booms and busts of the capitalist stock markets.

When Polič was brought before the disciplinary court of honor at *Sokol I* at Tabor because of his subversive activity I branded him with the word "depraved." That almost cost me my life. In the fifty years since that time Polič rose to the highest positions in the communist regime and, together with other communists, was directly responsible for loss of life and liberty and irreparable damage to many Yugoslavs. Finally, however, he unmasked himself before his compatriots who have now — fifty years too late — recognized and repudiated him as an enemy and gravedigger of liberty.

[75] Milovan Djilas: "*Wartime*" Harcourt Brace Jovanovich Inc., New York, 1977, p. 235.

World War II

The Yugoslav Tragedy Begins

Hitler's well-known plans for conquest and expansion of the German *Lebensraum* (living space) were gradually materializing. His plans were not directed only toward the East but also to the Southeast. Conquest of the Mediterranean would give Hitler a realistic basis for a world empire. His plans were being carried out one after another and it was clear that he could not be stopped by diplomacy — only force could arrest the German expansion. In Yugoslavia Hitler established an economic dictatorship through which he was dictating delivery quotas for the goods he needed. I have already mentioned the overbearing German interference in connection with such a minor matter as was the distribution of Rauschning's book. In brief, fear and expectation of Hitler's thrust toward the Mediterranean was in the air. Although the Yugoslav budget was dominated by enormous appropriations designated for defense it was too late even for equipping the army with motorized weapons. The resources were inadequate and there was not enough time. Civilian government works were limited, because of that every time a budget for civilian expenditures was approved, a government decree was issued reducing all appropriations by ten percent. Moreover, it was always necessary to go to great lengths before the appropriated funds were released. In 1940 it became almost impossible to make a private firm show an interest in contracting for government work. Persons subject to military service who were being called up for lengthy training did not receive appropriate sustenance on the grounds that they were not actually mobilized. Widespread uncertainty and Hitler's overbearing pressure caused stagnation of normal economic development. Public opinion was decidedly opposed to Hitler's ambitions.

The *Sokol* organization ordered special training at which members were instructed on possible means of resistance in case of enemy invasion and occupation. In the *Ljubljana Sokol District* I was in charge of such training.

The Yugoslav government was ever more intensely pressured by Germany to condition the public opinion in favor of it. German spies were tolerated, the fifth column was making preparations for the invasion, the German *Kulturbund* was not subject to the regulations to which Yugoslav associations had to conform and it was running out of control.

I have already narrated how the Germans demanded that Engelbert Gangl be removed as Deputy Presiding Elder of the *Sokol Federation of Yugoslavia* and the public reaction to the Prince Regent's subservience to the Nazis.

One of the numerous instances from which it might be surmised that the Yugoslav public opinion was resolutely opposed to the German overtures which had the intent of gaining a degree of control over Yugoslavia that would allow dictating economic terms or, under the guise of Yugoslav neutrality, advance the German aspirations for our living space and eventually conquering it. From the first of these it was possible to discern the Yugoslav position that any concessions to Hitler would be only of transitory nature and that Yugoslavia would in the end wind up on the side of the Western democracies. Yugoslav public opinion, however, repudiated a foreign policy of the kind that was conducted by Rumania.

In the spring of 1940 I attended, as vice president and chairman of the Ljubljana chapter, the general assembly of the *Reserve Officers' Association* in Belgrade. At the assembly delegates from cities near the border reported on an incredible tolerance of military authorities toward reserve officers of German nationality. They offered proof that those German nationals were already organized and associated with the *Gestapo* and were going on "vacations" or "recuperation" in Germany or Austria where they received training from the Gestapo. Those reserve officers were even promoted in the Yugoslav army. The general assembly of the Reserve Officers' Association which had some 60,000 members elected a deputation which was to call on the Minister of War Milan Nedić and recommend that those disloyal reserve officers be separated from the Yugoslav army. I was one of the members of the deputation in my capacity as the vice president. The minister listened to our detailed evidence and seemed inclined to investigate the matter. At that time the Germans had built an enormous transit camp for Germans from Besarabia and Rumania who responded to Hitler's summons to return to the *Fatherland*. The minister told us that he had been invited to inspect that camp but he declined because the place could not possibly be of any interest to him. The deputation thus left under impression that its mission was a success. The following week, however, we were surprised by a government *ukase* decreeing that the *Central Administration of the Reserve Officers' Association* be dissolved and its authority transferred to a commissioner appointed by the government. This was similar to the dismissal of Gangl as First Deputy Presiding Elder of the *Sokol Federation*. In case of the *Reserve Officers' Association* Yugoslavia's political leadership had again succumbed to German diplomatic intervention protecting the operatives of the fifth column.

A year later, in mid March of 1941, the government commissioner in charge of the *Reserve Officers' Association* called a general meeting for the purpose of electing a new administration to which he would then turn over his responsibilities. When the elections were about to be carried out one of the delegates stepped on a chair and indignantly reported that efforts to dissuade the government from joining the Axis alliance were failing. Prince Regent Paul, too, was impervious to any advice against him joining the proposed pact with the Axis. The delegate made a motion that the general assembly forgo the election of the new administration, leave the government commissioner at his post, and recess for an indefinite time as a sign of protest. The entire assembly rose, sang the national anthem, and disbanded instantly without a word of debate.

This, and similar incidents gave a clear and accurate picture to the government as to the disposition of the citizens. A few days later, on March 25, 1941 the pact with the Axis was indeed signed at the Belvedere palace in Vienna. A *coup d'etat* followed on March 27. The government was removed from power and king Peter II was declared to be of age which ended the rule of Prince Regent Paul. Although the new government headed by Air Force General Simović did not revoke the pact, Hitler, enraged, ordered a destructive attack on Yugoslavia.

War and Capitulation

On April 1, 1941 Yugoslavia ordered general mobilization. I left for Osijek in Croatia where the staff of the 1st pontoon group of the First Army to which I belonged was being formed. The staff was also in charge of the river navigation. On April 6, 1941 when Hitler without any declaration of war invaded Yugoslavia the staff departed for Novi Sad. Soon thereafter we received orders to withdraw to the South. After the army columns crossed the bridge over the Danube the staff had the bridge blown up and ordered that all navigable ships be sunk, then we joined the retreating army. The general staff of the first army had to withdraw from Bijelina to Mostar so that the Germans whose intelligence found out the location of its headquarters, could not attack it with the *Luftwaffe*. We thus lost connection with the command and continued an orderly retreat among immense military columns to the South. Our staff had one automobile, one truck and thirty horse drawn carriages. The automobile and the truck began to break down. In Šabac we intended to remain for a day to have our truck repaired. The automobile had to be abandoned on the road. Both vehicles had civilian drivers who probably sabotaged them, but there was nothing we could do about it. We were thus left with twenty four horses and their carriages. We had a long way to go to Mostar. As we were retreating through Srem the Hungarian militiamen appeared.

We were traveling by night and rested during the day away from the main road. In Šabac we survived the first artillery attack by the Germans who were forcing their way across the bridge on the Sava River. An incredible panic ensued. Soldiers were running out of the town with no idea where they were going, horse drawn wagons without drivers were racing along the chaotic throng. When we regrouped two of our officers were missing but the enlisted men were all there. We joined the immense column of the army which was retreating toward the Southwest. In moonlight we crossed to the left bank of the Drina Rriver and tried to find a place away from the road where we would wait for daylight. At that instant vigorous shooting burst out on the right riverbank. The Germans in armored vehicles were pushing south and were continuously firing their machine guns ahead of them although there was no organized resistance. From the direction of Zvornik we heard the detonation when our military blew up the bridge over the Drina River. As enemy planes began to overfly our positions we could not remain in place but set forth in broad daylight toward Zvornik. Midway on our march we were overtaken by a motorcycle with a side car. It was driven by an armed German officer. In the side car sat a Yugoslav officer with a white flag who informed us that the Yugoslav army had surrendered. The German declared that we were all prisoners of war and directed us to continue on our way to the suburbs of Zvornik. He spoke flawless Serbo-Croatian which he explained to someone who was interested where he had learned the language so well: *"I am a native of Dalmatia living in Vienna."*

We stopped in front of Zvornik. Germans in motorized vehicles were overtaking us and hurrying to town from where we heard rumbling, shooting and saw raging fires. As we were thus waiting in front of Zvornik a group of about hundred Germans gradually aggregated around us. One of them wanted to take my officer's bayonet "as a souvenir" which I resolutely resisted. Suddenly all the Germans vanished in the direction of Zvornik from where we heard the detonation when the bridge had been blown up and saw large conflagrations .The Germans began to occupy Zvornik where they had the right to loot according to their rules of war.

German Captivity Evaded

We remained alone as the main military traffic was still on the right bank of the Drina River. Our commanding officer observed with regret that we failed to take the opportunity in moving to a side road away from the western bank of the Drina River about a kilometer before the place where the Germans overtook us. I repeated my earlier advice pointing out that we still had that option and he ordered that we go back. It was already dusk when we reached a farmhouse where we were hospitably received. Our captivity had thus not lasted longer than four hours. We bivouacked overnight and the next day headed west toward the pass from where we moved on toward Brčko. At the pass we found out that people had never heard of any mobilization or German invasion. We tried to telephone to Mostar but we could not establish a contact with the First Army.

We spent the night in the abandoned worker's quarters of the Lopare mine. The following day we moved on to Brčko where German flags with the swastika were already flying on all buildings. The *Ustashi*[76] police seized our horses, wagons and equipment. Both bridges over the Sava River had been blown up but the Germans were going back and forth across the river in the two steamers, the *Slovenac* and the *Crnogorac*. These two steamers should have been scuttled according to the orders we issued in Novi Sad. Obviously our orders had not been carried out.

We crossed the river in a private boat and went to the railway station on the left bank of the River Sava where we found out that the railway traffic had not yet been reestablished but that there was a chance that a train for Vinkovci might be put in operation the next morning. Next morning, still at dusk, we were indeed on the train to Vinkovci. There the control had been assumed by the *Ustashi* who were ransacking our luggage and disarming the military personnel.

Eventually a train for Zagreb arrived from Belgrade and we boarded it without serious difficulties. In Zagreb the terminal was packed by an immense throng of soldiers. We had to get off the train but it was announced that the train for Ljubljana would depart the next morning.

All officers were required to report to the main office where a German officer was in charge of a group of former Yugoslav officers with *Ustashi* insignia carrying out his orders. The questioning was to establish only whether one was a Serb, Croat or Slovenian. Slovenians and Croats were allowed to return to the platform but the fate of the detained Serbs was foreboding nothing good. All of them were probably murdered.

Next morning the train to Brežice departed as promised. At Savski Marof, however, the train stopped and we had to get off. The Ustashi and the Germans had not yet established railway communications. Brežice had already been occupied by the Germans and we had to walk the ten kilometers' distance to get there. In Brežice a train was soon put together and took us to Ljubljana.

Homecoming

Although we were all in uniforms the Germans did not molest us and we were able to board the train without any tickets. My acquaintances on the train knew that in view of the new situation I was in grave danger. They warned me not to

[76] Ustashi — Croatian separatists who at German invasion declared independent Croatia.

go to Ljubljana but rather get off the train in Zalog, change into civilian clothes and then try to get home as a civilian. But I remained in the uniform with the insignia of my rank and the officer's short dagger. I even managed to keep my standard officer's trunk with my luggage and sidearm. Thus equipped I arrived to Ljubljana on April 18, 1941, twelve days after Germany invaded Yugoslavia. An enlisted man was kind enough to carry my trunk. Walking home I encountered my family which was going to wait for me at the railway terminal. The Italians had already been in Ljubljana for a week but apparently still did not quite know what to do in the unexpected role of a victorious occupying power. At the time of the German attack on Yugoslavia without a declaration of war the Italians assumed that Yugoslavia would carry out defensive operations also against Italy. Because the Germans, as was their custom, did not advise their allies of their planned aggression, the Italians had put out white flags in Trieste and throughout the Julian March. The Germans, on the other hand, had their plans for revenge worked out to the last detail. Those plans were prepared by the *Kulturbund* and their fifth column. Even in Ljubljana the Gestapo established itself in Hotel Metropol Miklič, opposite to the railway terminal. On the façade of the hotel was spread an ominous black flag with runic letter *SS* in white. The Germans were fast filling the prisons throughout Slovenia. The Gestapo had orders to immediately arrest hundreds of Slovenians who were destined either to be interned in the deadly concentration camps or shot outright by the Gestapo.

When I arrived home I was told that prior to my return messengers from my friends were coming to our apartment strongly urging my wife to destroy my archive. She had indeed done so and that to such extent that numerous mementos of my public service were destroyed.

Arrested by the Gestapo

Being a civil servant I immediately resumed my duties on the job. Coming home on April 21, I found our apartment invaded by a strange uninvited company: a Gestapo officer with a skull on his hat who told me: "*Sie sind verhaftet*,"("You are under arrest"), an Italian officer accompanied by a *carabiniere* as representatives of the authorities, representative of the *Kulturbund* Muehleisen by name, and some helpers. In front of the house was parked a taxi into which they were piling whatever they found in our apartment that could be suspicious or represent a threat to the existence of the German state. At that time numerous irreplaceable books and topographical maps disappeared from my library which the search party categorized in the above sense. My wife told me that they were at work several hours before I returned from office. The principal looter was the representative of *Kulturbund* who attempted to seize even the memorial badges commemorating various national events which, however, the Gestapo officer did not want. Later I discovered that they made off also with my collection of stamps issued at the time of the plebiscite in Carinthia and similar philatelic rarities. My library and what was left of my archive were thoroughly looted and, of course, no inventory of abstracted material was made. The car was not big enough to carry all that material and they had to drive it away in several installments presumably to a place at *Stari Trg*. Years later, the communists would attempt a similar expedition but they found my library so thoroughly looted by their *Gestapo* colleagues that they had to leave empty handed.

The Gestapo man allowed me to have lunch with my family before they would take me away. He watched me during lunch sitting on a sofa in the dining room. The lunch consisting of potatoes and beets was probably more meager than he expected. After lunch he affably asked the children, in German of course, if they had enjoyed their lunch. As I put on my overcoat, the Gestapo man wondered if I was really ready for prison. He expected me to take warmer clothing and some luggage. My wife then asked him how long would the interrogation take and the Gestapo man replied that it would certainly take several days before my turn would come. My wife then pointed out to him that I had barely come from the front and that I was still running a fever from the cold I caught at the front. She told him that I would really have to be in bed, that there was a telephone in the house upstairs and that they can always summon me to report when I get well. She also argued that being a good citizen in one's own country could not be a crime. The Gestapo man did not contradict her, he only said that Germany was in mortal peril, that it was fighting for its existence and as such had the right of self-defense (*Notrecht*) to ensure its security during the war. He tried to make excuses by pointing to the two Italians present who, he said, might also have security concerns. My wife, who is fluent in Italian, then turned to the Italian officer and told him the reasons why they should leave me at home. The Italian officer could not possibly care less whether I would be taken away or left at home, he seemed annoyed by the whole affair. In the end, the Gestapo man agreed that I remain, so to speak, under a house arrest. He took down the telephone numbers of the people who rented the apartment above us and of my office and the search party left. Two days later the flag of the *SS* disappeared from Hotel Metropol Miklič housing the *Gestapo* headquarters. The Italians finally realized that they were the victorious occupying power in Ljubljana and asked the *Gestapo* to leave. My arrest thus became moot as it was solely a project of the *Kulturbund*. The prisoners whom the *Gestapo* had arrested on the advice of *Kulturbund* were taken with them and on the way to Germany shot at Begunje in

Upper Carniola (*Gorenjska*). My friend Stanko Dimnik got hold of photographs of the murdered prisoners made by the Gestapo who were fond of keeping souvenirs of their macabre undertakings.

Although we shall never forget the German savagery, Ivanka, my wife, who certainly never had any sympathies toward the Germans, expressed a wish that the Gestapo man who left me at home would be able to return to his family in Baden where he said he had his home.

When the Germans were deporting the Slovenian intelligentsia from the areas they occupied and annexed to Germany in defiance of international law many refugees who evaded forcible deportation to Serbia came to Ljubljana. Soon also those who were deported to Serbia found their way to Ljubljana or to the region south of it which was occupied by the Italians. One of these refugees told me that the Germans had locked him up in a holding cell in Maribor. There all the prisoners could hear the reading of the order of the day for German agents which listed about one hundred Slovenians wanted by the Gestapo in the interest of German security. I was included among those, evidence that my arrest had been carried out by order of the Gestapo command on the basis of information supplied by the *Kulturbund* and German espionage. Those arrested were immediately taken to the prison in Begunje in Upper Carniola (Gorenjska) for a preliminary interrogation and sometimes a summary sentencing to death. Those who survived the preliminary ordeal were interned in the concentration camp at Dachau where they remained until the end of the war if they did not succumb to an epidemic or suffered a violent death. Colonel Glušič who was interned at Dachau described to me an example of Nazi brutality. If an inmate fell into a guard's disfavor, the guard would as the prisoner would be walking near the fence, kick the hat off his head so that it fell beyond the fence and then order him to pick it up. Whatever the unfortunate did was grounds for killing him: if he disobeyed he was instantly killed, if he stepped over the fence to pick up his hat, he was mowed down by the machine gun from the watchtower along the fence.

In 1963 my wife and I were traveling through Munich. We wanted to see the memorial to the victims who died there among whom was my brother in law, Dr. Stanko Sket. We found it extremely difficult to get information as to where the concentration camp used to be. We got the impression that Germans did not want to give out this information because they were probably ashamed of what was going on there. Whenever the German concentration camps are mentioned in the press and other news media there is no mention of the Yugoslav or Czechoslovak victims although there is extensive discussion of others. Publicity depends on who has control of the mass media. The Yugoslavs had to be satisfied by symbolic punishment of a few German war criminals but the majority of them went free. When we were withering in the refugee camps we found out, for instance, that some agents of the *Gestapo* and guards from the prisons in Ljubljana were in the police service of the British military government in Klagenfurt.

Salvaging Sokol Property

Once the *Gestapo* departed from Ljubljana, I could get my bearings and adapt to the new, sad circumstances. The *Sokol Hall at Tabor*, the pride and joy of my professional and *Sokol* activity was seized immediately by the Italians who installed in it their *Dopolavoro*, a fascist recreational organization. In my absence the association's administration managed to save all three association's banners which were then buried in a metal casing in somebody's garden. The member of *Sokol I* in whose garden the banners were buried feared that someone might inform the authorities about that and came to see me with request that I find another place for the flags. I went to the director of the National Museum and explained the situation. I asked him to accept the banners in his custody. The Director agreed on condition that I give him a letter antedating the Italian occupation stating that the *Sokol Association* presents the banners to the museum which was easily done and the banners were put safely away as planned. The archive of the *Reserve Officers' Association* was successfully saved and hidden in a warehouse of a factory in Ljubljana's suburbs.

Because of the haste with which the Italian *Dopolavoro* occupied the *Sokol Hall at Tabor* we could save only part of the *Sokol I* archive. Association's secretary Fran Vrečar and deputy presiding elder Miroslav Urbas thus unfortunately could not hide much of the Association's property. The Italians dissolved the *Sokol* organization and, of course, confiscated all of its property.[77]

Sokol I had no cash on hand because it used all its income for making payments on the debt incurred by construction of the Sokol Hall. The principal creditor was the *Czech Sokol Community* (*Česka Obec Sokolska*) but because Czechoslovakia had already been occupied by the Germans we stopped all remittances to Prague. Instead, Association's comptroller Lozej established contact with the Czech *Sokol* refugees and was sending the payments due to the *Czech Sokol Community* to the organization of Czech refugees who were coming in large numbers to the South of Yugoslavia. Prior to the arrival of the Italians Lozej succeeded in timely sending

[77] Note by editor: The dissolution of the *Sokol* organization by the Italians serves as a very convenient excuse to the current government of the Republic of Slovenia which, although it allowed a new *Sokol I* organization to be established, contends that the *Sokol* organization which owned the *Sokol Hall at Tabor* had been dissolved and the ressurected organization is not the same as the original owner and thus ineligible for restitution of the *Sokol Hall*. Thus the predatory Slovenian government continues the injustices of the Italian invaders.

the entire surplus funds to the Czech refugee organization. We were thus in the enviable position of not having any monetary assets to hand over to the Italians. In examining the books the Italian superintendent of the *Sokol Hall at Tabor* discovered entries documenting large payments dated at the end of March 1941 and interrogated Lozej, Flegar and myself as to the particulars but he could get no information from us. We insisted that those expenses were made prior to the Italian occupation and that we were not accountable to him for the administration of any of the *Sokol* property on the entire occupied territory.[78]

When, following the Italian surrender, the Germans occupied Ljubljana they changed the government administration so that all administration was left to the Yugoslav officials and civil servants, they only assigned to each government unit their own advisor (*Berater*) who had the final say on all matters of importance. The regional administration could thus decide how should the confiscated *Sokol* property in the entire province be administered and it appointed Miroslav Urbas, deputy presiding elder, as the administrator of the *Sokol* property. The Italian administrator of the Sokol Hall attempted to leave with an enormous amount of luggage containing articles with which he was trying to abscond, including even the puppets from the marionette theatre. Urbas had the power to stop him and he prevented the Italian to take anything that did not belong to him; the Italian was allowed to leave essentially only with the clothes on his back. In assuming the responsibility for the *Sokol Hall* Urbas found, in the former apartment of my brother-in-law Boris Sket I believe, all of the *Sokol I Association's* archive in disarray but still intact. Among the papers was found a bound copy of the transcript of record of the disciplinary proceedings against communist infiltrators at Tabor: Lubej, Polič, Porenta, Podgornik and their cohorts which the Italians never bothered to examine. This defused the lies spread by the communists to the effect that we had turned over that record to the Italians who then shot some of the expelled communists as hostages.[79]

Vrečar did have on hand the day to day management funds of about forty thousand Dinars. When the Italians were changing the currency we decided not to hand over that money to the Italians nor would we exchange it for the Italian *lire*. The money remained in Vrečar's safe at his office in the editorial offices of the daily *Jutro*.

---

[78] Note by editor:Under the *Hague Convention II* of 29 July 1899, *Convenntion with Respect to the Laws and Customs of War on Land* Article 46, it is unlawful for an occupying country to disturb the title to private property on the occupied territory. This international law was routinely violated by Germany and Italy. The confiscations of the invading countries turned out to be convenient to the government of Slovenia which contends that property confiscated by the occupying countries became property of nationals of those countries which were in turn subject to confiscation under the *AVNOJ* decrees and thus need not be returned to the rightful owners.

[79] See, for instance: *Franjo Lubej: Kot žrtve so padli (As victims they died)*, Naši razgledi, v. 16, No. 13 (372), 8. July 1967, pp. 343–344.

During the war I was in the *Sokol Hall at Tabor* only once — at the time when the Italian administrator interrogated Lozej, Flegar and me about the expenditures made before the Italian occupation. Under the German occupation, which followed the Italian, some kind of a military kitchen was installed in the basement at which time the original arrangement of locker rooms and showers for the gymnasts was probably changed.

Surplus funds of the *Committee for the Memorial to King Peter I and the Committee for the Memorial of King Alexander I* in Ljubljana were, as already mentioned, distributed among the professional military officers who were stranded in Slovenia. Thus the Committees were able to come to the rescue of many families in distress. The residual funds of the *Committee for the memorial to King Alexander*, that had not be expended was entrusted to me to be deposited in a savings account. At the end of the war the communists confiscated all my property and made away with that savings account, too.

In my capacity as Secretary of the *Association for the Construction of the Engineering Faculty Buildings* I also was entrusted with the administration of a residual deposit from the time when the *Association* constructed the first building for the Engineering Faculty on Mirje in 1920. Those funds also were confiscated by the communists.

Occupation

Sokol War Council

Early in May 1941 Engelbert Gangl made a special trip from Metlika, where he had moved after he was relieved as deputy presiding elder of the *Sokol Federation of Yugoslavia,* to call on me in my office. Deeply moved he embraced me and urged me to organize *Sokol* resistance against the occupation and a *Sokol* effort for liberation. The visit gave me a further impetus to continue organizing the *Sokols* that I had already started. A few weeks later we held somewhere in the outskirts of the city a meeting of the representatives of all five *Sokol* associations in Ljubljana. It was decided to set up a *Sokol War Council* as the supreme leadership of *Sokols* in the subjugated homeland. I was entrusted with the presidency of the *Council* and served in this capacity throughout the war. At the same time the *Sokol Legion* was established. The relationship between the *Council* and the *Legion* was the same as between a *Sokol Association's* governing board, the association's trainers and the membership. The difference was that membership could not engage in gymnastics but had to devote itself to the military tasks of the resistance movement. Members of the *Sokol War Council* swore the oath of allegiance upon their election, as did members of the *Legion.* Because of the state of war the *Sokol Legion* was organized into companies and battalions. The *Sokol Legion* was led by a *commander* who was a member of the *Sokol War Council.* The first commander was Pavle Vošnar whom the communists later captured at Grčarice, sentenced to death, and executed. The second one Janko Mačkovšek[80] who was interned at the concentration camp in Dachau where he died. The third one, Edi Antosiewicz, was also interned at Dachau where he was liberated by the Americans. Antosiewicz emigrated to the United States and never returned to Yugoslavia. The fourth commander, Rudi Marinčič was, as Vošnar, also captured by the communists at Grčarice and executed.

"Better in the grave than a slave!" was the motto affirming the unanimous refusal of the Yugoslav population to assist, even temporarily and for tactical reasons, the Axis in the execution of its plans for destroying human liberties and Yugoslav

---

[80] Janko Mačkovšek (1888–1944), professional engineer, technical staff of the *Municipal Bureau of Construction* in Ljubljana, author of numerous papers on Slovenian national minorities abroad, committee member of *Branibor,* secretary of the *Cyril and Methodius Association,* and co-founder of the *National Minorities Institute,* leading member of the *Yugoslav Nationalist party* (*JNS*), leading personality in the Yugoslav resistance under the command of General Dragoljub Mihajlović in Slovenia. Arrested by the Nazis and interned at Dachau where he died

independence. As often before at times of a national crisis, a venomous propaganda machine swung into action attributing all responsibility for the nation's calamity to the new government which had escaped abroad. The whispering campaign overwhelmed the naïve and shortsighted people and was, of course, abetted by the cautious and opportunists. That was the propaganda with which the communists were preparing for civil war.

At the time of German invasion the communists were still straddling the fence because the Soviet Union was Hitler's ally. When in July 1941 Hitler attacked the Soviet Union, the communists found a reason to organize their *Liberation Front*[81], the leadership of which they firmly held in their hands while the cautious and the opportunists formed a docile chorus of "useful dupes" exploited by the communists for camouflaging their total control of the so-called *Liberation Front*. In that front automatically appeared the former *Sokols* whom we had before the war unmasked as leaders of the attempt to infiltrate the *Sokol* organization by communists and whom we expelled. The communists gave to those expelled *Sokols* the task to recruit *Sokols* for the *Liberation Front* but they were rather inefficient in carrying out their assignment. They were recruiting members of all *Sokol Associations* to join the *Liberation Front* and they wanted declarations from them that they are supporting it as *Sokols*. For understandable reasons members of *Sokol I at Tabor* were not approached. Only one member of the governing board had been approached and he referred the recruiters to a representative of the association. The *Liberation Front*'s proselytizers put down the signature of one out of the fifteen members who at the General Assembly in January 1940 represented the opposition that lost out against a 400 member's strong majority. It was in this way that Jože Rus, Franjo Lubej, and Zoran Polič obtained their credentials as representatives of the *Sokols* within the communist *Liberation Front*. As a reward, Jože Rus was selected at the communist meeting in Jajce in 1943 for the dubious honor of proposing that the title of *Marshal of Yugoslavia* be conferred on Josip Broz-Tito and who was later made the president of the first Yugoslav communist controlled parliament. Franjo Lubej became a member of the communist party and secretary of the *Liberation Front*. Zoran Polič, the unmasked communist, became a lifelong financial expert under the communist regime. All communist terrorists had attained high positions in the communist government. Klemenčič, the youth who during the sessions of the disciplinary proceedings at Tabor attempted to interrupt electric power, had by the end of the war attained the rank of colonel of *OZNA*. It was a strange coincidence of fate that electricity became his nemesis as I related elsewhere.

81 When and why the communists established the so-called Liberation Front, is clearly apparent from their marching song: "Readily we grabbed the guns/and took to the hills/when the Soviet heroes/ called for joint combat"

In the fall of 1941 an intermediary brought an invitation for the *Sokol War Council* to join the so-called *Liberation Front*. The proposition was freely discussed by the *Council* at the conclusion of which I remarked that I, for one, was in no hurry to jump on that bandwagon. The invitation to join the *Liberation Front* was unanimously declined. The refusal must have been quite a blow to the communists because at the kangaroo court trials in Kočevje in 1943 they forced Pavle Vošnar, one of the victims, to report that statement of mine which was subsequently published in the book *White Guards* [*Belogardizem*] by Franček Saje[82] as the only specific allegation concerning my supposedly treasonable activities to which the communists ever publicly alluded. In any case, communist snoopers had been stalking me since as early as 1941 when they began to threaten me with murder. The communist surveillance was so thorough that they knew even in whose company I would be returning home from the *Sokol Legion's* meetings although the city streets were under a complete blackout.

The *Sokol War Council* began publishing a clandestine newspaper. On the advice of Engelbert Gangl, who, however, was not a member of the *Council*, the paper was called *Naprej Zastava Slave* [*Forward, Banner of Old Glory*]. The first editorial was written by Gangl whose style could readily be recognized. The paper was printed in a small pocket format. It was prepared by reduction of a typewritten copy to a size that just allowed the print to be read. For reproduction we obtained a hand operated printing press. The "printing plant" was located in the center of the city in a garden house behind the first courtyard where our printer lived. The editor was the late Dr. Vinko Blaško, an American citizen by birth, who was also the commander of one of the companies of the *Sokol Legion*. Vinko Blaško was a young idealist, a good athlete who exercised regularly at *Sokol I*, and an unusually capable organizer. He was among the legionnaires captured at Grčarice in 1943 when Italy surrendered and became a victim of the communist terror.

Following the publication of the first issue, however, Gangl began to draw away from the *Sokol Legion*. At that time the railways were still operating and I was able to visit him in Metlika where Gangl deservedly enjoyed the esteem of its most distinguished citizen. He was president of the Dolenjska Savings and had his apartment in the bank's building. His wife told me Gangl was annoyed whenever emissaries of Emilio Grazioli, the Italian High Commissioner for the Province of Ljubljana, called on him bringing "Greetings to my poet Gangl." The local Italian commander, in emulation of his leader the Duce, took fancy to addressing his troops and any civilian onlookers from Gangl's window on the numerous Italian festive occasions and was pressing his attentions on Gangl, who did not care for

[82] Franček Saje: *Belogardizem*, Slovenski knjižni zavod, Ljubljana 1951, pp. 259–260. That is the sum total Saje could discover about author's activities.

them. To avoid such annoying meetings, Gangl would withdraw to his bedroom, leaving word for the unwelcome visitor that he was "ill."

When I arrived, Gangl received me in his office. As I came in he invited another person, unknown to me, to be present at our meeting. It was a clear signal that Gangl no longer wanted to be associated with our cause because of intensive communist surveillance. I could do nothing but engage in small talk to pass the time. Gangl's withdrawal from active cooperation in the liberation movement of the *Sokols* was dictated by unfavorable circumstances, but his devotion to the *Sokol* cause became clearly apparent when he later came to the station to see me off. I sensed his regret that he could not share the burden of my responsibility with me.

Toward the end of the war the communists appointed Gangl president of the Red Cross. He probably accepted that appointment while still in shock of bereavement following the tragic death of his wife during the German bombing of the Novo Mesto hospital. When, after the so-called liberation, Gangl died, the communist newspapers carried only a brief notice omitting all mention of his prominent pre-war public service and glossing over his considerable works as a poet and author; all that was said about him was that he was "President of the *Slovenian Red Cross.*"

Communists Raid our Paper

When, in the spring of 1942, *Naprej Zastava Slave* published an article which made it clear that we were quite capable of thinking with our own heads, the clandestine communist paper "*Slovenski Poročevalec*" [*Slovenian Reporter*] was incensed and editorialized: "We have been silent so long as the *Naprej Zastava Slave* was peddling lemonade but what it is writing now is the limit!" The stereotyped cliché about lemonade indicated that Jože Rus was author of the editorial. Shortly after that, on a Sunday morning at around eleven an armed band of communists broke into the *Sokol* printing plant, ransacked the printer's modest apartment, seized the printing press and supplies, ordered the printer under the threat of death not to get out of the house, and carted away the loot with their car in broad daylight under the noses of the ubiquitous Italian policemen and security personnel.

We were thus left without a publication. All the work of the *Sokol Legion* was conducted unselfishly and without compensation. We sought no financial support from friends and sympathizers. We had enough personnel consisting of devoted *Sokol* members whom we could not readily include in the *Legion's* detachments. We could always fill vacancies by members who waited to be accepted in a *Sokol Legion* detachment. The communists, on the other hand, were maintaining a well paid throng of informers and a fanatical group of executioners who were carrying out assassinations in broad daylight often in plain view of indifferent occupation authorities.

The attack on our paper was one of the first terrorist acts of the communist conspiracy. Assassinations carried out in broad daylight followed. The victims were people who opposed the communists and were therefore sentenced *in absentia* to death by the communist kangaroo courts. Such a sentence was also passed against me in absentia so that I could at any time expect an encounter with a communist executioner. Those executioners have now attained the status of *people's heroes*[83] Franjo Lubej left his job as an elementary school teacher and the apostate trio Rus-Lubej-Polič became full time operatives of the communist terror. Never in the history of the Slovenian people have there been such acts of terrorism as were those perpetrated by the communists.

Vladimir Dedijer in his book *The Fight Stalin Lost 1948–1953*, revealed that Edward Kardelj as a partisan assumed the alias *Bevc*. Kardelj would use that alias at the communist assemblies as well as in signing death warrants. As a student at the teachers' college Kardelj used to live in one of the courtyard flats at *Rimska Road No. 5* separated by an alley from the house in which I lived. I would from time to time see him on the street when he was returning from school. One day my wife told me that Kardelj's mother had come over and requested that I intercede for her son with the police who had arrested him on political charges but that he was innocent. My wife told Mrs. Kardelj that I had no connections with the police and that it would be impossible for me to intercede on her son's is behalf. Still, I made some inquiries in the course of which I found out that the man was known as a communist already at that time. His cowardly usurping of my good name is characteristics for the hatred against everybody who did not submit to communist ideology.

Sokols Stand Firm

The sole purpose of the occupation authorities was to remain in control and suppress all vestiges of the Slovenian national identity and independence. Concurrently with this, the communists were carrying out a revolution whose objective was to seize all power already during the time of the Axis occupation. In the communist scheme of things the national liberation was only of secondary importance. As the developments have shown, the Slovenian nation was not liberated

[83] One of those heroes is Edo Brajnik, during the war commander of the *Security Intelligence Service (VOS — Varnostna obveščevalna služba)*, executive arm of communist terror used for murdering the Slovenians who opposed the communists,now lieutnant colonel oft he *Yugoslav People's Army*.

at all when the occupation powers left, it remains enslaved to this date. All that happened was that the occupation authorities were replaced by the communists.

The communists also attempted to attract the reserve officers. In mid May of 1941 the reserve officers held a secret meeting at which appeared a querulous character who was accusing everyone who was prominent in the public life of responsibility for the catastrophe that had befallen Yugoslavia. The only ones to whom he attached no blame were Hitler and Mussolini. It was clear that he had been sent by the communist propaganda department which through the spreading of such insinuations was poisoning the public opinion in preparation of seizing power after destroying all vestiges of the national authority. By his harangue the communist agent managed to block any attempt at creating unity among the reserve officers. He succeeded in breaking up the meeting which I left before it was over. The work of *Sokols* became more difficult because the occupation forces were joined by the inimical communist opposition of the fratricidal war.

As soon as the *Sokol War Council* was organized it became subject of relentless attacks by the expelled *Sokols* who have joined the so-called *Liberation Front*. The membership of the *Sokol Legion* including those who were active in the field always exceeded one thousand. The Legion drew its strength from the deep commitment of its members to the *Sokol* ideals. Commensurate to this commitment were the communist efforts to subvert the *Legion's* ranks and win over to their side its membership. However, instances in which an individual member of the *Sokol Legion* went over to the communist side were rare.

To the communists any means that might suppress the independence of *Sokols* was welcome. They even managed to plant one of their spies on the *Legion's* military staff where he acted as a courier. Fortunately, his duties were restricted so that he never could reach me. As numerous other communist spies that person lacked the intelligence to effectively carry out his mission.

The communist intelligence network was very tightly woven. Its spies were paid out of the ample funds at the disposal of the communists. Various "business circles" which before the war were loudly complaining against "intolerable taxes" contributed generously to the communist coffers. When merchant and factory owner Ivan Jelačin was immediately after the so-called liberation hauled before the communist court he managed to prove that he gave 13 million lire to the *Liberation Front* during the time of the Axis occupation. Such sums were considered enormous at the time.

When the Germans occupied the northern part of Slovenia in 1941, many *Sokols* came to Ljubljana as refugees. We were busily making our rounds to secure some resources with which we could provide assistance to those refugees. In the process of this fund raising activity a friend of mine and I called on Jelačin, among others. I happened to know Jelačin because he had been my former landlord. Jelačin received us in the presence of his son who was smirking at us as we were explaining the plight of our brethren — it may well be that the youth had found it fashionable to be a parlor pink at that early time and, since at that time the Soviet Union was still an ally of Germany, he had scant sympathy for the people who would not submit to the tender mercies of the Third Reich. Jelačin himself acted as if he were tremendously bored by our imposition on his precious time but he did give me a thousand lire banknote from his wallet. However, when I mentioned that we planned to buy tickets for meals in the *Workers' Home* (*Delavski Dom*) he asked me to give him back the banknote saying that he would buy the tickets himself. He probably intended to intimate to us, as a gratuitous insult, that he did not trust us with the thousand lire. Jelačin was well-known for his railing against the "unbearable taxes" which he supposedly had to pay; but someone who was privy to his financial standing told me that Jelačin could pay all his taxes with the proceeds of selling the plums shaken off a tree in his garden.

With access to resources of such magnitude as were extorted from Jelačin and other captains of industry the *Liberation Front* was able to maintain an entire army of paid informers and couriers. Although the reports produced by the communist surveillance network were of little consequence, the *Liberation Front* was thus in position to attract numerous people who were grateful for being paid for carrying out their unsavory assignments. A typical report produced by such informers would read as follows: "Sunday, 11 AM, in *Cafe Slon*...a man of athletic build sits under the mirror...etc.." By no stretch of imagination can a rational person divine what kind of "threat" to the national liberation could the athletic person possibly represent to merit his inclusion into the files of the communist security archive.

Neither the *Sokol War Council* nor the *Sokol Legion* ever received monetary contribution from its compatriots. The *Sokols* were carrying out their activities without compensation of any kind, their work was based on the principles and ideals of the *Sokol* organization and had as its objective the expulsion of the alien occupational forces from our country. Thus we remained beholden to no one. Whatever we did was the personal contribution of each of the members who had joined our ranks.

Main Weapon of the Communists

A lie repeated a hundred times becomes accepted as truth. This was the basic precept according to which acted the expelled *Sokol* apostates in the *Liberation Front*. I suddenly found myself represented to the public opinion as a "traitor." People who before the war used to seek my company began to avoid me.

I made calls on four persons whom I caught slandering me. The slanderers were: mediocre writer Fran Albrecht, secretary of the university's School of Divinity, Arnošt Brilej, city counselor, secretary of the chamber of commerce Šmuc, and widow Tepina, mother of two parlor pinks. The latter's sons signed in May 1945 the decree by which I was discharged from the government service and deprived of all vested rights. He was author of the proposed design for a monument to King Alexander I bearing the inscription: "*De mortuis nil nisi bene.*" When confronted with my complaint, the first three hypocritically swore on their honor that they never had the slightest reason for doubting the integrity of my character and, of course, could come up with nothing that would in any way support the slanders they helped to spread about me. Widow Tepina, whom I did not know personally, had seen me arriving and came to meet me on the staircase and inform me the person I wanted to see "was not at home." The following day, she sent one of her confidantes to me with assurances that she had said nothing offensive when she was talking about me, that the person who reported the matter to me did not actually understand what she had been saying and that what was reported to me was in fact not true at all. I realized that under the circumstances I could not fight the lies with legal action. Under the Axis occupation no paper would be able to publish a defense or retractions against a flood of lies nor would the courts, controlled by the occupation authorities, be the appropriate forum for resolving such matters. I could indeed take to court a number of the *Liberation Front* sympathizers who had slandered me but I would never be able to reach the source consisting of the expelled *Sokol* renegades within the leadership of the *Liberation Front* who orchestrated the assassination of my character. By resorting to such base means the trio Rus-Lubej-Polič was able to carry out their vendetta. This, of course, was only the beginning. Toward the end of the war the word got out that the communist kangaroo court had sentenced me to death. Having managed to evade assassination at the hands of some future "people's hero" my only course of action to avoid being murdered by the communists was to leave the country when the communist occupation was imminent. Since the time when the communists occupied my homeland to this date they never attempted to stage a trial *in absentia* against me at which they would judge me as a "traitor" although they had enough of fawningly subservient judges who by pronouncing a judgment "in the name of the people" could give legal sanction to communist lies.

Lie Exposed

When the Italian occupation authorities began large scale persecution of people all too many found themselves in prison overnight. No one could comprehend that the massive raids and deportation to Italian concentration camps were a standard method of subjugation employed by the fascists, therefore they were looking for the root causes of such terrorist acts. The *Sokol* apostates expelled from the *Sokol I Association at Tabor* invented and began spreading the lie that the governing board of *Sokol I at Tabor* deliberately allowed the Italian authorities to get hold of the record of the disciplinary proceedings against the seven members who were expelled. Each of the affected parties adapted this lie to his own case and then began spreading it as a sacred truth. Three of the expelled former Sokols, Janez Porenta, a communist proselytizer, Lubej's brother Lojze, treasurer of the *Executive Committee of the Liberation Front*, and Dušan Podgornik, member of the General Staff of the *People's Liberation Army and Partisan Detachments of Slovenia*, were actually arrested by the Italians in one of the frequent raids, held for a time, and eventually shot as hostages. The Italians did not know what functions those three held in the communist organization. The shooting of hostages was in retaliation for the assassination of Gabriela Rea, a woman fascist official, at the *Figovec Inn* in Ljubljana by the communists. Wife of Lojze Lubej thought of making an appeal to the authorities for the release of her husband but when Franjo Lubej, her brother-in-law, heard of her plans he forbade her to go to the authorities saying: "If you go I will kick you in the rear."[84] When another member of that group was picked up by the Italians, his wife in despair jumped off a balcony and killed herself. The communists were attributing those arrests to the Italians' reading of the record of the disciplinary proceedings and finding out who was likely to be a communist sympathizer. I bore the brunt of the hatred generated by this campaign which was directed against the *Sokol I* governing board.

When, on April 1, 1941, I was called up to join the Army, my duties as the presiding elder of *Sokol I Association's* were carried out by deputy presiding elder, Miroslav Urbas, who tried all he could to save the *Sokol* archive. Unfortunately, the governing board was able to save only part of the *Association's* archive because the fascist recreational organization *Dopolavoro* moved very swiftly in seizing the *Sokol Hall at Tabor*, as already mentioned. After I returned from the front, it was no longer possible to save the rest of the archive. It was out of question for somebody to take anything like the file of disciplinary proceedings out of the *Sokol Hall* occupied by the Italians and turn it over to the police.

[84] Actually Lubej, true to his style, used a more vulgar expression which would not be appropriate to quote here.

Soon after declaring — in contravention to all precepts of international law — the annexation of the occupied Slovenian territory to Italy, the Italians dissolved all *Sokol Associations* and seized their property. When, in 1943, Italy capitulated the German administration of the Province of Ljubljana appointed Miroslav Urbas as commissioner in charge of administering the *Sokol* property. At the *Sokol Hall* commissioner Urbas found intact the remainder of the *Sokol Association's* archive which we could not rescue in 1941. Included in that part of the archive was the record of the disciplinary proceedings which, contrary to the communist calumny, was untouched and unmarked. After the war the record was appropriated by Lubej who wrote in his memoirs[85] that the names of the communists appearing in it were underlined and that Pestotnik through Avgust Praprotnik pointed out those men to the Italian occupation authorities who executed them as hostages.

Chetnik Detachment Destroyed

In 1943 we finally saw the long awaited breakup of fascism and the end of the Italian occupation. Italian general Cerutti had long before that time conducted several meetings with Boris Kidrič and other communist emissaries in the surroundings of Novo Mesto. When the Italians surrendered they, on instructions of the Allies, turned over their artillery equipment to the communists and, contrary to the Geneva convention, even provided their own personnel to operate it. The artillery was used to overcome the resistance of the chetniks at Grčarice of whom the majority were *Sokols*. The communists guaranteed to the chetniks an uninhibited departure to the Adriatic coast where we, naively, thought the British fleet would land. As soon as the chetniks came out of their fortifications, however, the communists seized them, dragged them to Kočevje and tried them in night sessions in the communist court. The investigating judge at that court was Zdenka Kidrič, wife of Boris Kidrič and organizer of the *Security Intelligence Service* (*Varnostna obveščevalna služba — VOS*). Vinko Blaško who was captured at Grčarice was not put on trial as the communists probably thought that as a brilliant lawyer he would put up a formidable defense. The court meted out death sentences for all of the accused. Drunkard and artist (in that order) Niko Pirnat was the officially appointed artist whose task was drawing cartoons of the accused. In attendance were members of the British mission who were full of admiration for the communist proceedings and later broadcast their admiring reports over the BBC from London. The abuse of the accused did not bother the British who are fond of preaching that captured soldiers may not be ridiculed or exposed to degrading treatment — of course they mean that this applies only to their soldiers.

---

[85] France Lubej: *Odločitve*, Partizanska knjiga, Ljubljana 1980, p 182.

When the Germans chased the communists out of Kočevje they dragged along those of their prisoners whom they had not yet executed. When the retreat turned in a rout, the communists, unable to take the prisoners with them, lined them up undressed in front of an open ditch and killed them by either shooting them in the neck or bludgeoning them with picks. This happened at Travna Gora. Among those who were thus murdered was Vinko Blaško. On December 31, 1953, the tenth anniversary of the Grčarice tragedy, I wrote for *Domovina*, an article entitled *Wake me not* [*Ne budite me*] commemorating my friends. Because the article is relevant on any occasion, I am quoting it here:

"When early in the Summer of 1943, the Legionnaires were leaving for the front, where within a few month the fate would include them among the heroes of the national tragedy at Grčarice, the late Dr. Vinko Blaško explained to me his decision to go with them thus: "Actually, my turn to go has not yet come because I am needed as a leader of numerous legionnaires who are still waiting to be called up. However, it is clear that I must leave with this group. The national calamity has formed such strong ties between us that we are all of the same will to contribute equally to the national liberation."

"He left, never to return. Neither he, nor his legionnaires. The Grčarice company was attacked by the partisans who, being in possession of Italian artillery and gunners, were able to overcome its defenses. At the stage managed trial in Kočevje the communist justice was washing its hands in the innocent blood of some of the heroes who survived the fighting while the rest of the victims were spared the indignity and humiliation of a communist court trial. They were lined up in front of a common grave and killed."

"How fast the time flies! It is nine years since that time and yet I feel as if my friend has just shut the door behind him. He left, and with him the rest, a youth by years, yet a mature man deeply attached to the nation's ideals, firm in his decision to do his duty. As we, the parents, wives, children, brothers, sisters, lovers and friends gather at the final resting place of these brave men we again feel the agony of our grief and the cruelty of the crime that sent them to their grave. And we hear them implore: "Disturb not our sleep!"

"True, they have been resting for ten years already. They found their peace in eternity. Their achievements and their martyrdom had set them in front of us in life as well as in death. The only way to become their equal is to follow the path they themselves have found and shown to us."

"We regret to have disturbed their sleep even for a fleeting moment at this sad anniversary. But let us tell them by way of apology that we would rather be with them for all those years and forever after than bear the oppressing burden of our

responsibilities. In our present struggles and endeavors we look up to the shining example of these brave men whom we ask to include us in their dreams before they resume their eternal sleep. We shall not disturb their sleep until the love we bear for them compels us again to break for a brief moment their command."

"Eternal glory to the brave men of Grčarice!"

German Occupation

After the Italians left the scene in 1943 they were replaced by the Germans who, however, came to the Ljubljana Province clad in sheep's clothing. They even introduced themselves to the public with posters in the Slovenian language the use of which they had forbidden in Lower Styria and in Carniola. The *Gestapo* and its infamous spying and snooping organization established its headquarters in *Kazina*, while the German supreme commander Erwin Rösener[86] lived in a villa in the Mirje district near the Engineering Middle School from which he had evicted the owner.

I had no connections with either the German or Italian occupation authorities. General Leon Rupnik,[87] soon after he was appointed President of the Provincial Administration, invited me to his office and suggested I become one of his advisors. He said: "In this position I do not have good advisors, I need better counsel and I am inviting you to become my advisor. If you agree, you will have access to me at all times." As I was deeply involved in the Mihajlović's resistance movement of which the Sokol Legion was part, I did not want to work with Rupnik and, indirectly, with the Germans. On the other hand I thought it imprudent to turn him down outright and so I remained silent. This was my only wartime meeting with General Rupnik who took umbrage at my silent refusal of his invitation.

Rösener demanded that the Provincial administration (formerly *banovina*) where I was the senior engineering counselor set up along our roads and in cities throughout the province topographic and traffic signs in German. We were delaying the execution of this order and when Rösener sensed that he called in the Principal engineer Franc Rueh for an audience. We were reasonably sure that Rösener wanted to punish us for dragging our feet. Rueh did not want to go and he delegated the task

---

[86] Erwin Rösener, Obergruppenführer der *Waffen SS*. After the war he was arrested in Austria by the British and sent to the Yugoslav authorities in Ljubljana to be tried for war crimes. The authorities staged a trial at which he appeared together with General Rupnik and Police Prefect Lovro Hacin and sentenced to death by hanging. Because that trial was probably under some kind of international supervision it was conducted in a considerably more civilized way than the trials against *Sokols* and other political opponents.

[87] Leon Rupnik, General of the Army, Mayor of Ljubljana 1941–1943, President of the Provincial Administration of the Province of Ljubljana 1943–1945.

to my colleague Ciril Pirc and me. Rösener received us at the door and haughtily shook hands with each attendee. After calling the meeting to order he proceeded to read the list of attendees, checking if everybody was present. When he called Rueh's name I responded and said that Pirc and I were his representatives. That enraged Rösener and he shouted: "*Ein Deutscher General verhandelt nicht mit Vertretern! Heraus!*"(*A German general does not deal with representatives! Get out!*) and he pointed to the door with outstretched arm. We wasted no time in leaving but on our way out I looked back at the German proconsul and inquired what we should tell our chief. Still with a raised voice, Rösener shot back: "I shall let him know that soon enough!" Rueh was quite shaken when we reported this to him. He wasted no time in engaging all his political friends to save him; fortunately they must have had enough influence and Rueh was spared the concentration camp.

Before the war I had acquired a reputation of being able to help people get jobs. During the Great Depression as well as after it people who were unemployed and their relatives besieged me with their requests wherever they could find me: in my office, at home, in various clubs and associations, as well as on the street. Many of them found work because of my referrals. With the onset of the Axis occupation, however, my magic touch vanished because I eschewed any and all contact with the authorities. Nevertheless, following every series of raids and arrests, relatives of the arrested persons kept appearing with requests that I intercede with the authorities on their behalf. From the humanitarian point of view I would like to help everyone but under the circumstances it was not possible. Those visits and requests for help were rooted in the general opinion formed by the communist lies to the effect that I was a "traitor" and "collaborator" and as a consequence had influential connections with the occupation authorities. No such connections existed and I would be hard put even if I wanted to help my closest friends and relatives. Under the circumstances I regretfully had to turn down all requests for intercession with the authorities.

University professor and world class mathematician Josip Plemelj thus once called when I was not at home and asked my wife that I intercede on behalf of his son-in-law who had been imprisoned by the Italians. My wife asked him how he envisioned that I would be able to get a person released from the Italian prisons. Plemelj retorted that it was well-known that I had connections with the occupation authorities. On that account the visit terminated with a rather frosty parting. Logic of the world renowned professor was unfortunately limited by the narrow confines of his profession so that he could not understand how offensive his talking of my liaisons with the occupying authorities was.

Only in case of my distant cousin Josip Šest who was imprisoned by the Italians, I gave my personal guarantee to the authorities that he was not a communist. This was purely a formality which the authorities sometimes considered and for which, should it prove false, they might hold the guarantor accountable. The Italians released my cousin who soon after his release joined the partisans in the hills where he was killed in action. With this kind of record any further guarantee of mine would be rather ineffective.

In 1944, when my brother-in-law Dr. Stanko Sket, a brilliant attorney and linguist, was arrested by the police and we received word that he was being brutally beaten and tortured, the family expected me to intercede on his behalf by calling on the police prefect Lovro Hacin, whom I knew slightly from my various civic functions antedating the war. I would not go near the police headquarters because I knew that it was under communist surveillance. People coming in and going out of the police headquarters often were photographed by the communists. If I were seen there my presence at that place would be immediately published as yet another "proof" of my collaboration with the enemy. My wife then decided to go herself. Hacin received my wife but he was very arrogant, refused to be of any help and, in the end almost had her thrown out of his office.[88] I probably would not have done any better. My brother-in-law was sent to the German concentration camp in Dachau where, because of exposure and malnutrition, he became ill with meningitis of which he died with our family doctor Franta Mis, who was also interned in Dachau, helplessly looking on as the Nazis refused to provide the medicine necessary to cure my brother-in-law. Dr. Mis was sent to Dachau because he had protested the German seizure of a carload of medicine the *Swiss Red Cross* had sent to Ljubljana to be used for children. The Germans notified the Swiss that the Slovenian children are well provided for and arrested Dr. Mis. Fortunately he survived the concentration camp.

In spring of 1944 rumors were being circulated that extensive raids and arrests of the "liberals" were in preparation. My political friends wanted me to see General Rupnik about this. I categorically turned down that mission but agreed to see instead Colonel Krener, commander of the *National Guard*. Krener used to be the local commander in the Mihajlović's organization but when he resigned from his position and joined the *National Guard* we, naturally, broke off our relations with him. That visit was my only meeting with Krener who professed that he had no influence in such matters and that he knew nothing about the rumors of arrests. A few days later, however, Krener saw me in the park, stopped me, and gave his assurances that the rumors of the impending arrests of the "liberals" were without foundation. He did not tell me what was the source of his information.

[88] Lovro Hacin remained his own arrogant self to the end. At the trial where he was defendant along with General Rupnik and *Obergruppenfüher* Rösener he was aware that he was already with one foot in his grave. Therefore he responded curty and testily to his judges. When he was questioning about preparations for the defense of Ljubljana, the military prosecutor inquired from whom would Ljubljana have to be defended whereupon Hacin shot back: "From the Partisans, *whom else*." He was hanged.

Once I received an inquiry by telephone if I would receive deputy Marko Kranjec of Maribor. I agreed and Kranjec soon appeared and asked me to follow him without telling me the destination. We walked through the city to the Diocese which Kranjec entered and proceeded upstairs to the second floor. The door at the top of the staircase was locked. Kranjec rung the bell and the door was opened by Bishop Rožman[89] who dismissed Kranjec and invited me in. The bishop locked the door and took me to a large drawing room. After a short conversation he went to his office, returned with a bulky document and handed it to me. It was a situational circular of the *Yugoslav Army in the Homeland* with corresponding instructions to the principal national leaders which was being shown to me as a member of the organizational division of the general staff of the *Yugoslav Army in the Homeland.* The bishop told me to read it, initial, and return it to him. When I did that he took the paper back to his office. He had of course received the document through the Vatican the neutrality of which the Germans were still recognizing and where priests could travel without hindrance. So much for the calumny of the bishop's supposed collaboration with the enemy.

Friends and Personalities

The responsibility with which I was entrusted by the Sokol organization during the war had put me in touch with numerous people whom I had met then for the first time. I deliberately did not make notes and never insisted on identification by name. That practice saved me and my friends from the misfortune of revealing to the enemy the names and facts related to our clandestine organization under the duress of torture.

The number of associates with whom I collaborated was such that citing their names would require too much space. To those who are no longer with us as well as to the survivors I dedicate these lines with the most profound admiration for their work and sacrifice in defense of the *Sokol* idea. Holding high the banner of liberty has throughout history been the most difficult task of brave and dedicated men. The history repeats and teaches that brave men were sooner or later rewarded by acknowledgment and gratitude when the menacing terror of the struggle for achievement of their goals had faded away.

In my mind's eye I see the hosts of seemingly anonymous friends and coworkers each of whom I knew well: Vinko Blaško, Mirko Saks, Pavle Vošnar, Rudi Marinčič — I cannot enumerate all — so much of handsome youth, idealism and dedication who were destroyed by the international communist conspiracy, Nazism and facism.

---

89 Gregorij Rožman, Bishop of Ljubljana, 1883–1959, invisible spiritual leader of the *Slovenian People's Party*, died in Lamont, U.S.A.

At the end of the Italian occupation the communists declared a general mobilization. Among the conscripts was also *Sokol* and Olympic champion Leon Štukelj who was a judge in Novo Mesto where he moved from Maribor which was occupied by the Germans. Although an educated man he was conscripted as a private because the communists suspected that he did not have the appropriate enthusiasm for their cause. Štukelj was ordered to appear before the communist divisional Judge Advocate Jože Rus who greeted him with the ominous words: "*We hear that there is something wrong with you.*" When Štukelj came out of the interrogation his gun, which he had to leave outside, was gone and he was directed to a forced labor detachment assigned to clear the rubble from sites bombed by the Germans. Within a week or two, the Germans moved toward Novo Mesto and the partisans, who were guarding Štukelj and other prisoners, fled in a hurry before the advancing storm troopers, leaving the prisoners to their fate. The partisans knew better than to stand up to the SS. When the Germans came across the prisoners, who were clad in partisan uniforms, they lined them against the wall to be summarily shot. Štukelj was able to attract the attention of the German officer, explained to him that he was no communist, and showed him his identity card from the Olympic Games in Berlin in 1936 signed by Adolph Hitler himself. The German officer examined the document, decided that Štukelj must be some important personality and let him go.

The attempt of Jože Rus, later declared a "hero" by the Slovenian government, and the Slovenian communists to get rid of the "bourgeois" Olympics champion was thus foiled and Štukelj lived to see the golden old age and the Olympics in Atlanta. The same people who persecuted him before are now parading him before the world to gain prestige for their regime. But it was the unknown German officer, who thought more highly of the Olympic champion than his own compatriots, and the signature of Hitler that saved Štukelj from certain death.

We were fighting on three fronts: against the German and Italian invaders and against the fratricidal *Liberation Front*. The *Sokol Legion*, ready for any sacrifice in defense of the nation, persevered unwavering in the face of heavy losses it sustained in Grčarice because its ranks were always renewed by new volunteers.

I cannot name here the members of the *Sokol War Council* regardless of whether they may be at home or abroad. From my own experience I am aware that the Yugoslav secret police *OZNA* has not closed its books on the matter and continues to carry on its vendetta against the relatives of those who fought the communists. The *Sokol War Council* was guarding the ideological principles of *Sokol* members and made no compromises nor did it conclude any opportunistic alliances: it steadfastly stood on the principle of resistance against the Axis occupation. In accordance with this principle the *Sokol Legion* was allied with the *National Legion* and *Slovenian Legion*. All three legions were part of the *Yugoslav Army in the Homeland* ready to call up its members for military duty in the army. The clandestine military city command consisted of a commander and three members, one from each legion. I was one of the three members but for security reasons I cannot name the other two. I also was a member of the steering committee of the organizational department of the *Yugoslav Army in the Homeland (Jugoslovenska vojska u domovini — JVvD) — Slovenian Division*. The legions collaborated in harmony and friendship. Our friendship has endured and remained intact; it could serve as a model for relations between diverse political groups and parties.

The *Sokol War Council* and *Sokol Legion* have demonstrated with their resistence effort against the occupying authorities that there is no force in the world that could suppress the *Sokol* idea. Without subventions such as were received by the communists we maintained the highest number possible of members ready for combat. We sustained heavy losses, too many of which from the part of the fratricidal communist terror. *Sokol* apostates–communists had instead of resisting the occupation forces concentrated their "Liberation war" against their compatriots whom they wanted to rob of their *Sokol* ideals already during the war.

When the communists seized the power in Yugoslavia the *Sokol* apostates convened a meeting at Tabor where Franjo Lubej and Jože Rus presided. Those present at the meeting declared the dissolution of the *Sokol* organization and established a communist gymnast organization called *The Partisan*. In this way they were following the fascists and the Nazis, the third totalitarian regime that decreed the dissolution of the *Sokol* organization. They might have been able to destroy the body but they could not destroy the soul. The *Sokol* idea remains alive and well among us in the free world. We shall follow the *Sokol* idea long after there is no trace left of the agents of communist conspiracy which had for the moment triumphed in its base vendetta but what is a short moment of anguish compared to the bright future in which shines the indomitable *Sokol* idea.[90]

National Declaration

During the war the leading Slovenian political parties, the Slovenian part of the *Yugoslav Nationalist Party (Jugoslovenska nacionalna stranka — JNS)*, *Slovenian People's Party* (clericals) and *Social Democrats* maintained a joint

[90] In 1995, in the newly declared Republic of Slovenia, on the initiative of historian Matjaž Anžur, *Sokol I* was reestablished. The ressurected association included a few surviving members of the *Sokol I Association* suppressed by the communists. *Sokol I* demanded, in accordance with the law on restitution adopted in 1991 shortly after the declaration of Slovenia's secession from Yugoslavia the return of its *Sokol I Hall at Tabor*. The communists, however, maintain that *Sokol I* was after 1945 re-established by *Sokol* apostates — communists who voted that the association be dissolved and presented its property to the communist association *Partizan*. Communists, who have never created anything but were only destroying whatever they could are experts in looting the property which they could never create themselves.

committee in which was united by the idea of a free and indivisible Yugoslavia. Leader of the *Slovenian JNS* to which the liberals belonged was Albert Kramer, member of parliament and minister. Its core was represented by Janko Mačkovšek, Bogdan Ferlinc representing Ivan Pucelj, Otmar Pirkmajer,[91] Ernest Kalan and I. Following his retirement before World War II Pirkmajer was elected acting vice chairman of the Slovenian section of the *JNS*. When the government was considering plans for a civil service government, court deputy Stanković hinted that Pirkmajer was being considered for a cabinet post and Pirkmajer stepped down from his position of acting vice chairman of *JNS*. During the war he tried to make a comeback by revoking his earlier resignation but without success. He remained, however, member of the seven man directorate of *JNS* which, after the death of Albert Kramer, clandestinely caried on the affairs of the *Slovenian Democratic Party* and of which I, too, was a member. The directorate was chaired by Mačkovšek and, after he was arrested, by Bogdan Ferlinc. The directorate maintained contacts with the *Slovenian People's Party* and *Social Democratic Party* and was thus entitled to a share of funds sent by the emigré government in London via Vatican with the assistance of priests who were traveling between Ljubljana and Rome. The funds were distributed between the parties, the military and all who were in need because of the war. The young wing of the party, *OJNS — Omladina JNS* (*Youth of JNS*) was represented by Jože Rus,[92] Bogdan Drnovšek, Branko Vrčon, Andrej Uršič, Franc Juvan and Janko Koestl. The *Slovenian People's Party* was under a strong influence of the clergy whose unquestionable, albeit invisible, ideological leader was Bishop Gregorij Rožman. The *Socialist Party* was represented by Celestin Jelenc.[93]

As the end of World War II in Europe was approaching Dr. Marjan Zajec,[94] executive secretary of the *Slovenian JNS*, initiated discussions with the *Slovenian People's Party* for which he obtained my prior approval. The two parties concluded agreement on establishment of the *Slovenian National Committee* in which the *JNS*

---

[91] Otmar Pirkmajer, lawyer and politician, leading member of the *Independent Democratic Party* (*Samostojna demokratska stranka*) Lieutenant Governor, chairman of the state committee of *JNS*, chairman of the directorate of *JNS* during World War II, friend and collaborator of Janko Mačkovšek, publisher of *Dnevnik*. The *Gestapo* arrested him in 1944 and interned him at Dachau. After the war he remained in München as liaison officer of the *International Refugee Organization* in the American occupation zone.

[92] No relation to the communist judge and querulant gravedigger of *Sokol* Jože Rus.

[93] Celestin Jelenc (1884–1968), attorney in Ljubljana, leader of the *Socialist Party of Yugoslavia*, member of the *National Committee*, member of the leadership of *Progressive Alliance* and presidency of the *Slovenian Confederation*, member of the *Slovenian National Committee*. In 1945 he emigrated and lived in Argentina.

[94] Marjan Zajec (1905–1984), attorney in Ljubljana, executive secretary (*banovinski tajnik*) of the regional committee of *Jyugoslav Nationalist Party* (*JNS*), during the war coordinator of the detachment of the *Yugoslav Army in Homeland*, one of the founders of the *Slovenian National Committee*, emigrated in Argentina at the end of World War II.

would have six representatives and the *Slovenian People's Party* seven representatives. The *Slovenian JNS* gave one of its seats on the *National Committee* to the *Socialist Party* which was represented by Celestin Jelenc. Text of the agreement, which at that time was to be kept confidential, is quoted below.

> Political parties Slovenian People's Party (SLS) and Yugoslav Nationalist Party (JNS) as representatives of the Slovenian nation which are represented in the first National Committee for Slovenia based on the agreement between parties of September 1941 and the National Declaration of October 29, 1944, desiring that all Slovenian nationally and Yugoslavian oriented entities join in as close cooperation as possible in their common combat against internal and external enemies, hereby conclude the following

AGREEMENT

1. As long as it is necessary to have the National Committee for Slovenia the proportion of political representatives in the National Committee shall be 7 to 6, that is, the number of political representatives of the Slovenian People's Party versus those of other political parties shall be seven to six. The number of the members of the National Committee may be increased but its composition shall be such that the Slovenian People's Party always shall have one representative more than the sum total of representatives of the other parties.
2. The chairman of the National Committee for Slovenia shall be member of the Slovenian People's Party or a person delegated by the Slovenian People's Party to the National Committee upon prior hearing of the views of the other political parties represented in the National Committee. The vice-chairman shall be a person selected by other political parties again upon prior hearing of the views of the Slovenian People's Party. The secretarial affairs shall be conducted by two executive secretaries having equal rights of which one shall be selected by the Slovenian People's Party and the other by the Yugoslav Nationalist Party. Both executive secretaries shall be from the ranks of members of the National Committee.
3. The Plenum of the National Committee may transfer by authorization the discharge of its rights in part or in the entirety to another, smaller committee which consists of the members of the Presidium and three additional members of the National Committee of which two shall be from the Slovenian People's Party and one from other political parties.
4. The Chairman, the vice-chairman and both secretaries constitute the presidium of the National Committee. The presidium attends to current affairs, prepares agenda items for the meetings of the National Committee and carries out its decisions. Urgent matters may also be decided by the presidium alone, provided that the decision is unanimous. The presidium must present such decision to the National Committee for approval as soon as possible.

5. Proportions between the representatives of political parties agreed upon herein shall not be prejudicial to any other forums or combinations during or after the war.
6. If the National Committee should consider that the leadership of the military part should be entrusted to a military person that person shall not be included in the above proportion between the political parties constituting the National Committee. He shall have only a consulting vote.
7. On recommendation of the National Committee and by agreement between the parties signatories to this agreement new parties that recognize the National Declaration of October 29, 1944 and which have hitherto not been represented may be included in the National Committee. Such political parties shall in any case be considered to be other political parties with respect to the proportionality so far as the Slovenian People's Party is concerned. In such contingency the Slovenian People's Party shall retain the lead of one member more than the aggregate of members of all other political parties.
8. The National Committee shall adopt its decisions by simple majority. Where matters of important principles are considered votes of two thirds of the members present are required for a decision to be adopted. A proposal shall be considered to involve an important principle if it is declared as such by so many of the members present that their number corresponds to two thirds of all members of the National Committee. The decimal 5 shall be rounded upward. The restricted Committee also adopts decisions with a simple majority. A proposal shall be considered to involve an important principle if three members so state. The chairman of the Plenum or of the narrower committee shall have the right to vote. In case of tie his vote shall decide.
9. If a member of the National Committee for any reason should not be active in the National Committee, the party to which such a member belongs shall upon request of the National Committee appoint a new member to take his place. Membership of a representative in the National Committee shall terminate if a party signatory to this agreement recalls that representative. In such cases a new member shall be appointed to the vacated place by the party signatory to this agreement.
10. Decisions concerning changes of any item of this agreement shall be the responsibility of the parties signatories to this agreement.
11. The National Committee shall exist only until such time as the representation of United Slovenia charged with drafting of the constitution is elected by the Slovenian nation in free elections.
12. This agreement and the National Declaration of October 29, 1944 are strictly confidential in nature because of the existing situation and circumstances. They shall be published by a special decision of the National Committee and by a separate agreement of the parties signatories to this agreement.

13. This agreement has been prepared in two equal originals of which one shall be kept by the Slovenian People's Party, and the other by the Yugoslav Nationalist Party.
14. The National Declaration is prepared in four equal originals two of which shall be signed by members of the Slovenian People's Party and two by members of the Yugoslav Nationalist Party whereupon one signed original of each party shall be exchanged between the parties. Two originals with signatures of the respective parties shall constitute the complete document of the National Declaration of October 29, 1944.
15. This agreement shall be effective on the date of signature.

Dated on the national holiday this 29th day of October 1944.

On October 29, 1944, pursuant to their Agreement, all three Slovenian political parties signed the National Declaration and established the *Slovenian National Committee* of which I have been member from its inception to this date.[95]

The National Declaration appears on the next page.

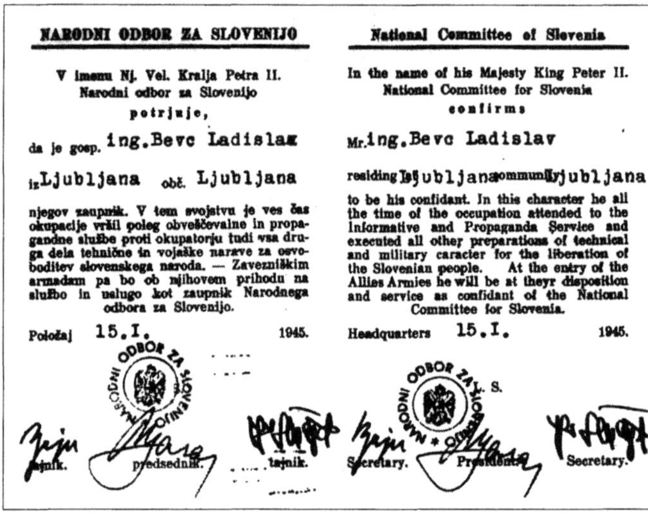

NARODNI ODBOR ZA SLOVENIJO

V imenu Nj. Vel. Kralja Petra II.
Narodni odbor za Slovenijo
potrjuje,

da je gosp. ing.Bevc Ladislav

iz Ljubljana obč. Ljubljana

njegov zaupnik. V tem svojstvu je ves čas okupacije vršil poleg obveščevalne in propagandne službe proti okupatorju tudi vsa druga dela tehnične in vojaške narave za osvoboditev slovenskega naroda. — Zavezniškim armadam pa bo ob njihovem prihodu na službo in uslugo kot zaupnik Narodnega odbora za Slovenijo.

Položaj 15.I. 1945.

tajnik. predsednik. tajnik.

National Committee of Slovenia

In the name of his Majesty King Peter II.
National Committee for Slovenia
confirms

Mr. ing.Bevc Ladislav

residing Ljubljana community Ljubljana

to be his confidant. In this character he all the time of the occupation attended to the Informative and Propaganda Service and executed all other preparations of technical and military caracter for the liberation of the Slovenian people. At the entry of the Allies Armies he will be at theyr disposition and service as confidant of the National Committee for Slovenia.

Headquarters 15.I. 1945.

L. S.

Secretary. President. Secretary.

[95] Note by editor: Written in 1984. In 1995 the *Archives of the Republic of Slovenia* allowed partial access to authors' dossier prepared under file number 4253 by communist Svetozar Frantar who declared therein that signing the National Declaration constituted treason. About a year after the Republic of Slovenia declared its secession from Yugoslavia the *Slovenian National Committee* voted to dissolve itself, a move that may well prove to have been premature

NATIONAL DECLARATION OF 29 OCTOBER 1944

Slovenians!

Representatives of the Slovenian public life have upon mutual consultation resolved to make the following declaration:

Based on the ethnic uniqueness of the Slovenian Nation and on the principle of nationalities we demand:

1. *Lawful governmental unification of all Slovenian national territory into a UNITED SLOVENIA, which must be in the geographic, economic, communication and strategic sense defined so as to ensure untrammeled national and economic development of the Slovenian Nation;*
2. *Federative Kingdom of Yugoslavia under the dynasty of Karadjordjes, based on democracy and social justice;*
3. *A federal government organized in such a way that the constituent national states as its integral parts will have such jurisdictional domain that the central federal authority will retain only those prerogatives as are expressly granted by the constitution and excluded from the jurisdiction of the national state.*

   The federal constitution of the Yugoslav state will be adopted by the constitutional assembly composed of lawful representatives of individual national states so that a majority of lawful representatives of each national state will vote for it.
4. *A fundamental social and economic restructuring which will destroy the rule of the capital over labor, recognize all working strata, families and individuals, their socio-economic function and guarantee a life worthy of the human person.*

To secure these demands the constituted National Committee for Slovenia as the supreme national authority is temporarily, until the normal state of affairs is re-established, assuming the rights of a sovereign and the highest administrative authority.

Dated on the national holiday,
this 29th day of October 1944.

Imprisoned by the Gestapo

In the summer of 1944 my friend Colonel Andrej Glušič[96] who was already imprisoned by the *Gestapo* sent word to me from prison by an intermediary that I, too, may soon be arrested. The intermediary met me on a street corner. He was soon arrested himself. Glušič was arrested when he went to a meeting in what he thought was a safe apartment in the Bat'a house in Ljubljana only to find the *Gestapo* waiting for him inside. If the *Gestapo* did not find out about that meeting place it is not impossible that an informer had been found who did this as a strike against the people of Draža Mihajović.

Andrej Glušič

I did not attend meetings in that place nor did I know about it. As president of the *Sokol War Council* and member of the local command and the organizational staff of the Slovenian part of the *Yugoslav Army in the Homeland* I was meeting with my collaborators elsewhere. Glušič maintained liaison with me through messengers. I, too, maintained liaison with the assistance of Etbin Bežek who was commander of one of the *Sokol Legion's* detachments, and also through the current leader of the *Sokol Legion*, the last of which was Edi Antosiewicz. Glušič once sent me a confidante who introduced himself as "Yuri" (George) and who was announced beforehand. Yuri was an older, well dressed gentleman whom I had never seen before or after. He could not find out from me more than what the Gestapo had already found about me through its espionage. My turn to be arrested had come according to Glušič's warning.

Early in December 1944, the *Gestapo* discovered in Ljubljana a weapons cache belonging to the *Slovenian Legion*. The *Sokol Legion*, too, had been collecting armament, particularly machine guns. The Legion was receiving limited funds for that purpose from the Yugoslav Army in the Homeland so that it could buy weapons from corrupt officers of the occupation forces. A machine gun, for instance, could be obtained for 6,000 lire.[97] The cache of the *Sokol Legion*, however, was not discovered. When the cache of *Slovenian Legion* was found Ivo Krmavner, commander of the *Legion* and a number of legionnaires were arrested early in December.

---

[96] Andrej Glušič (1905–1970), colonel in the Royal Yugoslav Army, specializing in cartography. Came to Ljubljana in 1942 and was interned by the Italians. Upon his release he became chief of staff of the command of the *Yugoslav Army in the Homeland*–the resistance movement of Dragoljub Mihajlović. In June 1944 the *Gestapo* arrested him and sent him to Dachau. Following the war he conducted covert actions against the communist regime from Germany. Emigrated to the United States in 1950 where he was a cartographer in the Pentagon.

[97] For comparison let me state here that montly pay of a relatively well paid government employee was 4,000 lire.

After the war questions were raised in certain circles whether the Slovenian clericals betrayed the leadership of the Slovenian Mihajlović's *Ravna Gora* movement. Insinuations that the clericals might have betrayed the leadership to the *Gestapo* is absurd for the *Slovenian Legion* which cooperated with the *Sokol Legion* and *National Legio*n belonged to the clerical *Slovenian People's Party.*

At about the same time the *Gestapo* discovered in Ljubljana a non-communist anti-Axis organization which included the *Sokol War Council* and the *Sokol Legion.* The arrests began gradually, apparently as information obtained from those arrested was extracted in interrogations. My turn came on December 23, 1944.

At around six o' clock in the evening as I was in the process of leaving my office, executive secretary of the *National Committee,* Marjan Zajec, intercepted me and handed me a draft law on defense and the military for review and comment. I was reluctant to accept such a document at such a time and place because I had to go home by streets darkened because of the air raids. Because Zajec insisted that it was urgent I returned to my office, read the document and hid it in a cabinet among plans and other files, and then went home. For the next day I had an appointment with the office driver to take me to a place where I was to pick up a spinning wheel my wife wanted for Christmas. The sale of the spinning wheel was arranged by Račič who was an expert on domestic arts and crafts. I asked the driver to take the spinning wheel to Ivanka and walked to my office. I came in about half an hour late.

In front of my office stood a man sporting riding boots who accosted me in Slovenian, introduced himself as Dolinar, agent of the *Gestapo,* and told me I was under arrest. He directed me to go to the office and hand over my affairs. From my office I telephoned Principal Rueh. When he came I told him what happened and requested that he notify my wife of what happened and give her 500 lire which I handed him. The *Gestapo* agent butted in saying that I would need the money for myself but I curtly told him that he was to leave that to me. Soon after that the commander of the *Sokol Legion* Edi Antosiewicz entered without knocking. He came for our daily discussion of *Sokol* affairs, as well as Etbin Bežek. Antosiewicz was surprised seeing us three in the office but I was able to ward him off with a meaningful stare which he understood and withdrew forthwith without uttering a word.

In the office I still had those highly sensitive documents which Zajec delivered to me the previous night which, if found, could send us all to the gallows. In addition I kept there a decree from General Mihajlović promoting me to the rank of Major in reserve of the *Yugoslav Army in the Homeland.* That promotion should have appeared in the regular edition of the *Official Military Gazette* but the German bombings of Belgrade prevented its publication. Fortunately Dolinar did not search my office, he only sealed it for future investigation.

The Gestapo agent took me to *Slavija*, a building owned by a Czech bank which the Gestapo appropriated.[98] He brought me before Chalupski, an interrogator known for his brutality and murders. Chalupski's favorite method of interrogation was to set his victim on a chair with a noose around his neck which was attached to the wall, then remove the chair so that the victim began suffocating. In the last moment he would replace the chair but sometimes he was not fast enough and his victim died.[99] As Chalupski interrogated me two Gestapo thugs stood behind me ready to commit any act of violence at his bidding. I was confronted with Ivo Krmavner, commander of the *Slovenian Legion,* from whom I believe Chalupski extorted the admission that we were in communication. Chalupski ordered Ivo to tell me: "Everybody has confessed" — "*Die Leute haben alles gesagt.*" Ivo began to tell me this in Slovenian but the *Gestapo* agent ordered him to say it in German. While the agents were emptying my pockets, Dolinar heaped profanity and abuse on me chortling with malicious glee: "No more professional engineer, no more senior counselor." He was one of those primitive types who find pleasure in abusing educated people whose social status they envy because they cannot comprehend their role in the society and the significance of their work. Because I did not "sing" interrogator Chalupski began shouting abuse using the foulest imaginable language. In the midst of Chalupski's harangue somebody in the adjoining room called Chalupski away and in a lowered voice forbade him to continue using foul language. Failing to get anything out of me they locked me in an unheated solitary cell furnished with only one thin tattered blanket. I was so cold that I could not sleep until the next night from the 24th and 25th of December when the heat was finally turned on. I remained in the solitary confinement for three weeks until the second interrogation. Trying to keep warm I was walking around the cell perhaps some 30 kilometers every day. Soon after that they arrested Stanko Dimnik and consigned him to solitary confinement, too. Gestapo agent Simon did not want to accept a blanket for him from his wife saying: "We must soften him up first."

The courthouse prison had three stories. The cells were built around an empty space in the middle which we called "the submarine." At about seven in the evening, when the *Gestapo* guards left for the day leaving only a domestic guard, a lively

[98] When the communists took over their political police *OZNA* moved into the same building. In addition to the fact that the place was already suitably equipped they no doubt appreciated that it already had tradition in the unsavory business of the secret police.

[99] The Gestapo let it be known that Glušič's coworker Kosta Domazetić had hanged himself in prison. Domazetić was a courageous young man, his wife was a concert pianist. He was by no means a suicidal type. I am convinced that his death was caused by Chalupski when he was trying to extract information out of him. After the war the communists also let it be known that Gangl's secretary Poharc had hanged himself in prison which is also unlikely: he was probably murdered by some disciple of Chalupski.

conversation between the cells began. When I was still in the solitary I overheard the following exchange which was reverberating throughout the "submarine."

Stanko Dimnik (member of *National Legion*): *"Ivo!"*
Ivo Krmavner (member of *Slovenian Legion*): *"I hear you."*
Dimnik: *"Where is Bevc?"*
Krmavner: *"I do not know."*
Dimnik: *"I want to tell him that they found that."*
Krmavner: *"What did they find?"*
Dimnik: *"Draft of the plan."*

Stanko Dimnik

I did not want to take part of that conversation because confidential things were not to be treated in this way. Dimnik's message perturbed me. It seemed that the Gestapo found the incriminating documents when searching my office. On sleepless nights I was thinking of how to field the questions concerning those documents when the interrogators would bring them up and of possible defenses but when the time came interrogator Wurnigg never mentioned those documents.

When we, the blue guards, were after some six weeks packed in a common cell and I saw Dimnik my first question was who had found "that." He explained that Bežek, who knew where I was hiding things, immediately after my arrest went to my office to remove the incriminating documents. The door to my office was sealed until such time as the Gestapo would come around to search it but fortunately there was another door to my office blocked by a filing cabinet which Dolinar did not notice. Bežek opened that door, pushed away the filing cabinet and retrieved the documents. He immediately buried them in his garden. Expecting that he, too, would be arrested Bežek asked a friend in the National Guard detachment at Postojna to hide him clad in a National Guard uniform. However, according to the Nazi procedure of "*Sippenhaft*" the Gestapo arrested his wife–his daughter had fortunately been married and safe away in Italy–and held her until Bežek at her request gave himself up. Dimnik would have saved me many hours of worry if instead of shouting "They found that" he had used the expression: "They put that safely away."

On my first days of solitary confinement I heard a persistent knocking on the wall which was coming from the adjacent cell. I approached the place from where the knocking seemed to be coming and heard a subdued voice: "Go to the window." The window was high above the floor and I had to climb up to it. From the cell next door my neighbor, an active military officer, instructed me to put my ear to the wall when I heard the knocking and he would speak into the wall on his side. Indeed I could understand everything he said. First he told me that former lieutenant governor Pirkmajer was inquiring if I wanted to receive meals from a nearby restaurant.

I gratefully turned down the offer. My wife had been sending, from time to time and to the extent that she was able to, some staples by bribing the prison warden Strohmayer with a carton of cigarettes so that he would deliver to me a package of food which would last me a few days.

My neighbor also was sending me reading materials by tying the bundle on a rope and throwing it from the window to me where I caught it. I was returning it in the same way. He was a courier of the *Yugoslav Army in the Homeland* whom the Gestapo had intercepted and imprisoned. I do not know what had happened to him.

A few times, at the very beginning, when I was in a cell looking over Cigale Street *(Cigaletova ulica)* I talked with my family who came there at night. The windows were blocked by screens installed by the Italians so that the prisoners could not signal out. It was necessary to climb up high and shout over the screen. One could hear very well in both directions. However, at the second contact of this kind, a Slovenian guard from the courthouse came up to my family and advised them to go away. He said that the *Gestapo* agent Simon would sometimes drive up in with his car, arrest the relatives of prisoners and take them to jail where they did not fare well at all. We thus abandoned that mode of communication and found better, safer ways to maintain contacts.

When my turn for the "trial" came the interrogation took four hours; it was preparatory for the sentencing which was to be done in my absence. From the questions at the interrogation I could infer that despite the numerous arrests the *Gestapo* was unable to discover the full extent of our organization and of my activities. The principal charge against me was that I was communicating with Glušič and collected for him reports from the three *Legions*. Fortunately I was not in position to tell anything more specific to *Gestapo* interrogator Wurnigg than the messenger's name "Yuri" who once called at my office. In concluding the interrogation Wurnigg said that he was under the impression that I had not told him everything but he could not come up with any other question with which he could extract something more out of me.

Antosiewicz had his own courier who posed as a *Sokol* but who was in fact a communist. Antosiewicz had used that courier to send the legionnaires a circular in which he was alerting them to be prepared for the final expulsion of the occupying forces from Yugoslavia. That circular fell into the hands of the *Gestapo*. Two weeks later Antosiewicz was arrested along with his courier. Prior to his arrest Antosiewicz went into hiding but the Gestapo imprisoned his wife and daughter and Antosiewicz gave himself up.

Antosiewicz or "Ante," as we called him, was sentenced to the concentration camp at Dachau and his courier to forced labor in Germany. In prison we learned that the police arrested that courier immediately before he had to board the transport to Germany. He was sentenced to death and summarily executed. In prison we could only guess at the reason; those who knew him held that he was a communist which fact the police apparently discovered. By infiltrating our organization he had done great damage, particularly because we lost Ante on his account. The full extent of the damage he had done to the *Legion* and how much of its workings he got to know remained unknown. The day before the transport left Ante and his courier were placed in our cell where they stayed overnight. Antosiewicz was very upset, incapable of conversation and sobbing all the time. The courier, on the other hand, was calm and composed. None of us had his own blanket, we only had some tattered covering. The courier, however, had a nice brand new quilt which seemed unusual to us. He was a young man, most likely without any profession. He approached me to "confide" something to me. He said that at the interrogations the *Gestapo* never found out that I was the chairman of the *Sokol War Council*. By instinct I distrusted him and did not let him draw me to a conversation. I had only known him as courier bringing Antosiewicz's messages when Ante could not do that in person.

I found a way for maintaining contact with my wife. Once when I had gone unshaven for three weeks they sent me a barber who was also prisoner. He immediately recognized me recalling that as a city councilman I helped him find employment at the time of the Great Depression. He offered to take letters to Ivanka as they would occasionally send him out on various errands. I gladly accepted and in this way we were regularly exchanging notes. With the help of *Gestapo* guard Strohmayer who always willingly accepted a few cartons of cigarettes, some other useful present or money, I was able to receive additional packages of food and underwear. Strohmayer even arranged for visits in the office of baroness Zois who maintained the card files of the *Gestapo*. This was quite illegal, as it was outside of the officially allowed visits which were always in the presence of guards.

When I was arrested Rupnik stopped my salary payments and approved only one-fourth of the amount of my salary as subsistence for my family.[100] Ivanka, however, did not have to worry. My friends Lenarčič, Žitnik[101] and Urbas had on several occasions given her 5,000 lire[102] as support, probably from their own funds as they were rich but it is also possible that the money had come from Mihajlović's relief funds.

100 Under the German occupation families of political prisoners were thus given at least a small allowance. When, however, I left the country as émigré the communists had formally discharged me from civil service and expressly decreed that no subsistence allowance is to be accorded my family.

101 Rudolf Žitnik (1901–1974). Sheet metal contractor, Slovenian liberal politician, vice-chairman of the Ljubljana committee of the *Yugoslav Nationalist Party*, active member of the Mihajlović resistance, member of the *Slovenian National Committee* from its inception. After World War II he emigrated to Argentina where he was active in the *Slovenian Democratic Party* and published the paper *Free Speech (Svobodna beseda)*.

102 My monthly salary amounted to approximately 4,000 lire.

I was soon taken to a medical examination without having asked for it. I concluded that I was selected to be sent to Dachau. They took me to the infirmary in front of which already waited Otmar Pirkmajer so that I could talk with him for some ten minutes. Pirkmajer was called in first, I was next. The doctor spoke Slovenian, next to him stood an armed *Gestapo* man. The physician suggested that I refrain from recounting any infirmities because that would only make matters worse. He probably knew that people who were in poor health were marked for immediate execution at Dachau which at that time was not yet universally known.[103] The physician dismissed me without even touching me.

A few days later we heard that Pirkmajer was sent to Dachau. At that transport the Nazis were particularly brutal; they might have known that it was one of the last transports because the railroads were badly damaged if not destroyed by the incessant American bombing. The Nazis robbed the prisoners of everything they had, watches, shaving kits, and the like. The last train with prisoners to Dachau left in February 1945; it was apparent that sending prisoners to Germany was becoming increasingly more difficult. The roads and railways were constantly under attack by the American Air Force which was concentrating on roads and railroads in Austria so that travel between Slovenia and Germany had become impossible. The Germans were already systematically withdrawing back to the *Reich*. That saved me from deportation to Dachau. We remained in the uncertainty of our prison, however. *Gestapo* agent Rechnagel, a young man of some twenty-five, came to prison and called some of us out to the corridor. He was offering work in Kočevje and he did persuade a few to accept. To me he was offering a professional assignment in Judenburg, Austria, which I declined preferring to remain in prison.

Details about the imprisonment, about my companions, cruel guards, interrogations, and physical and mental suffering cannot be narrated in this place because some of my companions are still alive and I could reveal something that could expose them to renewed persecution by *Gestapo's* successors. Later, in a refugee camp in Italy, I received word from Klagenfurt that both *Gestapo* prison guards and agent Dolinar were employed by the police of the British military government. About *Gestapo* interrogator Chalupski rumors were circulating that he was accepted in the service of *OZNA*, the communist political police where his professional experience was greatly appreciated.

---

[103] Once before the war we were on a steamship enroute from Sušak to Malinska on the Adriatic coast where we used to vacation. On the ship were two German women, probably Nazis, complaining that the Yugoslavs were not very hospitable while, so they said, in Germany everybody is received with hospitality. I could not help but telling them:"*Ja, besonders in Dachau*" (Indded so, especially in Dachau), at which the two Germans flinched as if I had slapped them across the face. Little did I think that in not too distant future German hospitality would be waiting for me at that infamous concentration camp known already before the war as an instrument of Nazi terror.

Otmar Pirkmajer became seriously ill at an epidemic which broke out at that time in Dachau. He was fortunate to recover. Following the German surrender he remained in Germany where he was employed by the *International Refugee Organization* in the key position of a liaison officer[104] between the various national refugee committees. He could not return to Yugoslavia because the communists in Ljubljana had sentenced him to death. When I was in America I was corresponding with him regularly. He sent me his report on the status of refugees in the Federal Republic of Germany.[105] In 1963 my wife and I were in Europe and at that time we went to visit Pirkmajer in Munich. The house keeper told us that Pirkmajer had a stroke and that his son came from Maribor and took him home. In Maribor he was occasionally seen in a wheelchair, paralyzed by the stroke. He died soon after that.

The overflights of the American Air Force were becoming more frequent. We counted as many as one thousand American bombers per day who were flying undisturbed and untouchable from their targets in Germany over Yugoslavia to bases in Southern Italy which had already been occupied by the Americans. Ljubljana, too, was bombed a couple of times. At an attack on the railway depot the projectiles broke the windows in our cell. The guards cowered in the air raid shelter while we had to remain in our cells without any cover or other protection. The Americans had discovered where the German headquarters were located and they laid a small "carpet" on Mirje where Rösener, the German general and proconsul for the Province of Ljubljana, and his staff were staying in villas from which they evicted the owners. My friend Vurnik, too, had been among those evicted from his house on the banks of Gradaščica. Unfortunately the Americans failed to score a direct hit; they missed by only some fifty yards and in the process killed sixteen innocent Slovenians. That unfortunate event was used by the Germans for anti-American propaganda. Rösener and his staff who divined that the Americans have found out where he was staying wasted no time to move to Hotel Bellevue under the *Šišenski hrib* where two tunnels had been dug as air raid shelters into the hillside under the hotel.

One day interrogator Wurnigg appeared for a "friendly" chat. He brought a few packs of cards and, if I remember correctly, some cigarettes. To me he mentioned that I had abused the trust of General Rupnik referring probably to Rupnik's offer that I become his adviser which I had turned down.

When Ivanka found out that I was sentenced to internment at Dachau she and Milena Pirc, wife of my friend and fellow prisoner Stane Pirc, went to see Rupnik himself with a request that he prevent the execution of the deportation sentence. Rupnik probably resented that I did not want to be his adviser and when the *Gestapo*

[104] *IRO US Zonen Verbindungsoffizier zu den Nationalkomitees.*

[105] *Die Lage der Displaced Persons und der andern von der IRO betreuten flüchtlinge in Deutschland und ihre Zurückgebliebenen Probleme nach dem Aufhörden der IRO.*

discovered the activities of the *Slovenian Legion* and the *Sokol Legion* he probably inadvertently mentioned that to the *Gestapo*. Ivanka and Milena were first received by Stanko Kociper, Rupnik's son-in-law and aide-de- camp, and then also by Rupnik. Both Kociper and Rupnik maintained that they had nothing to do with the arrests of Mihajlović's people. To this Ivanka observed that it was strange that the *Gestapo* agents said that I had abused the president's trust. Rupnik was embarrassed and tried to vindicate himself but he would not promise that he would try arranging our release. The two ladies thus left empty handed. On parting my wife said: "The tables will turn someday." Indeed they had for Rupnik but not to our advantage. Although the role of General Rupnik as chief of the provincial administration and inspector-general of the *National Guard* was purely administrative and in no way damaging to the Allied military operations the British arrested Rupnik in Austria and turned him over to the communists to judge along with Rösener and Hacin. According to some sources Rupnik could have evaded the arrest but he was so demoralized because of the mass murder of the National Guards that he no longer cared and was willing to go to a certain death. The communists sentenced him to death by shooting, not by hanging as Rösener, because he had supposedly regretted his anti-communist activities. Some time later rumors began circulating that Rupnik was mercifully shot unawares while he was taken to the place of execution. By spreading such rumors the communists possibly tried to convince people how understanding they were.

Ivanka was trying other ways to get me released from prison. She went to the *Gestapo* interrogator Wurnigg and told him directly that it no longer made any sense for Germany to keep political prisoners. Wurnigg pretended not to understand and asked for reasons why political prisoners should not stay in prison. Ivanka said that the war would be over soon to which Wurnigg retorted that this was far from certain because the Germans would bring out from their arsenal yet another secret weapon which would decide the outcome of the war. He said that the Germans possessed something which would produce a flash like lighting and annihilate all enemies within sight of it. Ivanka insisted and in the heat of the argument said that Germany would soon lose the war. She immediately realized that she had gone too far as she noticed that Wurnigg pressed a button under his desk. She left in haste his office and the *Slavija* building.

At that time Serbian general Damjanović with his *chetniks* arrived to Ljubljana and set up his quarters in Hotel Union. Ivanka went to see him, too. Damjanović kindly received her and told her that he was already negotiating with the Germans to trade several captured German officers for the imprisoned Mihajlović's men.

Soon after that we were visited in our prison cell by the boots sporting *Gestapo* agent Dolinar. He sat down at the table and announced that the Germans would release us — the Blue Guards. The murderous Chalupski, however, had dropped out of sight. Panic had already set in among our German guards. One of them, Hans Strohmayer from whom almost anything could be obtained for a suitable bribe, got drunk in the Bricelj's restaurant where he took practically all his meals at no cost to him, and drunkenly sobbed and wept over what would become of him when the Germans lose the war. His fears were unfounded, however, since as I already mentioned, the British gave him a good job with the police in Klagenfurt.

In the end I was released on April 10, 1945. Walking the streets after four month's in a prison cell almost seemed an unusual experience. Sixteen year old German youths, armed and in uniform, were presumably maintaining order on the streets of the city while the German army was retreating. Officers were walking around in some kind of raincoats and almost in mufti. The first day I spent alone with my family. Ljubljana had more frequent air raid alarms; we had a shelter in the cellar where we hid when the air raid alarm sounded.

The following day I reported at the office. The personnel manager issued instructions that I was to receive back pay for the entire time of my imprisonment which amounted to about ten thousand lire. I never received that money, probably because on May 3 martial law had been declared. I re-established my connections with the *Sokol War Council*, *Sokol Legion*, directorate of the *Yugoslav Nationalist Party* and the *National Committee.*

The *National Committee* put Ivo Krmavner, commander of the *Slovenian Legion,* who had in the meantime also been released, in charge of the preparations for the retreat across the border because the communists were already in process of cutting off Ljubljana, first in the south and then to the west from the direction of the Littoral. All this was news to me and the thoughts of leaving did not sit well with me. In retrospect I now regret that we were not informed of the political situation. If we had good information we could have withdrawn to Italy in time where the situation was better for the refugees and émigrés. It would probably also save lives of the thosands of National Guards who were murdered by the communists with British assistance.

At the end of April 1945, the *National Committee* invited General Rupnik to a conference I arranged under the auspices and in the presence of Bishop Rožman. In addition to the members of the *National Committee*, many other friends came so that the bishop's drawing room was packed. Rupnik with his retinue arrived last. At the door he noticed me and in a very friendly manner shook hands with me at which I recalled that the *Gestapo* interrogator told me that I was accused of abusing his trust. Perhaps his kindness was an attempt to make amends for the injustice he had done to me with his perhaps inadvertent remarks to the *Gestapo.*

The *National Committee* requested Rupnik to resign his position as President of the *Provincial Administration* and transfer his powers to the *National Commit-*

tee. Rupnik declined saying that the time for that had not yet come. He explained that he believed that the German *Operation Werewolf* would be activated and that it would eventually reverse the military situation to the advantage of the Germans. He said he would resign when the British tanks appeared on Tržaška Avenue (*Tržaška cesta*).

United Slovenia Proclaimed

Following that, on May 3, 1945, the *National Committee* called a meeting of a parliament of representatives of democratic Slovenians from all regions. Approximately one hundred and fifty representatives convened in the small hall of the *Sokol Hall at Tabor.* Jože Basaj,[106] chairman of the *Nationa Committee for Slovenia* was presiding. The Parliament acknowledged with approval the *Declaration* of the *National Committee for Slovenia* of October 29, 1944 with which it published its formation along with the entire contents of that constitutional document. The Parliament resolved to publish immediately the document with which it was established and declare the liberation of Slovenia as part of a free Yugoslavia, confirm the *National Committee* as the provisional government of the liberated Slovenia, approve all the measures hitherto adopted by the *National Committee,* and invite King Peter II to immediately return to the liberated homeland. The proclamation appears on page 173.

The short lived government of the *Slovenian National Committee* thus attempted to acquire for the Slovenians all the territory on which they lived including the Slovenian Littoral and Carinthia. That was considerably more than they could be offered by Tito's Yugoslavia, which with its hostile demeanor toward the Americans and the British, who were in any case traditionally disposed to favor Italy and Austria, lost any chance of getting their support of its ethnically justified territorial claims. Tito and Stalin comforted the Slovenians by telling them that after the final world-wide victory of communism all frontiers will be abolished. Fortunately there is every indication that the universal victory of communism will not materialize in the foreseeable future and Slovenians living outside Yugoslavia will in all likelihood sooner disappear from the scene as they already are disappearing in Carinthia. The newly proclaimed government tried to achieve reconciliation in a civil war but it had neither the military power nor diplomatic support of the so-called Allies to accomplish this.

Both major Ljubljana dailies, *Jutro* and *Slovenec,* were represented at the National Committee's session by their general editors, but it was only *Slovenec* that

[106] Dr. Jože Basaj, financier at *Zadružna zveza* (*Cooperative Alliance*) in Ljubljana.

TO THE SLOVENIAN NATION

In accordance with the Atlantic Charter and other solemn declarations of democratic Allies on the free determination and democratic government of nations

NATIONAL COMMITTEE FOR SLOVENIA

At this momentous historic hour for the Slovenian Nation

Declares

- All the territory on which Slovenians live is hereby declared to be the Nation State of Slovenia which is an integral part of a democratic federative Kingdom of Yugoslavia.
- The National Committee for Slovenia hereby establishes the Slovenian Government as a provisional highest administrative authority for the National State of Slovenia.
- The Slovenian Army has assumed the responsibility of maintaining peace and order in the nation state and will secure the frontiers of Slovenia. The Slovenian National Guard and other national armed units have placed themselves under command of the National Committee for Slovenia. They have already sworn the oath of allegiance as an integral part of the Royal Yugoslav Army in the Homeland. In the nation state of Slovenia each political party will in accordance with the principles of liberty have the opportunity to develop its power and realize its aspirations.
- The National Committee for Slovenia invites all Slovenians to universal concord and reconcilliation so that the modest and weakened forces of the nation will not be tied up in internal strife at a time when they are needed free and united for the achievement of the highest national aspirations. Accordingly the Committee calls on the partisan detachments to immediately terminate all hostilities against the Slovenian National Army and all actions against the peace-loving Slovenian population so that the terrible fratricidal blodshed may cease.
- In this crucial moment the National Committee for Slovenia calls upon all Slovenians to maintain their national discipline, that they act toward everybody with dignity, that they refrain from any personal vendetta and that they follow the directives of their National Government.
- Ljubljana, which will, thanks to the victorious Allied arms and our efforts, for the first time in history become the capital of United Slovenia must be a model of national pride, political maturity and civic discipline for all.
- Trusting in the Almighty God who guides the fate of nations and relying on the support of the great democratic Allies, the National Committee for Slovenia invites all Slovenians to cooperation for the benefit of the homeland, unity and reconciliation, as these are the only guarantee for the existence and progress of the Slovenian Nation.

Long live King Peter II!
Long live Federative Kingdom of Yugoslavia!
Long live the National State of Slovenia!

Ljubljana, this 3rd day of May, 1945.

published the proclamation. I still have a copy of it, it appears on page 192. The editors of *Jutro* preferred to remain circumspect.

The next day I was visited by Marjan Zajec, who was seeking candidates for the cabinet. He told me which portfolios had already been decided upon by mutual agreement and offered me the ministry of war. I turned it down because I would have preferred the portfolio of public works which would be in my area of professional expertise. We left the decision pending because the events were unfolding rapidly and we had to prepare for leaving the country. The cabinet posts offered by Zajec would not last long even if we succeeded in establishing a free Yugoslavia.

Bogumil Remec

Franc Kolterer

Albert Kramer, who was a minister several times in the pre-war Yugoslavia once mentioned that he was a minister *with* the government but not *in* the government. Similarly some members of the *National Committee* were under the impression that we were at the end of the war with the government but not in the government. To justify that observation let me cite one incident as an example. As I recall, in the morning of May 5 at the session of the *National Committee* in *Alojzijevišče* (St. Aloysius institute) Zajec introduced a man named Paul Duscha whom I had seen for the first and last time on that occasion. Zajec told us that Duscha had been a chief or something like that at the *Gestapo* in Ljubljana and that he was offering the services of his organization — the Gestapo! — to the *National Committee.* Duscha did not speak, obviously he was one of the rats leaving a sinking ship. I exchanged glances with Franc Kolterer[107] and Bogumil Remec[108] both of whom were, like me, astonished.

---

[107] Franc Kolterer (1888–1964), physician, senior sanitary inspector of state railways, senior medical counselor at the regional administration (*banovina*), chief of Rogaška Slatina health resort, founder and presiding elder of the *Sokol Association in Rogaška Slatina*, emigrated in 1945 and settled in Klagenfurt, Austria.

[108] Bogumil Remec (1878–1955), politician, professor of mathematics and physics at the college preparatory school in Kranj, founder and principal of *Commercial School* (*Trgovska šola*) in Ljubljana, president of *Cooperative bank* (*Zadružna gospodarska banka*), vice-president and director of *Mutual Savings and Loans* (*Vzajemna posojilnica*). Emigrated to Argentina in 1947 where he was a leading member of émigré associations.

German Occupation Ends

The Germans immediately banned all further publication of newspapers and decreed a martial law. Members of the *National Committee* decided to avoid possible arrests by spending the night away from their homes; I went to sleep at Dimnik's. Jože Basaj, chairman of the *National Committee*, however, was on May 5 surprised at his home by two SS officers. Thinking that they had come to arrest him he slid down a rope attached to the railing of his balcony on the second floor in the process of which he had badly burned the skin off his palms so that he had his hands bandaged for some time after that. It turned out, however, that the two officers did not come to arrest him; they only brought an invitation from Rösener requesting that the *National Committee* come to a meeting with the Nazi general at his headquarters in *Kazina* to discuss the situation.

This was to be the second time that I saw Rösener. I have already told of my first encounter with him in connection with the road sign construction. Present at the meeting with the *Slovenian National Committee* were Rösener, Rupnik and various police officials. Rupnik, who had been in constant consultations with Rösener during the recent two weeks, spoke first. He said he was resigning his position and transferring his functions to the *Slovenian National Committee,* but he would retain the command of the *National Guard*. Rösener then listened to the demands of the *National Committee* which, among other things, demanded that the Germans transfer all authority in Slovenia, including Lower Styria (*Štajerska*), to the *SNC*, give the *SNC* control of the Ljubljana Radio Station and release all imprisoned non-Communists. Rösener replied he had no authority in Styria and therefore could not meet that demand; he refused to give up the radio station and agreed only to consider the release of all non-Communists. It was apparent that Rösener was only trying to transfer the *pro forma* administrative functions to some civilian authority without actually giving up control until the Germans departed. On the other hand, it was also clear that the Germans intended to leave and were trying to accomplish an orderly transfer of authority. The Germans were leaving because they knew that their government would surrender to the Allies within a few days.

Tine Debeljak later reported[109] that Basaj, Remec, Žitnik and Zajec were in agreement that Rösener did not turn over his authority to the *Slovenian National Committee.*

In theory the *National Guard* who had already placed themselves under the command of the *National Committee* would have to disarm the Germans but in that situation at that point in time there was no way of doing so. It is clear, however,

[109] Tine Debeljak, Slovenian writer and editor in Buenos Aires, Argentina, translated Dante's *Divine Commedy* in 1959, in private communication dated November 8, 1977.

that the Germans were ready to turn over control to the Yugoslav government which would happen automatically at the time of German surrender. They left Slovenia not one minute sooner than they found it expedient and they would have left regardless of whether the Partisans existed or not. It is clear that the war did not end even five minutes sooner because of the partisans — the Germans left Ljubljana when the Allies forced them to surrender at Reims.

All the above events transpired on the morning of May 5. Zajec was summoned early in the afternoon to the governor's palace *(banovina)* where General Rupnik was to turn over the reins of government to the *Slovenian National Committee*. When Zajec arrived, Rupnik was already in his limousine and on his way to leave the country. Rupnik said to Zajec: "I am turning the government over to you (i.e., *Slovenian National Committee*); there is no money except in the Tobacco Factory. I am retaining command of the *National Guard*." With that Rupnik departed. According to some reports a Feisseler Storch standing by on Javornik's meadow took him directly to Carinthia by air. Some have interpreted this farewell as the transfer of the government to the *SNC*. The *National Committee* never exercised its power as a government. Leopold Eiletz,[110] however, confirmed that he assumed the powers of the government in Kranj in the name of *SNC*.

The National Committee appointed Dr. Hubad, chief in the department of internal affairs in the provincial administration, as the acting chief of the provincial administration. On the same day, May 5, most of the members of the Slovenian National Committee left Ljubljana and went into exile.

The end of the war separated Ladislav and Ivanka Bevc for ten years

[110] Leopold Eiletz, 1894–1981, politician, regional governor for the at Maribor, right bank.

Emigration

Way to Freedom

Early in the afternoon on May 5, 1945 I left Ljubljana. Žitnik notified me that members of the National Committee were assembling in the courtyard of the *Realka*. I asked him to take me with him in his car, a fairly large Mercedes, but his car was already full. He could only take me home to pick up my luggage and return me to *Realka*. At home I always kept my things packed for just such an eventuality so that all I had to do is pick the luggage up and go. While I was taking leave of my family Žitnik was nervously leaning on the horn. When we were finally pulling away he said to my wife: "We'll be back in two days." In the courtyard of *Realka* the governor's car driven by a chauffeur was available and I got a place in it. Other passengers in that car were Bogumil Remec and his wife, and if I recall correctly, Albin Šmajd.[111] I asked that we pick up Dr. Pestotnik so that we were five in the car. We headed toward Ljubelj, at that time the only direction that was still open. As we were riding along on *Gosposvetska* road Remec began to suggest to his wife to remain at home to which she only reluctantly agreed and stepped out of the car. The district was full of partisan sympathizers and Žitnik advised that we move along as fast as possible to avoid the malicious onlookers. However Mrs. Remec did not stay at home after all. She went to Carinthia later with other refugees, was intercepted on her way by the partisans, managed to evade them and after a few days joined her husband in Lienz.

Žitnik obtained from the police the password which we had to give to the border guards at Šentvid. We were on our own and, contrary to the communist insinuations, had no connection whatever with the retreating Germans. Tržič through which we drove was decorated with Slovenian flags — still without the repulsive red stars. Late in the afternoon we reached Ljubelj and pulled into the entrance of the tunnel connecting Carniola and Carinthia part of the way up under the Karavanke Alps. We slept in the car which, incidentally, became inoperative because a saboteur had put sugar in the gasoline. A German guard was kind enough to drain our gas tank and replace the fuel. Next morning we drove via Ferlach to Klagenfurt

[111] Albin Šmajd (1904–1946), attorney in Radovljica, leading member of the *Slovenian People's Party*, one of the founders of the *Slovenian National Committee*. After leaving Yugoslavia he was kidnapped by the communists in Trieste and dragged to Ljubljana where, after interrogation, in the quaint parlance of Slovenian historians he "disappeared without trace."

where we stopped in front of the diocese and had a picnic with the few refreshments we still had. The diocesan cook came to the doorway and tried to ward us off. She was concerned that the diocese would be compromised if the partisans whose arrival seemed imminent should find us there. As our car kept sputtering out and breaking down I asked Žitnik to take me in his car. He did and I rode with him through Villach, Spital and Oberdrauburg to Lienz which we reached on May 6. We took up quarters in Hotel Sonne; Miroslav Urbas and I shared a room.

On May 8, 1945, the British arrived in Lienz. We drove to the American demarcation line near Lienz and tried to obtain permission to enter the American zone of occupation in South Tyrol but the guard would not allow us to pass. That was the last trip with Žitnik's car which the British simply took away from him without giving him a receipt of any kind.[112] The British decreed that no one may remove himself beyond ten kilometers from Lienz. Food ration cards were being issued by the Nazi committee which the British left in charge of those affairs. We received the lowest class of ration cards. A weekly ration that these brought might be enough for one meal. Urbas and I had to comb the countryside around Lienz for food and we were able to come by some bread and milk at reasonable prices from the farmers. The farmers used to bake several months' supply of that bread and kept it on open shelves in the pantry. The taste of that bread was similar to that of bread found in Austrian military rations.

The number of Slovenian refuges was gradually increasing. One of them had a radio over which we could hear, in addition to news from home, transmissions of triumphal rejoicing of the communists. A woman greeted Pestotnik with: "Doctor, have you heard that you have been sentenced to death?" In fact, according to the news, Pestotnik, Pirkmajer, and Vindišer were all sentenced to death because the Credit and Loan Company (*Kreditni Zavod*) on the board of directors of which they served was transacting business with Italian clients during the war.

In the midst of such excitement that punctuated the daily tedium arrived a message from Klagenfurt — Viktring where part of the *National Committee* with Basaj, its President, was located. The *National Committee* sent word by a special emissary, Pater Odilo, that the British had decided to transport all Slovenian refugees to a better camp in Palmanova, Italy. Pater Odilo had been to America and spoke English. He came in a small truck which he drove himself. The refugees were to be transported in four groups on 27th through 31st of May 1945. The *National Committee* suggested that two of its members staying at Lienz come to Viktring so

[112] In this respect the British were different from the Germans. The Germans, for instance, had seized my friend Mačus's car but they gave him a receipt. When Mačus reached the Americans in Trieste he took the receipt to the American command and requested that they give him an equivalent automobile from among those that they had taken from the Germans. The Americans agreed and allowed him to choose a suitable car from those they held. Mačus then used the car for his travels around Italy.

that arrangements could be made for them to accompany the refugees to Italy and act as their spokesmen.

It was decided that Zajec and I would go to Viktring. I was to accompany the last group of the refugees to Italy and remain with it as a representative of the *National Committee.* On May 28, 1945 we drove through Spital, Villach and Klagenfurt to Viktring; Pater Odilo was informing the pesky British sentries on the way that he was taking two convalescents to Viktring.

In Viktring the *National Committee* was in permanent session. Zajec and I immediately joined them. Presiding was Dr. Basaj with Celestin Jelenc as vice-president. Also in attendance were Bajlec, Bitenc[113] and Drčar[114] who, however, were not members of the *National Committee.* Bitenc was most of the time in another room where the military leadership held conferences. He came several times out of that room to me at the session of the *National Committee.* First he told me that the military command had reservations in promoting Franci Slapar to the rank of lieutenant because it was felt he was not reliable. Slapar had earlier been a member of the *chetnik* group under Bitenc's command but soon left that group and joined the genuine *chetniks* in Dolenjsko; he had told me that it was one of Bitenc's idiosyncrasies to distribute religious emblems and rosaries among his men and expect them to carry those on their persons at all times. I told Bitenc that Franci was my brother-in-law and that any reservations as to his loyalty were out of question. Soon Bitenc came back to tell me that Franci Slapar had been promoted and that I, too, had been promoted to the rank of major. I responded that I had already received a regular promotion to that rank. I had submitted my application for promotion at the military district's office in Ljubljana in 1940. At that time communist agent Milan Lemež, mobilized into the service and assigned to the district office, received my application with a smirk on his face. My promotion was approved and should have been published on April 6, 1941 but the printing plant of the *Official Military Gazette* (*Vojaški uradni list*) had been bombed by the Germans. Later on Colonel Andrej Glušič, using the files that had been saved from the occupation authorities, managed to put that promotion through: it was approved by the then minister of war Mihajlović. After a while Bitenc returned

[113] Mirko Bitenc (1898–1948), professor of mathematics and physics at a college preparatory school in Celje, Major in reserve of the *Royal Yugoslav Army.*

[114] Ivan J. Drčar (1911–1982) Graduate of Military Academy in Belgrade, officer in the Royal Guards, deputy coach at *Sokol V Senjzak* in Belgrade, interned by the Italians in 1942, after collapse of Italy organized clandenstine resistance against the Axis occupation under the command of Colonel Andrej Glušič and served as liaison officer to the *Sokol Legion*, joined *Slovenian National Guard* in 1943 where he was Chief of Staff. At the end of the war he withdrew to Austria and thence emigrated to the United States where he lived in Cicero, Illinois, employed as engineer in machine design and construction. Member of the general and executive council of the *Slovenian Democratic Party* in exile 1974, member of the *Slovenian National Committee*, 1979, active in the restoration of the *Yugoslav Sokol in the Free World* and in establishing the *Sokol Herald* (*Sokolski vestnik*), member of the *Association of Friendship of Nations of Central and Eastern Europe.*

with a message that I had been promoted colonel. I felt that pursuing the matters of rank and promotions a time when we were confronted by a national calamity was inappropriate and indicative that Bitenc was out of touch with reality.

In Viktring I had to sleep in the open but the food was ample; the refugees were preparing it themselves. I wondered where all the meat that was served with cabbage had come from. Later I found out that the refugees were slaughtering the horses and cattle they brought along because they had no fodder for them. Before I left Lienz Urbas gave me 1,000 Lire for his son whom he asked me to find in Viktring with the *National Guard*. I found his son and was able to give him the money and his father's greetings. He was with the group that left on May 29, 1945.

British Treachery

At the conference in Yalta (February 2 to 12, 1945) where Roosevelt, Churchill and Stalin met, the British promised the Soviets that they would forcibly return to them all Soviet subjects who would not want to be repatriated voluntarily. In his wartime memoirs Churchill wrote that Stalin said: "We shall shoot all the officers." Churchill objected to that and when Stalin saw his reaction and disgust he quickly added that he had said that in jest. Churchill noted in his book that he was not altogether sure that Stalin had been joking. Roosevelt was already very ill at that time and it was rather strange that his doctor allowed him to go on such a long journey to a strenuous conference. But Roosevelt had such fervent desire to see Stalin and fulfill each and every one of his wishes that he went anyway. Two months later he died. In any case the results of Yalta conference were a success of Stalin's policies and an agreement that will live in infamy. The consequence of theat agreement between Roosevelt, Churchill and Stalin, whom the two western statesmen thought would be willing to share with them the control of the rest of the world, was that Stalin did what he pleased on the Balkans and in Eastern Europe. Churchill simply dictated to the Yugoslav government in exile and even to King Peter that they had to leave Tito free hands for introducing a communist regime.

In our immediate vicinity–some ten kilometers from Lienz in the direction toward Klagenfurt — was a camp of Cossacks and Byelorussians. These were Russian soldiers whom the Germans had organized in the *Russian Liberation Army* under the command of former Soviet General Vlasov which fought alongside the German army. After the breakdown of Germany the Vlasov soldiers made the said camp in which they lived together with their wives and families. The British prepared everything for their transport to the Soviet Union and by surprise overpowered the Vlasov troops who resisted when the British began to drag them to the prepared trucks and trains. Some of those unfortunates succeeded in fleeing to the woods thus escaping a certain death. By eyewitness accounts a great deal of blood was spilled, the despair of the unfortunates was indescribable and many committed suicide. The British took them to the nearest ports from whence the ships, which were always available for such purposes but never for the transport of refugees to the countries of their choice, went to Odessa where they were officially accepted by Soviet authorities. The Soviets were shooting repatriated refugees right on the piers The British knew full well that they would do so. Women and children also were delivered to the Soviets as were Byelorussians and even Russian émigrés from the time of the revolution although these were not included in the Yalta agreement.[115]

On May 29, however, the second day of the shipments of Yugoslav refugees, a *National Guard* officer managed to return from Rosenthal with the news that the British were handing refugees over to the communists who were waiting for them concealed in railroad cars in Rosenthal. Basaj immediately went to see the British commander Lieutenant General Keightley in Klagenfurt to find out what was going on. After considerable delay Basaj was eventually admitted to Keightley's presence only to hear indignant denials that anything like the handing over of refugees was occurring coupled with threats of severe punishment of anyone who might spread such rumors. Tolstoy's paper, however, reveals that Keightley, the British officer who most likely by direction of Harold Macmillan, then the British resident minister for the Mediterranean at the Allied Supreme Command Headquarters at Caserta, arranged for the treachery, was lying.

While Basaj and his deputation were cooling their heels in Keightley's antechamber a certain "Major Jones" was bustling in and out of the general's office arranging, no doubt the details of what the general would shortly vigorously deny on his honor as a British officer. Apparently he was the same Jones whom the Partisans used to show off at their meetings during the war as evidence that their cause enjoyed full support of the Allies. During the war Slovenian papers did not want to report the presence of such British observers as Major Jones in the partisan detachments lest the public be demoralized; the papers even expressed doubts of the authenticity of such British emissaries. Thus for instance it was well known that Vito Kraigher, one of the leading communists, had posed as an Englishman when the *National Guard* surprised him at Bloke. He spoke only English but the *National Guardsmen* recognized him and arrested him. At his arrest the man who recognized said to him:"Quit fooling around!"[116] Major Jones, formerly a British observer with the partisans, was at Klagenfurt a member of the staff of the British commander Keightley. From such observations one can draw inferences concerning the sinister forces that played the decisive role in the betrayal of the refugees.

[115] Nicholas Bethell, *The Last Secret*, Basic Books, New York, 1974.

[116] Vito Kraigher was shot shortly before the end of the war near Turjak.

In 1983 Nikolai Tolstoy,[117] descendant of the Russian writer count Lev Tolstoy, was able to obtain access to official documents, which were under the British law made available to the pubic in a suitably sanitized form, and established that the principal culprits for the delivery of the refugees to the communists were general Keightley in Klagenfurt and Harold Macmillan, minister for the Mediterranean attached to the staff of general Alexander in Caserta. The British now maintain that Macmillan and Keightley were not authorized to deliver the refugees to the communists but such statements appear to be the customary diplomatic evasions which lack credibility.

At the same time another terrified fugitive from a "shipment to Italy" showed up with an identical report to that given earlier by the escaped officer. The *National Committee* still was not able to make up its mind what to do and continued its session through May 30, 1945. At around noontime, on that date, however, Dr. Janez Janež[118], a young military surgeon, who had been sent with the first transport on May 29, appeared, still in full uniform. He was exhausted because he had come to Viktring from Bleiburg walking day and night without rest, evading the British sentries who, naturally, also guarded the bridges on the Drava River. He reported that he was assigned to go with the group to be transported as a medical officer in an ambulance. When they arrived at the railway station at Bleiburg the refugees were mustered out of the train by the British. In front of the railway station Tito's partisans were already waiting for them. The British soldiers, who knew what would happen to the refugees, robbed them systematically of their meager personal belongings such as watches, shaving kits, and the like, before they handed them over to the communists. The communists, in turn, made the refugees undress so that they could take their uniforms leaving them, for the time being, their own lice infested rags. The military surgeon managed to avoid the muster by heading toward the nearby inn, ostensibly looking for a restroom. A British guard followed closely behind him but under his charge's reproachful look he eventually stepped aside when the surgeon entered the facility. As soon as the latter was sure he was no longer being followed he threw himself into a wheat field and lay still and unobserved. He heard the frightful cries of the refugees whom the partisans were loading into boxcars which were then locked. The night fell but he still heard the plaintive cries from the sealed cars. From the position of the stars he estimated that it was midnight when the train with its wretched cargo departed and silence fell on

---

[117] Nikolai Tolstoy, *The Klagenfurt Conspiracy, War Crimes and Diplomatic Secrets*, Encounter, v. 40, No.5, pp. 24–37; Nikolai Tolstoy, *The Minister and the Massacres*, Century Hutchinson, London 1986. The latter book is banned in Great Britain but is available in the university libraries in the United States and elsewhere.

[118] Following these events Dr. Janez Janež emigrated to Taiwan where he continued to practice medicine.

the station and the inn. He rose from the field and began to walk away from that place. It was still dark when he reached a farm and raised the farmer. Fortunately the farmer was not a communist sympathizer and he advised Janež how to reach the Drava River. The farmer also warned him about the British sentries on the bridges and told him where he could find a ferryman who would take him across. The surgeon found the ferry, crossed the Drava River, and without rest or sleep continued on to Viktring.

The *National Committee* thus had an authentic report about the infamous treachery by which the British were delivering the refugees to the Partisans. It was decided at once to warn the refugees advising them to disperse immediately and save themselves as best they could; from then on it was every man for himself. Later we learned that some refugees, notwithstanding the clear warning they received, boarded the last transport train on May 31, 1945, and went to their deaths. It was that last group with which I was to go as a representative of the *National Committee* to the "new and better refugee camp in Palmanova, Italy". It is a pity that the *National Committee* did not believe the first reports of those who fled. Thousands of lives could be saved.

Pater Odilo was ready to take both Zajec and myself, as well as Celestin Jelenc, — if I remember correctly — back to Lienz at once. We left on May 30, 1945 and arrived at Lienz on the same day, Pater Odilo was again informing the British sentries on the way he was transporting convalescents. The dreadful news spread instantly and it was universally agreed that we must disappear as soon as possible into the American sector. Urbas wept all night over the loss of his only son who left with one of the transports.

Later, in a refugee camp, I read a sworn statement by a refugee who succeeded in escaping after he was handed over to the partisans at Bleiburg. He gave the statement to the American Military Government in Trieste after he escaped from Yugoslavia. The victim reported that on arrival at the territory under the communist control the Partisans would stop the train at larger stations and drive the captives from boxcars through the gauntlets of crowds that spat on them and abused them in any way they could. When they reached the concentration camp at Teharje women and children were separated from men who were sent to a separate concentration camp. Guards were coming throughout the night taking the victims to natural underground caves and gullies. There the victims were lined up with their hands bound along the edges of the pits and shot in the neck. The young man in Trieste who bore witness to these massacres, however, had not been mortally wounded. After a while he regained consciousness and began feeling around. Finding the bloodied corpses all around him he tried to undo the wire with which his hands were bound. Suddenly he noticed another man rising and sinking again between the dead walking over them toward him. When the man reached him, he untied his hands. Both men waited inside below the rim of the pit until the executioners finished off another group, then they climbed out and ran in different directions. Although it was dark the sentries around the pit noticed their escape and began firing at them. Fortunately, our fugitive was not gunned down. He got away and managed to reach, still at night, a farm where the people dressed the wound in his neck. He moved on, mostly at night, and eventually reached his home somewhere in Dolenjska. There he remained until his wound healed and he recovered from his ordeal. When he got well his relatives advised him to leave the country lest the communists find out about him.

Another staging camp for mass executions, described by Tolstoy, was St. Vid near Ljubljana. Some victims at that camp were summarily executed there but the majority was taken to Kočevje, the Slovenian Katyn. There the communist executioners in their systematic way prepared a large clearing. The victims were bound in pairs, tied to a line and taken to a common grave where they were mowed down with machine guns. Then the communists finished off those that were still showing signs of life and threw hand grenades in the pit for good measure. Those mass graves were only superficially covered and local communist authorities, later complained on the earth heaving up and uncovering the partially decomposed dead. Thus Anton Gliha, a friend of ours in Krško, noticed that a foul smell of decay filled the air as soon as he began excavating on his property where he wanted to build a fence. As soon as he started the work a person came from the Local People's Committee forbidding him to dig further and ordering him to fill in the excavation. Spelunkers, exploring the various subterranean caves in Charso also often came across skeletons of people who were executed and thrown into the pits.

Some 13,000 Slovenians were brutally murdered in this way. Tolstoy says indications point to Lieutenant General Keightley and Harold Macmillan as persons most likely responsible for these crimes and conjectures that the payoff for sending the Slovenian refugees back to Tito and their death may have been the withdrawal of communist troops from the portions of Carinthia they occupied. Thus the communists sold the Slovenian soil to the British and eventually to Austria in order to gain revenge over their political opponents.[119] At the same time the diplomats at the founding of the United Nations at San Francisco were mouthing high flown oratory on human rights and fundamental freedoms.

[119] Charles Zalar, *Yugoslav Communism, a Critical Study,* prepared for the Subcommittee to Investigate the Administration of the Internal Security Act and other Internal Security Laws of the Committee on the Judiciary, United States Senate, 87th Congress, 1st Session, October 18, 1961, U.S. Government Printing Office, Washington, 1961. Contains a detailed index and comprehensive bibliography. Viktring is mentioned on pages 114 and 115. The book is no longer available from the Superintendent of Documents; the author obtained a copy by courtesy of the Chairman of the Senate Committee on the Judiciary.

A year or two after these events a boy was recounting to my son how he, as a young partisan camp follower, participated at such murderous orgies. The Partisans on purpose did not tie-up all their victims and let some escape from the pits so that other communists, including the boy who was telling about this, hidden some distance away in the forest, could amuse themselves by shooting down those trying to run away. The boy was not much older than my son who was fourteen when these heroic deeds were narrated to him. The communists made no effort to conceal what they did at all in those days, on the contrary — they used to describe these executions in morbid detail while they watched the listener to see how he would react. It was the same with concentration camps. Regimes that maintain concentration camps do not try to cover up the atrocities committed there. On the contrary rumors of such atrocities serve the regime as an essential element of its reign of terror. Tales that people in leading positions of such regimes were not aware of what was going on in concentration camps are nonsense. If somebody was punished because he was publicly talking about the horrors in the death camps that was only a pretext so that he could be imprisoned but in no way an attempt to stop spreading the word about what was going on in concentration camps.

Soon after these events both the British armed forces as well as Whitehall disclaimed any intentional complicity in the handing over of the Slovenian refugees to the Partisans. I observe, however, that while diplomats of most countries as a rule receive precise instructions for action from their governments Whitehall does not issue specific instructions to its senior diplomats, it only provides them with all the information that is deemed necessary for them to act. Thus, actions of the resident minister for the Mediterranean were for all practical purposes actions for which the British government must assume full responsibility and any disclaimers appear disingenuous. In response to condemnations of those heinous acts the British were quick to point out that it was not they who pulled the trigger. True enough, this is how the British fight the wars of their Empire: by letting others do the shooting.

We may justly ask whether crimes of such proportions perpetuated by an Allied government are not equally or even more deserving of prosecution as that of the defendants at Nüremberg or those Nazis that were apprehended more recently? In the case of the latter, the intervening time appears to be no bar to prosecution Indeed, while one could argue that the trials at Nüremberg had their legal foundation only in the generally perceived limits on the conduct of governments and the conscience of mankind, the actions of the British officers and diplomats were clearly illegal under the laws of their own country and the covenants to which it was a party, namely the Geneva Convention and the Atlantic Charter.

As the unfortunate defeat of Serbs at Kosovo in 1389 gave the impetus to the Serbian nation to concentrate its efforts toward the eventual liberation from the

Islamic invaders so should the majority of our nation oppressed by communists remember the Yalta conference as our national calamity equal to that of Kosovo. And bearing this in mind we should concentrate in our endeavors to regain liberty and the ideals of human rights and fundamental freedoms. Yalta — our second Kosovo.

Flight from the British Zone

When on May 30 Pater Odilo brought us again back to Lienz news of the infamous British treachery and delivery of over ten thousand Slovenian and two thousand Serbian refugees to the communists instantly became known to numerous refugees who had reached Lienz before that. Notwithstanding the roadblocks and checkpoints on the road to Italy which was in the American zone of occupation the refugees found ways to bypass the roadblocks and gathered in Anras, a mountain village to which Bishop Rožman had removed himself from Lienz. The bishop still had his car which was chauffeured by his chaplain. As narrated above, the car belonging to my friend Žitnik in which we traveled was seized by the British who did not even bother to give Žitnik a receipt. Members of the *National Committee* decided to follow the refugees to Anras where we would find out how to get around the checkpoint on the border of the American zone at Silian. The chaplain-driver agreed to load our knapsacks in his car and wait for us some distance away from the roadblock at Lienz. We were to get around it on foot and rejoin the waiting car. On June 2, 1945 this plan was carried out without difficulty. There were five of us in the group: Bogumil Remec, educator and financier, writer Tine Debeljak, his son-in-law, Josip Porenta, civil engineer, Miroslav Urbas, and I. In Abfaltersbach Urbas got out and headed for Gail Tal where he had relatives; finding his kin inhospitable, however, he soon moved on to Klagenfurt where he remained until his death. We remained in touch as long as he lived.

Tine Debeljak

The rest of us in the car continued uphill to Anras. On recommendation of the rector who was an old acquaintance of Bishop Rožman the farmers in Anras were very hospitable towards the refugees and willingly provided us with food and shelter. I was pleasantly surprised to find there Andrej Uršič, a political friend of mine, and his friends. We slept in a larger room, each on his own bed. The hostess fed us well as a matter of course. On Sunday, June 3, 1945, Bishop Rožman, officiating at the mass in the local church, conducted the confirmation of refugee children. The following day we started on our way to Silian to carry out our plan for reaching the American zone of occupation. The people in Anras told us where along our way we should stop to get detailed information for this purpose. The bishop gave me a letter of recommendation to all ecclesiastics requesting that they give me help in case of need. This is what he wrote:

> *"Litterae comendatitis*
>
> *Episcopus Labacensis omnibus Rev. is Sacerdotibus et ordinibus eclesiasticis virum eggregium Ladislavum Bevc quam optime comendat et rogat ut eum opportunis consiliis et informationibus adjuvent*
>
> *In itinere datum 1. Junii 1945.*
>
> *+Gregorius Rožman*
> *eppus Labacen."*
> *SEAL: F. B. Pfaramt in Anras, Ost Tirol*

The chaplain again drove Remec and me from Anras. On our way we picked up Debeljak and Porenta. In Absfaltersbach we got out and continued on foot through the villages of Strassen, Tassenbach, and Panzendorf to Silian. The trip took two days. The best we could do for the night was to stay in a shelter.

Outside Silian we received precise information on the movements of the patrols near the roadblock and instructions how to avoid them. We arrived safely to Silian where we called on the dean who welcomed us and invited us to dinner. We could only offer him the few ration cards we had left from Lienz as recompense. The dean promised to get a guide for each of us the next morning who would take us to the Austro-Italian border in the part of Tyrol occupied by the Italians. The border was being patrolled by Italian guards but the locals knew at what time the guards returned to their barracks.

That night before crossing the border I had a dream which I could never explain. In my dream I saw four youths whom the dean had assigned to us. The youngest was about fifteen and looked very much like my son; in the dream I was observing that he seemed a little older than he. Imagine my surprise next morning when the guides showed up as promised. They were four strong youths aged 15 to 18 years — exactly like the youths I had seen in my dream. Because my knapsack was the lightest the youngest guide was assigned to me. The guides were thoroughly familiar with the route and the timetables of the border patrols. When we reached the place of the crossing we paid each guide his fee. They pointed out the direction of mountain pass Windibach at an altitude of 1,900 meters where we would find a large farm. We set out across the snowfields and slush. Remec, who was the oldest among us and who had the most luggage was moving up the trail with considerable difficulty. From time to time we had to decide which turn to take, and each time we decided for the steepest path.

It was already getting dark when we sighted the farm. We were received without objections and given rooms, two persons to a room. Our host was the only man on the farm but there were several women who, in the absence of men who had gone to war, had to do all the work on the farm. We were invited to an opulent dinner after which we turned in to rest after the exhausting trip. In the morning Remec arranged for a porter to carry his luggage on our descent toward Dobbiaco. We were surprised to be treated to a glass of cognac at breakfast; our host explained that the military in the valley had abandoned their supply stores which the local population was quick to empty out.

Already back in Lienz we had decided that each group of the *National Committee* members would independently try to find a suitable route to Rome. Our group had lost at least two days at Anras. The first group, however, did not wait around and so it was able to get on its way in the desired direction without impediment.

The descent into the valley was considerably less demanding than the previous day's climb. In good time we reached a tourist inn on the outskirts of Dobbiaco where the porter left. Continuing on toward Dobbiaco we stopped on a meadow and agreed that I should go to the municipal offices and seek information about transportation to Rome; my companions would wait there until I returned.

I readily found the mayor's office. The mayor, a German appointed by the Italian Government, was in and he carefully listened to my request. He told me that all civilian traffic was suspended and that all transportation was reserved exclusively for the use of the military. The only suggestion he could make was that I might be able to get further information at the Italian military command post. I returned with this information to my companions and we all went to the Italian military command where they were flattered that anyone would come to them for advice but could only suggest that we go to the carabinieri for further information.

The carabiniere was very reserved and questioned us closely. I was the spokesman for the group because I was the only one who spoke Italian. When the carabiniere ran out of questions he barked that we were under arrest and, in keeping with the time honored Italian practice with which we became acquainted back in Ljubljana, demanded that we hand over our pocket knives. I was prepared for this and handed over a cheap small penknife which I had bought for just such an eventuality as Italian military authorities in Ljubljana used to be fond of staging raids in public places at which they would confiscate pocket knives. Thus I managed to keep my better knife with scissors, a can opener and the like which was a gift brought to me from Paris by Mr. Perhauz, father of a youth whom I had been tutoring as a student. Naturally, I was running some risk for the bully could decide to frisk us. Fortunately, however, the thoroughness with which people are nowadays frisked and searched in America was not yet the practice with the carabinieri at that time. After he took away our knives the carabiniere shoved us into a cell ordering us to leave our luggage outside. The cell was small, without windows or other sources of light; in place of beds there were some kind of planks without bedding arranged along the walls leaving hardly enough space for a person to get up and carefully move toward the door when it was open.

Confined to such close quarters all we could do was lie closely together. We were locked up and spent the following three days in that cell. Every morning the carabiniere unlocked the door to let us out where we could wash a little, drink some water and take a few crumbs out of our packs. We were also issued one American canned meat ration per person whereupon the carabiniere would again lock us up in that windowless cell. Our only fear was that we would be returned to the British in Lienz.

On June 7, 1945, the third day of our imprisonment, the carabiniere opened the door of our cell and told us we would be taken to the American command in Brunico for interrogation by the OSS. We were received by a captain who questioned us through an interpreter. Again, I was the spokesman for the group. The interpreter was a Jew who had not yet become naturalized. When I complained about the way the carabiniere treated us the captain said that the carabiniere had no right to detain us. He ordered that we be put up in a hotel where we also were to have meals. Because civilian transportation was still suspended he would at the first opportunity send us to San Candido (Innichen, elevation 1,175 meters) to the depot of the *22nd American Military Government Evacuation Camp* where we could stay and have food and shelter until he could arrange for the transportation to the interior of Italy or wherever we might wish to go. He sympathized with our plight and expressed hope that some day Tito might yet be put in his proper place by the United States. The captain instructed his interpreter to give us at least some food rations immediately but the latter could come up with nothing but cigarettes.

During the interview Remec discovered that part of his supply of gold coins disappeared. He carried some gold coins in his underwear tied at the ankles with a ribbon that became loose during the ride to Brunico to the pleasant surprise of the dishonest finder. Unfortunately, Remec did not wish to confide this to the American representative, he merely observed to us that he had enough hard currency left for all four of us to live in comfort for several years.

Travel to the Interior of Italy

In a few days, the American commander sent a car that took us to Innichen. The quartermaster at the American depot there was an Italian who showed scant enthusiasm at our arrival but he had to comply with the orders. The morose quartermaster's face bore a resemblance to the pompous features of Mussolini whose portraits used to hang at railway stations. Paying little attention to him we found some living quarters and although there were no beds available we managed to improvise quite acceptable bunks with bedding. We took our meals with the employees of the depot. In town, where we were allowed to go at will, we found a laundress and other conveniences.

Within a week a few more refugees showed up, among them Prelog. They asked us to pretend we did not know them. Beyond Silian they had been intercepted by the Italian police who were taking them to an interrogation. Thus we could not associate with them.

On June 12, 1945, the American commander was able to carry out his promise. He sent a truck with which we left Innichen through Dobbiaco, Brunico, Muhlbach, Brixen and eventually arrived at Bolzano. The driver took us to the *24th A. M.G. Evacuation Camp* and went to report our arrival. We were still on the truck when a Slovenian came to look us over. Still out of sight, we heard him asking as he was coming around the corner "Where are those hooligans?" On seeing us, however, he realized what the situation was and shut up. He was one of the Slovenians who were returning to Yugoslavia from the various internment camps. These returning Slovenians acted in a rowdy and boisterous manner similar to that of the communists in Ljubljana whose carousing we could hear over the radio. Naturally, we told the Americans that we had nothing to do with those people.

We then went to the barracks where we stayed for two days without anyone disturbing us. In town we looked for transportation further to the south. We came across a deacon dressed in white with the Vatican yellow insignia; he turned out to be an employee of the Vatican repatriation assistance. I introduced myself showing him Bishop Rožman's letter of recommendation which the deacon attempted to read but, not having been conversant in Latin as is the case for the lower clergy in Italy, he could not understand. Almost immediately a small crowd of curious onlookers gathered around us peppering the deacon with questions as to the contents of the letter I handed him to which the deacon responded: "It says here that he is a good man." [*Dice che è un buon uomo.*] I asked him to take my letter to the commission's office which he did but he soon returned and told us that the commission was there to help only Italians.

Eventually we found a truck that was going to Verona and thence to Modena. We climbed on it along with a sizeable crowd of others and, standing all the way, arrived at Modena and the *24th A.M.G. Evacuation Camp*[120] where we stored our belongings. The city was swarming with people returning to their homes. We stayed overnight; Remec and his son-in-law found quarters with a family that was renting rooms while Porenta and I spent the night in an abandoned house. The next day, June 16th, we climbed on an already overloaded truck full of people who had enormous amounts of luggage: bundles of bedding and clothes, cages of geese and all kinds of things they were able to salvage when the fascist regime shipped them off as laborers on loan to Germany. Enroute to Rome, however, we learned that the road was closed because a bridge had been destroyed. The truck then took the route through Bologna to Forli where it stopped at the *23rd AM.G. Evacuation Camp.*

In Forli we found quarters in a vacant house where we could store our packs and sleep on the floor. A Royalist Yugoslav and *Chetnik* camp was nearby and the following day I visited it and called on General Andrej (Prezelj)[121] and Chief of Staff Skalar (Valentin Benedik)[122]. My brother-in-law Franci Slapar also was there as a staff officer. The *Chetniks* lived in tents which they had put up: They invited me to stay with them but I declined because I wanted to see my companions safely to their destination.

I recalled that daughter of my good friend Etbin Bežek who had married in Italy was living in Forli. At the city recorder's office where I inquired about her address the clerks knew about her as her marriage to Count Canestri-Trotti represented quite a social event in Forli. I found the Canestri-Trottis without difficulty. Tatjana was with her infant son Giorgio whom her mother-in-law came to watch during our visit. Visiting her was a Slovenian friend married to an Italian colonel from Forli who dropped by for a visit. I told Tatjana that her father intended to leave Ljubljana the same day as I but that he probably changed his mind. During the conversation Tatjana's visitor was expressing sympathies for the communist regime in Yugoslavia; apparently she had been spared the opportunity to experience its benefits first hand without the protection of her Italian passport. I expressed a desire for finding transportation to the nearby refuge camp at Riccione on the Adriatic coast and Tatjana asked her husband to arrange the transportation for us

[120] The manuscript gives the same number as that for the camp in Bolzano.

[121] Ivan Prezelj — Andrej (Račič) (1895-1973), Colonel of the general staff of the Royal Yugoslav Army. During World War II coworker of Major Karlo Novak and Colonel Vladimir Vauhnik, succeeded Novak in the position of commander for Slovenia of the Yugoslav *Army in Homeland*, at the end of war withdrew with chetniks to Italy.

[122] Valentin Benedik(1898–1978) Juris doctor, University in Ljubljana 1925, member of *Sokol I* at Tabor, district judge at Brdo, Kranj, and Ljubljana, During World War II member of supreme command of the Yugoslav *Army in Homeland*, emigrated to the United States, presiding elder of the Sokol Association in Cleveland.

He readily found a driver with a car who for 500 lire took us to Riccione where we arrived after dark on June 18, 1945. We found acquaintances who told us to report the next morning to the camp administration. It was already dark when we got to a villa and entered a second floor room packed full of refugees lying on the floor; some were already asleep. There was no light in the room. As we entered the man in charge protested our intrusion at such late hour. Remec then responded that two of us were members of the National Committee. We literally had to squeeze ourselves between the people lying on the stone floor without covers or other bedding.

My flight from the murderous communists and their British accomplices was over and the long wait for emigration to America began. I was to spend two years in that refugee camp.

SLOVENEC

Združena Slovenija vstaja!

LJUBLJANA, 3. maja. Na nocojšnji zgodovinski seji prvega slovenskega parlamenta je bil z navdušenjem sprejet naslednji oklic:

Proclamation of the Slovenian National Committe, May 3, 1945

Refugee Camps

Camp Riccione

Riccione turned out to be a seaside resort twelve kilometers from Rimini to the north of Ancona. The occupation authorities did not bother to ask the municipality or anyone else for permission to use these buildings for the refugees. They simply established a refugee camp at the location of the seaside resort. The villa where we spent our first night there happened to belong to Mussolini's son-in-law Galeazzo Ciano.[123]

Early in the morning on June 19 we reported to the camp administration. The spokesman for the refugees in this camp was a journalist from Gorizia, Kemperle by name. Most of the refugees were people from the Julian March who fled from the communists who had temporarily occupied that territory and who even after the withdrawal of the Yugoslav troops continued to terrorize the local population. We were accepted as the camp's residents with few formalities. I was billeted in *Hotel Zanzani* which also housed the camp headquarters. Some friends and acquaintances of ours were already there: Boris Mihalič,[124] Anton Komotar, Božidar Golob,[125] and Vlado Gorazd; others came later. Kacin from Gorizia was also there with his family. Remec could afford to rent private quarters but he would show up at lunch to be in our company although he could not stand the food we had to eat. He bought the material for a suit and found a tailor to make it. At the fittings I acted as his interpreter. I was quite concerned how I would be able to replace my own clothes and the sole pair of shoes when they wore out. Initially no political divisions or animosities seemed to exist among the refugees, particularly not among the people from the countryside.

The camp administration had a vast supply of dried peas. Those peas, hard as stone, were on the menu day in and day out. The cooking of this staple started early in the morning and went on all day; we were served a bowl of peas at lunch and

---

[123] Galeazzo Ciano, Italian Minister of Foreign Affairs, executed by the fascists in Verona in 1944.

[124] Boris Mihalič (1898–1985) judge in Maribor and later in Gorizia, defender of the northern frontier, leading member in the *Sokol* organization, member of the resistance movement of Dragoljub Mihaljović. After World War 2 he was living in Trieste as member of the legal department of the directorate of the Allied Military Government and coworker of Radio Trieste. When Trieste was annexed to Italy he returned to Yugoslavia.

[125] Božidar Golob (1923–2005) professional engineer, member of the *Sokol Legion*, emigrated to Canada in 1948

another bowl at supper, as the camp administration wanted to use up as much of them as possible. I soon became ill from that diet. The camp doctor, a consumptive Turk who would soon die, showed no understanding or concern for my complaint and I had to find a private physician who took kind interest in my complaint and prescribed as my regular diet a broth of rice boiled in water. That doctor was kind enough to treat me without charge. I worked out an arrangement with an elderly couple, Tomaso by name, who were innkeepers near the camp. They would prepare my meals in accordance with the doctor's instructions and with their kind help my health was eventually restored. Teresa Tomaso, the wife of the innkeeper, told me that she was a *Slava* who came to Italy from Dalmatia but that she could no longer speak her native language. In *Hotel Zanzani* I shared a small room with two other refugees. The room had a sink with running water which unfortunately was unusable. A restroom for the use of at least 200 people who lived in the hotel was on the floor below our quarters. Our room was equipped with three bunks with a straw mat and one blanket on each; there were no pillows and no linen. After some time we were issued another blanket and eventually we managed to wheedle even a third blanket from the uncaring administration. A nondescript stand of a sort served as a table, I do not remember if there had been a chair in the room or not: we used to sit on the bunks most of the time anyway. The kitchen, where we had to go three times a day to get our meals, was outside of the hotel. Once the supply of peas was exhausted the fare consisted of unsweetened tea in the morning, some kind of thin soup with potato or vegetables for lunch, no meat of course, and again unsweetened tea and a tiny piece of bread for dinner. That piece of bread was the entire ration for the following day. From time to time we were regaled with meat: the so-called "*spam*" from American C-rations. The contents of one can from these rations were cut into sixteen pieces each no larger than a thimble. On those occasions we would buy white onions at the town market. White onions are mild and we used them as a substitute for fruit. A regular loaf of bread could be obtained in town for ten lire.

The camp commanders were initially English officers Chase and Carrigher. The latter in particular was very poorly qualified for dealing with the refugees and his superior officers. Following Carrigher's reassignment the camp administration was taken over by two Americans: Major Brown and Captain Sampson who were true gentlemen and very favorably disposed toward the Slovenians. The British officer Chase was a younger man who spoke only English, his Italian secretary acted as an interpreter. He was inaccessible to the refugees and apparently made it a point not to shake hands with anyone. The cook's name was Samohod; his wife was Slovenian, a former hotel chambermaid in Ljubljana. Samohod spoke only Italian and English, which he learned in America. The camp doctor, Bertha Aschner, a Viennese, whose mother was Jewish, was particularly interested in Samohod's background. Once, when Samohod was in the clinic for a medical examination, Dr. Aschner asked him to take a deep breath in German: "*Tief atmen!*" Samohod took a deep breath immediately and the doctor thus found out that Samohod understood German. From this Aschner inferred that Samohod might have been a German from the *Sudeten* region in Czechoslovakia who had lived in America from where he returned to the German *Fatherland* when Hitler recalled all the overseas Germans back to the *Reich*. Samohod's service to the *German Reich* took him to Ljubljana where he met his wife. At the end of the war he, too, became a refugee. Because he could speak English the British employed him as a cook. Naturally, Samohod always had access to the camp commander because of his ability to speak English. On the other hand, the refugee committee had no access to the camp commander, let alone to Samohod, and could never present its requests and suggestions for improvements.

The camp medical officer, Bertha Aschner, as I already mentioned, was originally from Vienna. After the *Anschluss* — fusion of Austria and Germany — she found a man in Yugoslavia who was willing to marry her by proxy for a valuable consideration. As a wife of a Yugoslav citizen Aschner could obtain a Yugoslav passport, leave Germany and move to Yugoslavia. Without ever having seen her Yugoslav husband she settled in Dalmatia but when Hitler occupied Yugoslavia she and her mother moved on to Italy where they lived inconspicuously in the countryside, eking out a living by giving English language lessons. Aschner also spoke Serbo-Croatian and Italian. Although Mussolini did officially promulgate anti-Semitic laws no one in Italy persecuted the Aschners because of their Jewish ancestry. At the end of the war the allied authorities employed her as a medical doctor in several refugee camps in Italy. Her last assignment was Bagnoli, the camp from which the refugees were finally, after prolonged waiting, allowed to leave for various countries of the free world to be resettled. Aschner eventually came to New York, where she was employed in various state hospitals for the mentally ill, the only professional employment the *American Medical Association* would allow European doctors to hold. On reaching retirement she returned to Vienna. We remained in touch by exchanging holiday greetings. In 1963 when the Engineering University in Vienna honored me by conferring upon me a *Golden Diploma* in recognition of fifty years of professional activity, Aschner represented me at the ceremony and accepted the diploma on my behalf because I could not attend in person at the time.

When the British camp commander was, on a rather short notice, transferred to another assignment he was replaced by a Major Carrigher, another Englishman of morose military bearing, who, as a rule was inaccessible and aloof. With his administration came a flurry of appeals to the refugees exhorting them to return home so that the British would be rid of them. At one time we were all herded together to listen to the prattle of a confused British colonel who told us that he had just come from Yugoslavia where he had seen that everybody was happy and that everything was well. Anyone who should chose to return to Yugoslavia, prattled the colonel, could go about his business without interference from the communists and would never be persecuted. Two of the refugees actually believed him and went back. One of them was Mayor Hafner of Stara Loka who was arrested as soon as he crossed the border and kept in prison for a considerable length of time.

Sometime in August 1945 it was announced that the British would disband the Riccione camp. The refugees were instructed to either leave the camp and return home or find a refuge elsewhere in Italy or in the Italian part of the Slovenian Littoral. People began to leave the camp. Among those leaving were primarily the people from the Julian March and the regions of the Slovenian Littoral occupied by the Allies.

I wrote to my wife's friend Countess Paola Di Trento, née Slokar, who lived in San Giovanni al Natisone near Udine. She replied forthwith regretting that she could not be of help because the situation there was still in flux. Later she told me that at that time an armed partisan woman was marauding around the countryside, probably looking for refugees. Next I wrote to Milan Lenarčič, a colleague and friend of mine, asking him to give me shelter until I should be allowed to go to America. During the first half of the war, Lenarčič, who lived in Vrhnika would frequently come to Ljubljana and often call on me. He was a *Sokol* and an unwavering supporter of Yugoslavia. He was a nephew of Franc Kotnik, a great benefactor of the *Cyril and Methodius Association.* At one time Lenarčič asked me to review and comment on his last will and testament. The will listed all the income from his properties with estimated value in gold French franks from which it was obvious that he was immensely rich, quite likely the richest or at least one of the richest persons in Slovenia. He owned properties in Vrhnika and in its surroundings, and a lumber mill and a quarry in Pohorje. His only son had died in a fatal accident early during the war. In his testament Lenarčič willed all his estate to a trust fund, governed by a board of trustees, which would provide scholarships for students. His widow would be chairman of the board of trustees among whom would be included our colleague Janko Mačkovšek and I. During the time of the Italian occupation Lenarčič bought in Preganziol, a small town on the route from Venice to Treviso, a mansion *Vila Pace* along with some land settled by four sharecroppers. In accordance with the ancient feudal system the sharecroppers worked their farms and gave part of their harvest to the landlord in lieu of rent. At the end of the war Lenarčič thus had a place where he and his wife could stay and were thus saved from the fate of the refugees and perhaps from even a worse fate at the hands of the communists who had, naturally, confiscated all his property in Yugoslavia.

I asked Lenarčič if I could stay with him until my immigration visa for the United States would come through. At that time I had no inkling that there would be great difficulties and delays in securing that visa. Lenarčič regretted that he could not oblige because he was under a barrage of such requests from many refugees. His mansion was by far too small to accommodate the hundred and more requests that he was getting. He asked me to wait a little longer until the pressure wore off whereupon he would welcome me with open arms.

After a few weeks, however, when the camp was already alarmingly empty the British changed their mind, allowed the remaining refugees to remain in the camp and filled it again with a new contingent of displaced persons. By that time Lenarčič was no longer besieged with requests for shelter in his villa and he invited me to come and stay with him for good. Because I had been among the first who had applied for a visa to the United States, I expected that it would be forthcoming soon and decided to remain in the camp where I would qualify also as a refugee. Lenarčič, on the other hand, persisted and kept sending me invitations to join him, he would even send me invitations by wire. However, I never made that move.

After the British announced they would continue maintaining the Riccione camp, Kemperle, a journalist from Gorizia who was president of the refugee committee, decided to leave. Before his departure he had, with the approval of the refugee committee, turned over his duties as president to me. As I had already mentioned, the refugee committee had neither access to nor any influence on the British camp commander. However, when the former camp commander was transferred to another assignment, he called me in to take his leave and even shook hands with me which for him was something extraordinary.

Under the administration of Major Carrigher, the new camp commander, a few refugee youths staked out Samohod as he was returning to his room after "dinner." The youths snatched out of Samohod's hands a bucket filled with fat. When it thus came to light that Samohod was appropriating for his own use the provisions that should have been used for the preparation of the refugees' meals, the refugee committee authorized me to call on the commander and protest against Samohod's thievery and the fact that our meals were much worse because of it. Carrigher, however, did not listen to me at all. With no explanation at all he told me that he was dissolving the refugee committee but that I was to remain as the chairman. It was clear that Samohod had gotten to him first.

New refugees were coming to the camp after it was reactivated. A larger, well organized group, came from another camp south of Riccione. Their more prominent and eloquent spokesmen were: professor Kessler, journalist Mirko Javornik,[126] Zupan and his cousin who used an alias. I called a meeting of the refugee committee and proposed that we elect a new committee. We held an assembly of all refugees so that new candidates could be nominated. At the assembly Zupan's cousin

[126] Mirko Javornik (1909–1986), writer, journalist, editor of *Slovenski dom*, author of *Črne bukve*. Emigrated to the United States after World War II and lived in Washington.

immediately took charge of the floor and actively participated in the discussion. The primary thrust of all his proposals was directed at preventing the possibility that I might be reelected although I was not particularly interested in continuing as chairman. Zupan's cousin promoted Kessler as his candidate and he eventually won the election defeating candidate Vlado Gorazd.

With his sister's help Kessler began to straighten out the situation in the kitchen and through his efforts the quality of food was in fact improved. Provisions for the camp were being supplied by the British quartermaster's store in Rimini, eight miles from Riccione. The camp had several trucks operated by the Jewish refugees who were billeted separately from us and had a separate kitchen. The camp administration was fond of soccer and soccer players enjoyed the privilege of having separate quarters and a kitchen of their own.

The irregularities in the distribution of food supplies were subject of a great deal of discussion among the refugees. Thefts of provisions and supplies were organized so that on the days when supplies were being transported, accomplices were stationed at suitable places along the road between Rimini and Riccione to whom sacks of flour, rice, sugar, meat and the like were thrown from the trucks. The thieves and their accomplices then divided the loot among themselves.

We found out what was going on in the kitchen from young refugees detailed by the camp commander to take turns in helping Samohod with the chores. One of them was Janez Žitnik, who shared the room with me. After his turn he was able to tell us about the advantages of working in the kitchen: the helpers could drink as much coffee with as much sugar as they wanted. Once Janez put so much sugar in his coffee that he could not drink it and he poured the entire jug down the drain, a terrible waste in those days. At ten in the morning Samohod's wife would call the boys in and treated them to a *gulyasz*.

Javornik obtained a subsidy from the camp commander for publishing a refugee newsletter. The subsidy consisted of a free supply of printing paper and the use of the duplicating machine. Soon, however, Javornik brushed aside the etiquette which, in view of the catastrophe that had befallen all the refugees, dictated that in the camps all political disputes be suspended. First the paper took to task the "*liberals*" and affronted us with the insult of dubbing us the "*liberal OZNA*." Kessler called on me and, in an interminable diatribe, delivered a vitriolic criticism of Bishop Rožman. No one reacted to the insult about the "*liberal OZNA*" nor were any steps taken to lodge a complaint about it. Next the paper directed its attacks against the *Slovenian National Committee* by repeating the well-known recriminations about the concordance of the political parties. That, however, must have been the proverbial last straw because the *International Refugee Organization* (*IRO*) in Rome, where Miha Krek, the former minister of the Yugoslav government in exile, had some influence, directed the camp administration to withhold any further support of that paper.

On September 6, 1945, one of the priests invited all the refugees to a small church located nearby to celebrate the birthday of King Peter II of Yugoslavia. The small church was packed with refugees. Upon conclusion of the services late in the afternoon but still some time before the sunset we were returning to the camp. On our way home we saw a number of refugees from the "*Serbian House*" who were going to the church but were obviously too late for the celebration. The same night, when the lights were out and we had retired to our bunks, the *carabinieri* came to the camp and, shining their torches here and there, went from room to room. No one was questioned; the *carabinieri* were merely picking up our shoes and examining the soles. After some time they left without having talked to anyone. In the morning we learned that the shopkeeper from the "*Serbian House*" was murdered in her own house and robbed of one million lire. In the "*Serbian House*" the carabinieri noticed a bloody stain on Zupan's shirt. He explained that he had cut his finger and wiped off the blood with his shirt. The *carabinieri* took his shirt to have it examined in the laboratory and placed Zupan under arrest until they would obtain the results of the laboratory tests. Zupan's detention lasted a long time and several refugees called on me requesting that I ask the camp commander to intercede with the authorities and urge an early release of Zupan. Of course I did what I was asked to do although I had no official standing in the camp at the time. Eventually the tests came out negative and Zupan was released from custody. Soon after that the "*Serbian House*" began to empty out as the young people were leaving in droves for the Trieste region where they were appointed primarily as schoolteachers in the Slovenian ethnic territory occupied by the Allied Military Government. Others, including Zupan, simply left the camp.

Initially, prior to the election of the new Italian government, every camp commander would routinely secure for the refugees in his charge the so-called *soggiorno,* that is, a permit to live in Italy indefinitely, but the camp commander in Riccione did not want to secure those permits for us. In this way I never obtained a *soggiorno* in Italy and when we were being moved around we, the first refugees from Riccione, were the only ones who never got our residence papers.

We had a considerable problem with the children and youths who had nothing to do and were left to themselves. One of the refugees took the trouble to get his son admitted to a nearby Italian secondary school (*gymnasium*) but this was a rare and isolated case. Camp commander Carrigher did show some interest for the education of the youth. He called me in and recommended that I organize a school. The camp administration would reimburse us for any expenses incurred in purchasing the textbooks and other instructional materials. The children were of various ages and at various levels of education. I arranged for some refugees to take on the responsibility of providing instruction in arithmetic, languages, geography, history, physical education, and generally in subjects considered to constitute a basic curriculum. We acquired a sufficient number of books and materials and began with the instruction. The initial response and attendance were good but soon the pupils began to drop out and, as we had no authority to require attendance and maintain discipline, our efforts at conducting classes soon came to an end.

Through the Red Cross I succeeded in establishing contact with my sister Josephine who had been living in the United States since 1905. When she learned of the misfortune that had befallen me and my family she began to send me a few dollars every month which sufficed for most of my basic needs. A young university student agreed to give me instruction in the English language for a modest fee. My sister sent me several packages of clothing and shoes when she found out what sizes I needed. My eldest niece Angela Schneider also sent me packages of clothing and dried fruit from her ranch on Mt Veeder Road in Napa, California.

In time I was able to establish contact by mail with my wife: I was sending her postcards under the address of Tatjana Canestri-Trotti composed as correspondence from a fictitious friend. My wife would then reply to the Canestri-Trotti address. Such correspondence was not entirely without danger. Colonel Glušič's wife Marija, was arrested and spent some time in the communist jail being accused, among other things, of "carrying on illegal correspondence with her husband". Colonel Glušič was conducting intelligence operations against the communists from Austria and Germany but, fortunately, he would have nothing to do with the British, a precaution which probably saved his life.

My wife found a courier who was routinely crossing the border and surprised me by sending me several shipments of clothing, underwear, shoes and even my winter overcoat. In this way I became independent of the camp administration's handouts. Once the camp commander opened the store and summoned all the refugees to come so that he might give them such clothing and other supplies as they might need. I, too, went but my reason for going was to make the attendance as complete as possible because the allocation of supplies depended on the number of people who needed them. In the process of doling out the goods the commander noticed me standing on the side and he pressed on me a pair of shoes which I accepted so as not to rebuff his courtesy.

All those shipments of clothing and supplies improved my situation so that I could buy a can of olive oil and some other nonperishable staples which I gave to the courier to take to my wife. The courier took the mail and the packages to Trieste where someone else who had a border crossing permit took it along to its destination; the return mail would travel by the same route in the opposite direction. That particular package, however, never reached my wife. The courier was able to trace it as far as Postojna, which at the time was still in the Yugoslav zone of occupation, where any trace of it disappeared. Later, I was able to send small food packages through the regular mail but the Italians soon put an end to such shipments on grounds that it was prohibited to send foodstuffs out of Italy.

My letters sent via courier were marked with number nine and only the courier and the person who traveled across the border knew the address of my wife. Certain precautions had to be taken in view of what had befallen the wife of Colonel Glušič. Once Dušan Lajovic, who is now a prosperous industrialist in Australia, took one of my letters along but before he could hand it over to the person who would take it across the border he was arrested by the British military police in Trieste. At his interrogation the British waxed hysterical in their efforts to discover the meaning of Code Nine which they suspected to be a cipher involving some sinister conspiracy. Peter Horn[127], who was staying in Trieste at the time, wrote me and suggested that I clear out of the camp as soon as possible because the British were about to arrest me on account of the cipher. I ignored his warning because I did not care if the British actually would go to such extremes in trying to find out who was behind Code Nine. The *Intelligence Service*, however, never showed up.

An acquaintance sent me a message that the British Military Government in Klagenfurt, Austria, employed Hans Strohmayer, a former *Gestapo* prison guard in Ljubljana in 1944 and 1945 when I was imprisoned by the *Gestapo*. Strohmayer had actually been very venal and helpful to the prisoners. In consideration of suitable bribes such as cigarettes, liquor, and other commodities, he was willing to perform almost every favor for the prisoners short of unlocking the prison door and letting them go. Toward the end of the war Strohmayer was becoming very concerned about what would happen to him after the war. It appears, however, that the "Allies" had treated him well, at least initially, and in fact much better than they treated the Yugoslavs. Of course, we may never know how many gold sovereigns may have changed hands for the British to be so understanding. When I thus mentioned Strohmayer's new situation in a company of friends which included Ashner who liked to visit with us socially, the doctor made a beeline for the office of the British *Intelligence Service* agent in Riccione and reported the matter. Shortly after that I was summoned to that office and interrogated in German by a Jewish employee who worked there. When I confirmed what the doctor had told him my inquisitor flew into a rage and began to chastise me for failing to report the whole affair to him earlier. I let him know that it was not my duty to serve as an informer and suggested that he direct his indignation at the British authorities in Klagenfurt.

---

[127] Peter Horn emigrated to the United States and was teaching Slovenian at the *Defense Language Institute* in Monterey, California.

Rudolf Žitnik and Marjan Zajec, both members of the *Slovenian National Committee*, came to the Riccione camp from Rome. They reported that there was no chance of finding employment in Rome although some refugees did manage to get employed as guards with the American Military Government.

Franjo Mačus and his wife, on the other hand, did remain in Rome. Mačus used to have an automobile at home in Maribor which he had to turn over to the Germans who, fortunately, gave him a receipt for the car. In this the Germans were more mindful of the legalities than the British in Lienz who had quite unceremoniously seized Žitnik's car in which we were traveling without giving him a reason or any kind of a receipt. Mačus took his receipt to the American authorities in Italy and the Americans allowed him to select a car from the multitude of the German vehicles they had seized. Miha Krek in Rome, gave Mačus credentials which stated that he was a lieutenant in the Royal Yugoslav Army although Mačus had never been in the military service. Wearing a suitable uniform Mačus could travel around Italy by car and obtain meals and lodging at the allied military facilities. On one of his travels he visited us in Riccione. At that time Mačus was headed north and he agreed to go and see Paola Di Trento who gave him some dried fruit to bring back to me from his trip. As soon as it became possible Mačus emigrated to Argentina.

In Argentina the Slovenian priest Hladnik obtained an audience with President Juan Peron to whom he explained the plight of the Slovenian refugees whereupon Peron magnanimously granted *Libero sbarco* (*Permission to immigrate to Argentina*) for ten thousand Slovenians. Large numbers of refugees applied for emigration to Argentina and I was busy preparing the lists of those who wanted to go. Argentina has shown its generosity to Slovenians at a time when other countries, the United States included, had no understanding and no place for them.

Miha Krek also visited Riccione and he received me together with several others. Unfortunately we could not have the kind of political discussion which I hoped for because we were joined by the late career officer Andreé[128] who, being extremely nervous, was monopolizing the conversation with his interminable complaints so that it was impossible to introduce any other topics.

Bogumil Vošnjak lived in Rome; for a while he was staying with Vekoslav Bučar who was married to an Italian baroness. They lived in an apartment commandeered from a high ranking fascist. Vošnjak, however, could not remain with them indefinitely because Bučar and his wife had no servants and could not provide the care Vošnjak needed.

Vošnjak was well-known already before the World War I as the publisher of *Veda* (*Science*) and a promoter of the Yugoslav union. During World War I he lived

[128] Slavko L. Andreé (1908–1971), Major off the General Staff of the Royal Yugoslav Army, in 1942 interned in Italy, 1944–1945 officer of the Royal Yugoslav Army in Italy, 1945 employed by the Office for Displaced Persons of the American military Government in Trieste, collaborator of Radio Trieste.

in Washington. He was able to do that because he had considerable personal resources. Together with Thomas Masaryk[129] he worked for the liberation of the Slavic nations from the Austro-Hungarian domination. In those days, when the automobile was not yet the dominant means of transportation, Vošnjak and Masaryk often rode on the horseback through the countrysidde near Washington. They gained access to President Woodrow Wilson and helped persuading him to proclaim his *fourteen points* on the self-determination of nations. These points triggered the disintegration of the Austrian army at the front and strengthened the resolve of the Czech and Yugoslav national committees to demand independence of the Slavic nations from the Austrian domination and the reduction of the Hapsburg Empire to the Austrian republic. Masaryk was the unquestioned leader and advocate of the right of Czechs and Slovaks to their own state while Vošnjak was the secretary of the *Yugoslav National Committee* of which Ante Trumbič was the president.[130] Vošnjak and Trumbič traveled on board of a French cruiser to the island of Corfu where the Serbian government had retreated under the protection of allied navies, and signed on behalf of the Slovenians and the Croats, respectively, the historical *Corfu Declaration* by which the Croats, the Serbs and the Slovenians declared that they wanted to unite in a common state. Vošnjak wrote a diary of the negotiations leading to the Corfu Declaration.[131] Following World War I, Vošnjak was elected a deputy in the National Assembly, served as the Yugoslav Ambassador in Prague after which he became a professor of political science at the University in Zagreb. Vošnjak's wife was killed in the German air raid on Belgrade on the Palm Sunday in 1941. During World War II he served in the *Yugoslav Army in the Homeland* under the command of General Dragoljub Mihajlović where he participated in the drawing of the plans for the post war restructuring of Yugoslavia and he naturally had to emigrate when the communists took over, this time as a penniless refugee.

Vošnjak still carried his diplomatic passport which served as his identification. Once, on the way to Riccione, he made a stopover in Ancona where he forgot his attache case in a park on a bench; when he returned to retrieve it, the attaché case was gone and with it his precious diplomatic passport. The police, of course, could not find the thief.[132]

---

[129] Thomas G. Masaryk, professor of philosophy, Czech politician, founder and President of Czechoslovakia.

[130] Ante Trumbič, Deputy Leader of the *Croatian National Party* in the Dalmatian Diet, Mayor of Split, Deputy for the District of Zadar in the Austrian Parliament, and Minister of Foreign Affairs of Yugoslavia.

[131] V. Bevc, *The Corfu Diary of Dr Bogumil Vošnjak*, The South Slav Journal, v. 9, No. 3–4, (33–34), 1986, pp 25–32.

[132] Something similar happened to King Peter II at his visit in Washington where he forgot his passport in a taxi. When he inquired about it, the taxi driver reported that he had taken it to Tito's embassy in Washington which returned the passport only upon the intervention of the United States Department of State.

Carrigher showed kind interest in Vošnjak when he made a courtesy call on him. Vošnjak used to come to Riccione often and it was then that we established our friendship. When Vošnjak tried to go to Washington he ran into formidable obstacles posed by the strictures of the new immigration law. Somehow he obtained passage to Cuba from whence he managed to reach the United States without too much difficulty. Once in the United States he sought out his old connections and influential acquaintances; he found several senators who held key positions during World War I when he worked with them. After some time in Washington, Vošnjak was appointed lecturer in political science at the University of California at Berkeley where he knew Professor Joseph Kerner from the time of the Paris peace conference in 1919 where they both were advisors to their respective governments.

The administrative organization of the camp included *welfare officers* whose assignment was to look after the welfare of the refugees. Welfare officers were mainly females who as a rule were not very effective and there was a considerable turnover in these appointments. Some of the welfare officers focused their activity on the administration supplies and distributed chocolate, butter, sugar, peanuts and other, perhaps less desirable, items. The better part of these provisions and supplies was intended for the infirm but generally the way in which they were distributed failed to improve the lot of the refugees. Younger refugees were often employed as auxiliary personnel in the warehouses where they had the opportunity to note considerably larger quantities of supplies than ever trickled down to the refugees.

Personnel changes were sometimes to our advantage. A younger and apparently better educated Englishwoman, Gotubed by name, was one of the better appointments. Gotubed started an English language course which was attended by many refugees. I asked her to help me prepare a resume which I felt would be well to have on hand in seeking suitable employment when I immigrated to the United States and she readily obliged. I sent a copy of my *vitae* to Vošnjak who made and sent out a number of copies and kept some for himself which he used when he came to California in trying to secure a suitable employment for me. Vošnjak was successful in this search: he found a position for me with a construction company in Los Angeles the owner of which was a Serb but later developments precluded me from going to work for that company.

At that time the American immigration law was very unfavorable for the prospective immigrants. The law assigned an annual quota of persons that were allowed to immigrate to the United States which was based on the number of the inhabitants in the United States whose origin could be traced to those countries. For Yugoslavia this quota amounted to 933 persons per year, less than arrive illegally from Mexico every day. The quota was based on the number of immigrants from Serbia and while the number of immigrants from Austria-Hungary remained intact for the benefit of the immigrants from the rump states of Austria and Hungary, no one took in the account that following the World War I the population of the defunct Austro-Hungarian empire was redistributed between Yugoslavia and other successor states. The matter was compounded by the fact that many Slovenian and Croatian immigrants listed their former nationality as Austro-Hungarian. On the other hand, Northern European countries like Germany and England had immigration quotas on the order of 45,000 and 65,000 per year which were largely unused. This unfavorable quota system was the reason that I had to wait four and a half years for my quota number to come up. Vošnjak also had difficulties in his second immigration to the United States; he observed that among his political acquaintances the descendants of the more recent immigrants were those who were the least favorably disposed toward the newer immigrants.

Welfare officer Gotubed decided to organize some trips for the benefit of the refugees. She secured a large truck and invited the students of her English class on a trip. Her fiancée, a partly disabled military officer, also joined the party. On March 4, 1946 we left for San Marino taking the mountain road that branches off the highway at Rimini. San Marino is an independent republic whose foreign affairs are administered by the Italian Ministry of Foreign Affairs. The republic consists of four municipalities; its main product is wine. From the pre-war times I recalled the frequent newspaper advertisements of the municipality *Saravallo* for its wines. The republic is headed by two presidents. The most notable buildings in the city are the Parliament, the Post Office and the cathedral. The latter is housing the statute of the republic's founder — a saintly monk who many years ago arrived from Dalmatia, some kind of an early Yugoslav émigré whose name the republic bears. The government of San Marino apparently collects no taxes, its revenue is mainly derived from the sale of stamps of which new versions are issued as needed. The main street has a whole string of businesses which specialize in the philatelic trade. I had several occasions to visit San Marino and each time new issues of stamps were available of which I sent some to my family in Yugoslavia. The postal service is, naturally, busy with the processing of the tourist correspondence. Apart from these there were no other attractions. In taking in the sights of the diminutive city we came across the city jail holding a lone prisoner who had been jailed for larceny. We also saw a cannon which, however, could not be used for defense in modern times. We were told that in the autumn of 1943, when Italy capitulated and declared war on Germany, San Marino, too, joined in declaring war on Germany who had up to that time been Italy's ally and comrade in arms. The declaration was delivered to the German garrison in Rimini which dispatched a sergeant and some men in a few trucks to San Marino. As the Germans were approaching the Republic all the able bodied men scattered and hid themselves in the vineyards on the slopes of the mountain. On arriving to his destination, all that the German sergeant had to do was to declare that the republic was occupied. That occupation did not last long as the Germans began to retreat to the North. At the time of our visit San Marino had not yet signed a peace treaty with Germany. Miss Gotubed treated all of us to lunch in a restaurant. The trip was a success and we enjoyed the sightseeing and the relaxing drive through the Italian *campagna.*

Krener's Extradition Foiled

In Camp at Riccione we had only one incident of the British trying to arrest someone and hand him over to the communists as requested by the communist regime in Yugoslavia. One day a Slovenian refugee who worked as a clerk in the camp administration office overheard that the British were coming to arrest Colonel Franc Krener and promptly reported the news to me. Krener was initially the commander of Mihajlović's forces in the Ljubljana region. I opposed his appointment because his spouse was a German and he had reportedly declared himself a *German national (Volksdeutsche)* in 1941 when captured by the Germans in Priština where he was the commander of a Yugoslav artillery division. When Krener later became the commander of the *National Guard,* he could no longer remain in the Mihajlović's organization. Colonel Andrej Glušič maintained that Krener was supplying information to the Germans both directly as well as through the *Slovenian Information Service* which was headed by the *Gestapo* agent Ivan Kregar, a former captain in the Yugoslav Navy. In 1944 Lieutenant-colonel Peterlin[133] alerted the Slovenian command of Mihajlović's forces that Krener was informing the Germans of everything he knew. When Glušič was later arrested and interrogated by the *Gestapo* he learned that Peterlin's warnings were well justified. After the war, in 1946 in Trieste, Albin Šmajd told Žitnik that when the *Slovenian National Committee* demanded of Krener that he, as the commander of the *National Guard,* attack the Germans, shoot the German regional commander Rösener and remove Rupnik, Krener turned pale and, visibly shaken, turned down the request saying that he would never act against Rupnik or the Germans.

In Riccione Krener was staying in private quarters outside the camp and was regularly coming to the beach. When I learned what the British were up to, I immediately sent Janez Žitnik to find Krener at the beach and warn him of his impending arrest. Krener wasted no time; he went home, packed, and the same day left for

133 Ernest Peterlin (1903–1946), Lieutenant-colonel of the *Royal Yugoslav Army,* interned at Gonars in 1942, member of the clandestine war council in the Mihajlović resistance movement, worked with colonel Vladimir Vauhnik, joined the *Slovenian National Guard* (*Slovenski Domobranci*), arrested by the *Gestapo* in 1945 and interned at Dachau, upon his return to Yugoslavia in 1945 he was arrested and executed by the communists.

the Vatican, a neutral enclave where the British writ did not run. For once at least the British perfidy was frustrated. Whatever Krener's role during the war might have been, the communists had neither legal nor moral authority to sit in judgment over him. Krener may not have been a friend of mine but as a free man I was duty bound to save him. He who saves one life saves the world entire.

Communist Reign of Terror

While we were still in Lienz, Austria, we could listen to the communist carousing and celebrating their victory on the radio radio. The festivities were interspersed with the communist kangaroo court proceedings at which the accused were as a rule summarily sentenced to death. If the victim was in their custody, the communists carried out the sentence immediately. My friend and colleague Stanko Dimnik, who was in the communist prison at the time, described later to my son how those executions were carried out. The victim would be taken out of the cell into the corridor and the executioner would shoot him with a revolver at point blank range into the head or in the face. There was no time as yet to construct such elaborate tiled execution chambers as the Soviets had in their *Lubyanka* prison, the corridor had to do. The bodies would later be loaded on trucks like so many logs and hauled away to undisclosed burial grounds.

The first edict that the partisans issued as soon as they came to Ljubljana was that all privately owned cars were confiscated for their use so that the communist *comrades* could drive around. The confiscated villas and houses were bestowed on the comrades according to their seniority and their contribution to the communist victory. Some had selected their villas already during the war; they certainly knew what they were fighting for. Although the communists maintain that the concept of private property will become unimportant because everyone will have everything necessary to satisfy his needs they are now very zealously striving to acquire as much property as possible. Because private property is the foundation of political independence the communist state cannot tolerate that its potential enemies or even the public at large should be able to do what they want with their property.

Soon after the arrival of the partisans and when my family had not yet been evicted from our apartment Vinko Vrhunec, father-in-law of Marjan Tepina who had submitted to the *Committee for the Memorial to King Alexander* his miserable work of art with the insolent inscription "*De mortuis nil nisi bene*", accompanied by preparatory school teacher Pregel appeared at our apartment and demanded access to my library to confiscate my military books which the partisans could use in training their army until the time when the "Allies" would be able to train them properly. When he found nothing in my library my wife informed him that the

Gestapo had in 1941 beaten him to the loot. The sarcasm of the remark was not lost on Vrhunec. For a good measure, Ivanka added that Vrhunec, who before the war always wanted to weasel himself into the leadership of the *Reserve Officers' Association,* can well imagine how I will appreciate his visit when I would have learned about it. The looters had to leave empty handed.

Soon it was our apartment's turn. It was located in a villa at Erjavčeva cesta number 26 (now No. 28). A communist by name of Anton Košak arrived from the *Administration of People's Property* and ordered my wife to leave within twenty-four hours, he said all our belongings were confiscated, and it was no concern to him where my family went. The procedure, Košak said, was considerably more lenient than that used by the Germans who used to give only ten minutes to the people whom they were evicting from their homes. He told the story of his sufferings at the hands of the Germans which, he said, became so unbearable that he attempted a suicide by slashing his wrists. Obviously he was one of those victims of violence who acquire the thinking and conduct of their tormentors. Košak placed a guard in front of the house lest my family take something out but my wife was not aware of it. She found this out when our friend Edit Dimnik who lived nearby came by and my son tried to pass my binoculars over the fence to her. The guard readily intercepted him, seized the binoculars which he retained as his personal loot. Then he came in to scold and threaten my wife for having attempted to salvage the binoculars. In the ensuing argument my wife maintained that the guard should leave the binoculars in the house because, being unfamiliar with the ways of the communists, she was not aware that outright looting was the order of the day. The guard then spotted my briefcase which was made of soft black leather and wanted to "trade" the binoculars for it. What appeal that briefcase might have had on the primitive communist partisan is somewhat of a mystery. He might have thought that people would take him for one of the commissars or higher officials if he paraded around with my briefcase in hand. My wife refused the trade and decided to keep the briefcase which she eventually brought with her to America and I still have it.

Into our apartment moved Dragutin Gustinčič with his mistress Fedja Šturm who posed as his secretary. Gustinčič demanded that my wife unpack some of her linen and put it on the beds. Later he wrote and demanded the key to the piano which my wife locked hoping that the communists would soon be chased away by the "Allies." Gustinčič was a person who barely managed to squeak through the school of agriculture and later posed as an agricultural engineer although he had no right to the professional title which he had no ability to earn. I had known him since the time of my studies in Vienna and later when he was a communist councilman in Ljubljana. Gustinčič was not much sought after for his engineering expertise, therefore he could spend most of his time on the promotion of communism among the workers. I heard that they were buying him lunches, consisting of simple bowls of beans in exchange for his proselytizing activities. Among the workers who were paying for his lunches were my cousins Nace and Franc Gorenc.[134] Those endeavors eventually took Gustinčič to the Soviet Union where he had reportedly been arrested and imprisoned immediately upon his arrival and it took some time before the communist comrades cleared him. He reportedly also saw action in Spain, on the republican side, to be sure. When the war, which Gustinčič sat out in Kuybishev far away from any front, was over and Stalin had his protégé Tito installed as a communist dictator of Yugoslavia, the Soviets wasted no time in getting rid of Gustinčič and he was sent back to Yugoslavia. According to Gustinčič, he was to establishing university department of agricultural science where he was the only member of the faculty. It is highly questionable what he could contribute to the Yugoslav agriculture as he could not have learned much in the starving communist Russia. Ivanka's friend Vida Gorjup who was employed at the university as a translator reported that Gustinčič, who sported a Stalin-like mustache and spoke with an affected Russian accent, was openly treated with contempt even by those faculty members who supported the regime. Unfortunately for Gustinčič, the old maxim that the revolution eats its own children soon showed how true it was in his case. In 1948, when Stalin expelled Tito from the Cominform, Gustinčič, declared himself as a supporter of Stalin which act landed him in prison. He was released only after he publicly recanted his former views on the second page of *Ljudska Pravica*[135] (*People's Justice*). In his recantation Gustinčič explained that because he had been brought up in the iron discipline of the Soviet communist party he had acted almost reflexively when he heard the clarion call of comrade Stalin. The kind and understanding comrades of the Yugoslav political police had, however, great patience with Gustinčič and with gentle urging and kind persuasion showed him the error of his ways and helped him to recognize that the only true road to communism was the one scientifically discovered by comrade Tito and the Yugoslav communist party. Notwithstanding his recantation, Gustinčič was no longer allowed to return to our former apartment and enjoy the stolen property because our furniture had in the meantime been removed and the apartment assigned to Miha Kambič, a state secretary of the Slovenian People's Republic who as a "leftist

[134] In 1920 Franc Gorenc was fatally shot when the police fired on a group of demonstrators who were marching on the city along the Zaloška Road. The brutal act of the police was inexcusable but the ultimate responsibility for those deaths rests with the communist agitators like Gustinčič.

[135] *Ljudska Pravica* means "People's Justice" and is thus different from the Soviet *Pravda* which means "Truth," although many people in Slovenia tended to associate one with the other in their abject propensity to emulate everything Soviet. The actual association of the two names, however, came from the Orwellian perversion of their basic meaning.

Sokol"[136] had made great, albeit perhaps not very well known, contributions to the communist cause. Upon his release, Gustinčič was seen about the town walking with a limp supported by the arm of his wife with whom he must have made up after his affair with the torrid comrade Fedja Šturm; the limp might well have been a souvenir of one of those sessions with the kind and understanding comrades of the *UDB* at which they so solicitously revealed to Gustinčič the ideological errors of Stalinist ways.[137]

After the communists evicted my family from our apartment in their own barbarous way, Ivanka found temporary lodging in the Delcott villa in the Poljane district. Marin Pavlovič, a Croatian count and former Croatian consul in Ljubljana who had married Hermina Delcott, was quite happy to rent two rooms to her. Hermina, sister of the late Hugo Delcott, a schoolfellow of mine, had already been exiled to Austria because she was of German origin although her family had ancient roots in Ljubljana. The two rooms rented to my family used to be occupied by Boris Popov, a Russian opera singer who was on Soviet orders arrested as soon as the communists arrived because of his connections with the army of the Russian anti-communist General Vlasov. Shortly after his arrest word was received that he had died in prison, not likely of natural causes. In the Delcott villa also lived our relative Slavica Novak, wife of Ciril Novak, postal superintendent in Celje whom the Germans had shot in the early days of their occupation. Her sons Bogdan and Slavko had also escaped across the border before the communists took over. Pavlovič hoped that it would be more difficult for the communists to occupy the entire villa if he filled it with people who were more to his liking. In the basement lived family Križanič with their mother-in-law Štimec who were fierce communists.

My wife and family were for some time getting lunches and dinners at the nearby convent, Jožefišče where her relative Lenčka was Mother Superior. That arrangement soon ended because the communists established strict control over the institution where young women from the countryside were taught sewing. Lenčka was then forced to wash heavy linen sheets in the laundry where everything was done by hand. Advanced in age she once fell down at work, injured her spine and became paralyzed. Shortly after that she died.

The ground floor of that large villa which was full of Australian aborigine artifacts, was occupied by the offices of the communist district committee out of which

[136] A "leftist *Sokol*" is, of course an oxymoron which anyone familiar with the Sokol ideology and the principles enunciated by its founder Miroslav Tyrš will immediately recognize. Here it means a *Sokol* apostate who became a communist.

[137] Note by Editor: As mentioned earlier, Kardelj is accused of plagiarizing Gustinčič's work on the Slovenian nationality question. Such claims shed an interesting light on the internecine relations within the communist party and show that even the intellectual property rights of party comrades are not respected.

the communists conducted surveillance of the population according to the Soviet model. A few rooms were occupied by supernumerary members of some partisan brigade with whom the communists apparently did not know what to do. Those were under the command of a major Hribar, a former stable boy who had risen to that rank while with the partisans. Among those partisans was a young man called Srečo from Lower Styria who was the only decent person in the entire group. Once when Srečo saw a picture of King Alexander in an old calendar he remarked to my wife: "The things would not have come so far had he lived." For a partisan this was a remarkable statement meaning that not everyone wearing their uniform was a committed communist.

The partisans in the house had nothing to do and so they were mainly loafing around during the day and drinking to excess at night. At times they would also shoot in the air. The count's maid Ančka was cooking for them. Ančka was an energetic woman from Trebnje whom the communists were not able to intimidate. The partisans usually ate in the kitchen. One of them once shot a squirrel and cooked it for his commanding officer to remind him of the delicacies they had to eat in the woods. But the commander told him to go to hell with that squirrel. On one occasion my son was told that a partisan who came to dinner had murdered governor Marko Natlačen. Natlačen's villa was in the same block in the Strossmayer Street as Delcott's. The governor's family had prudently left before the communists arrived. They left behind their dog, a German shepherd called Lord. The dog, with an open cut on his leg, took refuge in the Delcott's villa where he was adopted by my son Vlado who took care of his wound and spent a lot of time with him. At first it was not known to whom the dog belonged but because it had a tag Vlado in time found out at the city offices who the dog's owner was. The dog did not like the partisans; there must be something about the sixth sense instinct people attribute to animals. Whenever somebody would shoot the dog began to howl. Probably he connected the shot with the death of his master. One day on the Poljanska street he chased a partisan riding a motorcycle who with his riding pants ripped at the thigh swerved onto the sidewalk and very nearly hit a pregnant woman. The people's militia investigated the incident and when it discovered to whom the dog belonged decreed that the dog must be put in the care of a young married couple who were no doubt politically reliable.

One of the partisans at the house was Ivan Bobič, a never-do-well horse trader — some would say a horse thief — from Šentjernej, known as communist sympathizer before the war. My wife had rescued her bicycle, an almost brand new *Neumann*, from the confiscation. Soon uniformed communists began to appear at the Delcott villa demanding that my wife hand over the bicycle which they were "confiscating." Although by then robbery and confiscation had become established synonyms my wife still demanded to see a written order of confiscation which none of those types could produce. On one occasion the resident communist partisans even sided with her and suggested to the would be bailiff to either show his papers or get lost. Finally Bobič asked my wife if she would lend him the bicycle so that he could ride it to Šentjernej. She did so and from then on Bobič and her bicycle disappeared. Ivanka went through all the motions by reporting the theft to the police who could not possibly have cared less although bicycles had to be registered with the police and there would be no problem finding it. Ivanka asked his commanding officer, the debauching "Major" Hribar as to Bobič's whereabouts and he responded with: "Where is your husband?" When Ivanka said she did not know, Hribar said: "Oh, we have him already." Fortunately she had received a letter from me via courier the day before and Hribar's answer did not upset her. After a long time and only after Ivanka wrote to the commanding general of the *People's Army* in Ljubljana that such conduct reflects adversely on the *People's Army* was Bobič's current address released to her. In a totalitarian state where everybody's residence and temporary movements must be registered with the police it is an absurd claim indeed that a discharged officer of the army cannot be found. Ivanka took Bobič to court — the authorities would not prosecute an outright theft as a criminal act — and Bobič was required to appear. When she asked him where her bicycle was he said that it was confiscated and taken from him. Bobič was treated with extreme leniency by the communist court. He was not required to return the bicycle — all that the court decreed was that he must pay Ivanka 5,000 Dinars at the time when a used bicycle sold for 35,000 Dinars.

Even in those modest circumstances the communists would not leave my family alone. Once when my wife was not at home the bailiffs of the *Administration of People's Property* (*Uprava ljudske imovine*), the office for plundering and confiscations, came and ransacked their rooms. At that occasion they also came to my sister-in-law Tončka who lived near Tabor where my wife had stored some bedding and silverware all of which they confiscated. The janitor at our residence on Erjavčeva had reported that some of our belongings were taken to Tončka. However, when my wife succeeded in proving in court that those furnishings were her personal property the *Administration of People's Property* which kept the items in its storage at Mestni Trg and which had not yet found a suitably meritorious communist to be regaled with them actually returned the items to my wife. That was accomplished through the kind help of Stojko Tuma who was an Administrator of the looted property and Ivanka's friend from the time she lived in Gorizia. Tuma's father had been a communist at a time when hardly anyone had heard of communism. We still have the silverware here in the United States. That, however, was all we ever got back from the communists.

Once, in the summer, a train loaded with ammunition exploded at the Ljubljana railway terminal. The explosions followed one after another for several hours and no one knew what was going on. The communists were frightened because they feared that the Western powers had attacked Yugoslavia. Everybody ran to the bomb shelter in the basement where it was interesting to observe how the communists who had been harassing Pavlovič suddenly became exceedingly polite toward him. Influenced by anti-Western propaganda they actually thought that Tito's Yugoslavia might be threatened by the Western powers which at that time still did not know what to do to show their obsequious friendship and admiration of Tito and his Soviet allies.

Soon Pavlovič was arrested and imprisoned because he happened to be a diplomatic representative of one of the countries that lost the war. Before they took him away he asked my wife to take care of his daughter. One night, soon after his arrest, communists armed with submachine guns barged into my family's rooms pointing their guns at my wife and children. They were looking for Pavlovič's daughter Majda. When they found her asleep in the Pavlovič rooms — those that were left to them — they in their characteristically brutal manner pulled her out of bed and took her to an unknown destination. That was on one of those nights when they were staging raids on Germans or persons suspected to be of German origin because of their surname. It was on one of such nights that they had deported my sister-in-law Milena Slapar, which is reported below.

To everybody's surprise, however, Majda showed up at five in the morning on the same day. She said that the communists intended to deport all persons they picked up to Austria where she would find her mother Hermina. Before they loaded her on the truck, however, she was taken before some higher ranking communist probably because she had a serious speech impediment and was constantly trembling from birth at which the incompetent midwife Štimec was in attendance. That communist decided that it would not be a good idea to send the girl in such a condition to Austria and his British allies because it could reflect adversely on the "esteem" in which the British held Tito and his partisans.

After some time Pavlovič was released and his wife Hermina was permitted to return to Yugoslavia. Pavlovič soon after that left for Split where he devoted himself almost exclusively to his sailboat. He did not take his family with him, Hermina and Majda went to live in Višnja Gora where they lived in a house owned by a postal employee who also was staying in the Delcott villa. Vlado once hitchhiked through Višnja Gora and happened to see Majda and Hermina in front of a house looking forlorn and lost.

In winter the communists decreed that my family must move four kilometers out of town to *Ižanska Road* to a relatively new house they appropriated from a once prosperous farmer Vidmar, known also as Mokar, near Črna Vas in the middle of the Ljubljana moor (*Ljubljansko barje*). Because *OZNA* wanted my family's rooms in the Delcott villa *OZNA* simply moved them. The driver who came with the moving truck was our former neighbor's janitor, who as a member of the *Security Intelligence Service (Varnostno obveščevalna služba — VOS)* had undoubtedly kept my neighbor Hrovat and me under surveillance during the war. He was very hostile and arrogant in carrying out his moving assignment. Vidmar-Mokar also operated a pub called "At Mokar's". His new house was situated at the junction of *Ižanska* road with the road leading to *Črna Vas* opposite the elementary school and not far from the church in *Črna Vas* designed by Plečnik. My family got a room and a kitchen on the second floor. The other apartment was occupied by *OZNA* captain ("kopetan" in the communist vernacular) Jevc. There was a communal bathroom for both apartments as the owner had originally planned the house to be used by his daughters but because the Jevcs did not know how to use such facilities they used the bathroom as a laundry until it became unusable. Vidmar-Mokar was pleased to have my family there and was selling them half a gallon of milk each day which at the time was hard to obtain. Jevc was responsible for the expropriation of a sizable part of Mokar's property which was intended for building a farmer's cooperative house along the lines of the Soviet *kolkhoz*; therefore Mokar could not stand the sight of him. He recounted that Jevc used to spend time with the Salesian monks at Rakovnik where he was guarding a replica of Don Bosco's sepulcher. Actually Jevc had been with the Salesians in some kind of a rehabilitation program for wayward boys — a juvenile authority delegated to the Church.

My family had two bicycles which the children used to ride to school weather permitting; otherwise they had to slog along in the mud or slush. Sometimes drivers of horse drawn carriages would give them a ride in the back of the cart. The infrequent and irregular bus service which was available twice a day was not of much use. Jevc of course had an automobile, a Ford Eifel. He would sometimes give rides to the children if he happened to go their way. Thus he once gave a ride to Breda, then about twelve. During the ride he asked her if she liked the partisans to which Breda replied that she did. Jevc then inquired further as to why and Breda wisely responded that because they fought the Germans. Once Vlado waved at him in town and Jevc gave him a ride home. During the ride Vlado overheard him saying to the other man in the front seat that the owner of the car was making noises about getting it back. Shortly after that Jevc began riding a motorcycle. Apparently the owner managed to get his car back which was quite a feat or it could have been an indication that Jevc did not rank very highly at the *OZNA*.

Next Spring a detachment of antiaircraft defense of the *Yugoslav People's Army* moved into the house. Jevc let commander Lieutenant Luk Stupar, a Serb from

Bosnia, have a connecting room of his apartment, other personnel was billeted in the unfinished attic and in the rooms on the first floor which were originally designed to house an inn and a small shop. The house and the overloaded facilities were diligently destroyed by the soldiers. In winter, for instance they would dance the *kolo* in the entrance hall below so that the entire house was shaking from the pounding of their heavy boots. On such occasions Vlado would sometimes cause a short circuit so that the lights went out and the dancing was interrupted. However the soldiers soon wised up to this and the commissar came to harangue my wife with all kinds of threats. My friend Bežek talked to Vlado and seriously advised him to forgo such pranks in the future.

Vlado used to play chess with the soldiers a great deal and in this way got an insight into the day to day workings of the *Yugoslav People's Army*. Once the commander, who at times tried to be friendly, took Vlado to the optical ranging equipment at the antiaircraft battery located out in the field but it seemed that he did not quite know how the equipment is used. The commander left momentarily on some business and promptly the commissar's aide, who must have been tailing the two, showed up and chased Vlado away from the instrument. Of interest to Vlado were also the army cars and trucks consisting mostly of captured German equipment. For a long time after the end of the war the German prisoners of war served as drivers and mechanics. Now and then they would volunteer interesting comments on the mechanical competence of the Yugoslav army. Sometimes the drivers would allow Vlado to move a truck.

My family had to live out there in the sticks for three years. At times they would be visited by friends like the family of the late Mirko Saks, murdered after the infamous Kočevje trial, and wife and daughters of Stanko Dimnik who was imprisoned at Ig where they could visit him from time to time. *Sokol* friends from Tabor, Franko and Albinca Krajger and their daughter Jelka, often came to visit by their boat on the Ljubljanica river. One Summer Franko let Vlado have his boat for a good part of the vacation so that he could row as far as Podpeč. On the river Vlado regularly saw Ante Tavčar who was rowing along in his *skiff*.

One day Ivanka was visited by a stranger who introduced himself as Peter. Later we found out that he was Peter Zorman. He appeared worried because he was armed and there were soldiers in the house. He brought a message from the wife of of the *National Guard* lieutenant colonel Ivan Drčar. Peter said he as a courier often crossed the Yugoslav northern border and helped people escape from Yugoslavia and that it was he who had taken Drčar's wife across the border. Ivanka asked him how much he charged for taking people across and Peter mentioned a substantial sum which he justified with the fact that he was risking his life and also provided armed cover to his clients in case they should be intercepted by border patrol. Peter said he was armed and acted so furtively that he would be likely to attract attention of an alert observer. However, the troops in the house took no notice of him. Ivanka had a feeling that something was not in order with the man and did not trust him.

About a month later the communists staged a big show trial against Črtomir Nagode,[138] Franjo[139] and Ljubo Sirc[140] and their group in the course of which "courier Peter" was also mentioned. Peter, however, was not a defendant at the trial; he was freely walking about in Ljubljana. Colonel Glušič explained to me that Peter Zorman took Drčar's wife across the border to Austria so that he could infiltrate Glušič's organization but Glušič was very careful and Peter did not succeed. When Peter took the first letters of Nagode's group to Austria, *OZNA* had been watching his group for a long time already. Then Peter tried to arrange for a meeting with someone of higher rank than Drčar, namely Glušič. Peter promised that he would bring from Yugoslavia professor Hribar who had reportedly just escaped from prison, wife of *National Guard* Lieutenant Skvarča and yet another man to that meeting. Under admonition to exercise extreme caution Glušič sent two of his officers, Janez Grum[141] and Rado Slanc,[142] to reconnoiter. The two officers found that Peter came across the border with thirteen members of *KNOJ*[143] armed to the teeth who had on the night preceding the planned meeting kidnapped in a nearby village a young man who was opposed to the communists. Thanks to a timely warning by a woman who told them

138 Črtomir Nagode (1903-1947), civil engineer, secretary of the *Association of Friends of the Soviet Union* (*Društvo prijateljev Sovjetske zveze*), leader of *Stara Pravda* (*Ancient Rights*) during World War II. After the war he attempted to organize democratic opposition to the communist regime. At the so named Nagode Trial he was sentenced to death and reportedly executed in a manner involving torture; the Slovenian government refuses to disclose where his remains are buried.

139 Franjo Sirc, industrialist, founder of textile industry in Kranj. Arrested by communists in connection with Nagode's political trial, imprisoned, and died in prison.

140 Ljubo Sirc, economist, lecturer in economics at Universities of St. Andrews and Glasgow, Scotland, author of numerous economic papers and books. During the World War II he succeeded in leaving for Switzerland from where he joined the overseas detachment of partisans. When the communists occupied Yugoslavia he jointly with Črtomir Nagode attempted to organize democratic opposition to the communist regime. The communists arrested him and following a show trial sentenced to death and then commuted his sentence to seven year's imprisonment. Upon his release he succeeded in escaping through Italy to Switzerland and obtained his doctorate at the university of Fribourg. He is one of the most active Slovenian émigrés, member of the *Slovenian National Committee, Liberal International* and other anti-communist organizations. The British government decorated him as *Companion of the British Empire* (CBE) for his contributions to the development of democracy in the world

141 Janez Grum, lieutenant in the *Chetnik* unit commanded by colonel Glušič, now professor in the United States.

142 Rado Slanc, member of the organization of colonel Glušič, businessman in Villach, member of the Principal Committee of the *Slovenian Democratic Party.*

143 *KNOJ — Korpusi narodnog oslobodjenja Jugoslavije* (*Corps of Yugoslav People's Liberation*) semimilitary formation, executive arm of the *OZNA* or in later time the *UDB* (*Uprava državne bezbednosti — Administration of state security*) used similarly as was the *SS* by the *Gestapo.*

what was waiting for them a hundred yards away in the forest Grum and Slanc just barely evaded the ambush. Subsequent to that Peter sent word that the people who came to the meeting had returned to Yugoslavia because Glušič's men did not show up. No one in his right mind would believe that persons Peter said he would bring across would return to Yugoslavia voluntarily.

Ivanka was at first supporting the family by knitting various garments for people who brought their own wool and who paid for the finished product. My sister was sending her a great deal of clothing for the children including some very nice dresses for my daughter so that the village girls looked with envy at how Breda was dressed. Because Ivanka did not need so much clothing she was able to exchange them for food.

Captain Jevc next door kept pressuring Ivanka to integrate into the socialist economy by finding a job. She told him that she could not get a job because she did not have the right to vote. Taking away the right to vote is of course an integral part of the communist program with the objective to ensure that the communist candidates — there are no other anyway — get elected with a hundred percent majority.[144] She eventually decided to apply for the right to vote which was granted after prolonged bureaucratic procedures not the least of which were rebuttals of laboriously penned reports of local informers that she was supporting the "*White Guard*." Slava Saks then helped her to get employment in the bookkeeping department of the enterprise *Tekstil-Obutev* (*Textiles and footwear*) which was later renamed *Železnina* (*Hardware*). Kovačič, the enterprise's director slighted her at every opportunity nad shortchanged her at benefits to which employees of the socialist country were entitled. She was regularly required to show up with both children for "voluntary" shock worker assignments such as destroying forests or moving logs at a railroad freight terminal and similar inanities. The point of all this was not so much as to eke out the last calorie of production but rather to ensure universal regimentation of the subjects of a socialist state.

The good side of Ivanka's employment was that once a year she was entitled to a substantial discount for traveling to a vacation resort. My family was thus able to go vacationing at the seaside again. The first year they went to Strunjan near Pirano where the enterprise maintained a vacation home. Next year they went to Malinska on the island of Krk where we used to vacation before the war. There they found out that the owner of Hotel Zagreb where we used to stay was expropriated, that he had bought a fishing boat with which he was able to evade the numerous OZNA patrol boats and successfully escape to Italy from whence, he returned to the United States

[144] Note by editor. Even today the newly proclaimed Republic of Slovenia, invoking communist legislation denies the right to vote to Slovenians who had renounced their citizenship of the Socialist Federal Republic of Yugoslavia and thus faithfully preserves the principle of controlling the electorate introduced by the communists.

where he used to be a permanent resident. On hearing this Vlado began to fantasize about me sending him an outboard motor — he even specified the brand "Johnson" which was particularly silent — for a boat which he would have built in Ljubljana. For this purpose he went to see my friend architect Lado Kham and inquired if he would design a low silhouette boat, almost a submersible which he would then have built. Kham told him that he certainly could design the boat but shipping such a boat from Ljubljana to the seaside by rail would be too much of a signal for *OZNA* which would intervene before the boat would touch the water.

Another vacation taken by my family was at Crikvenica. They invited my mother-in-law to go with them. The next year they went to Tučepi near Makarska where a luxurious hotel had been built on a very lovely beach and which was initially intended for members of the communist party only but was later opened to the ordinary mortals. Finally, the last year my wife spent in Yugoslavia, they went to Split where they stayed with Antonieta Franetovič, my wife's friend and former colleague from the time when she was teaching in Split. At that time they took a steamboat that plied the waters along the Adriatic coast and sailed all the way to Dubrovnik. When they visited island Lokrum off the coast at Dubrovnik they were told that it was prohibited to stay on that romantic island overnight because people were taking boats and fleeing to Italy which was more than sixty nautical miles away. There were cases when people accomplished the feat in a kayak. However, those people were probably picked up by some steamer by previous arrangement.

For the children it was extremely difficult to avoid conscription to the shock work *youth brigades*. These were sent to construct railroads on relations such as Brčko-Banoviči and Šamac-Doboj in Bosnia and the highway *Autoput* Zagreb-Belgrade. The point of those brigades was not so much the physical labor witch the communists extracted from the youths but the regimentation and indoctrination associated with those projects. The communists did not make any pretensions about that. Vlado had often been told: We do not care how much you do there, although your efforts will certainly be appreciated, but what we really want is that you share our company at campfires and develop the spirit of camaraderie befitting a citizen of a socialist country. The communists were constantly making threats that those who fail to join the brigades would not be admitted to the university.

At the departure of the first youth brigade from Ljubljana to Brčko many of those who enrolled to get the communist organizers off their backs simply did not show up and since it was then too late to round them up the brigade had to leave without the no shows. Next year the communists organized the departure so that those enrolled were required to report a few days in advance to the gymnasium at their respective schools. Seeing the parents who brought refreshments or simply came to see their sons and daughters awaiting on the other side of the fence transport to Bosnia was reminiscent of the times the same gyms were used by the Italians to collect people picked up during the numerous raids for internment at the island Rab, Gonars, and elsewhere in Italy. Enforcement squads were dispatched to bring in the recalcitrant "volunteers." If parents had reservations about sending their children to distant Bosnia or even dared to resist outright the aspiring political commissars of the People's Youth organization would threaten them with a loss of their job or worse. When everybody that could be rounded up was in the "volunteers" were marched off to the railway depot to board the train for Bosnia. Even so quite a few students managed to desert enroute to the depot and they could not be apprehended before the train left.

Some people resorted to other means of evading the working brigades. My son's friend whose father was a physician and professor at the medical school had to volunteer for a brigade otherwise he would not be accepted by the medical school where political persuasion was the principal factor for admission. Thus he left with the brigade. When the brigade reached the destination he injected in his leg some substance his father had given him. Immediately an ugly eczema appeared over the entire leg and he had to be released from the brigade and sent home. At home his father gave him another injection which cleared up the skin and the entire family left for a seaside vacation.

Vlado tried to evade this nuissance by always finding a summer job in some engineering field. In the last grade at the college preparatory school however when volunteering for the brigade was critical for admission to the university he devised a unique way out of this predicament. He suggested to a schoolfellow who was a first category chess master that they should exchange lessons. Vlado would give him lessons in mathematics which he would reciprocate by giving Vlado lessons in chess. The schoolmate agreed but requested that at the mathematical lessons his buddy should also be present. That fellow was the class president of the *People's Youth* organization, in other words a communist agent who monitored the students and prepared entries for their dossiers. Vlado had expected, in fact counted on that because he knew that that schoolmate was spending most of his time with the class president for whom he did many homework assignments and, he fancied, even taught him a little mathematics. When the time for "volunteering" for the brigades came, the class president said to Vlado: "I hear that you expressed a desire to go as a leader to *a pioneer* camp. I can now tell you that your request has been approved."

The *pioneers* were elementary school children and they could go to summer camps if they wanted where they were supervised by the elder comrades, members of the *People's Youth*. That would, of course, be equivalent to volunteering for a working brigade. A person in the free world would find it incredible what power those communist youths had in shaping one's destiny but here is one example in black and white for the record. The young communist added that he had already spoken to Nace Gorenc — *Nacek* who at the *City People's Committee* was in charge of *People's Youth* affairs and that Nace had approved his recommendation. He did not know that Nace was my nephew, son of my first cousin of the same name with whom I was always on very friendly terms. The junior Nace had in fact often gone sledding with Vlado on Rožnik, a hill near Ljubljana, and I was the godfather at his confirmation — before the communist era, of course. After a while, however, Vlado was notified that he does not have to go to the *pioneer* summer camp either. The communists no doubt had reservations about letting someone with his background and fairly well known antipathy toward everything in any way connected or related to the communists having the opportunity to be a role model to young citizens of the socialist homeland.

Nace at the *City Committee* resolved the matter so that he shoved Vlado's signed enrollment form for the working brigade to the bottom of a drawer where it got conveniently "lost." No one was then pursuing Vlado to join a working brigade and he was admitted to the university in the Fall. Professor Venče Koželj, dean of electrical engineering, who had been one of my best students when I was lecturing at the university in its early years, told him several years later that he preparatory school gave a special recommendation that Vlado should be admitted. The recommendation was probably made on the advice of Miroslav Adlešič, professor of physics.

In keeping with the communist principle that people whom they don't like should be constantly terrorized, my family was constantly being threatened with further evictions and moving. They were subjected to constant "investigations" at which the bailiffs measured those few square meters of living space they were allowed to have to make sure that they had not a square centimeter of superfluous living space. Once the communists decreed that my family must move to a decrepit house on the banks of the Gruber canal opposite to the botanical garden allegedly because their flat was needed by the army which already was occupying enough rooms in Mokar's house. It turned out that they did not coordinate their edict with the army because, to Ivanka's great relief, two days before the eviction the soldiers picked up their things, field batteries and all, and went back to the barracks in the city. But the communists would not let go. They came up with another notice of eviction. Ivanka's brother Janko kindly invited her to share his family's small apartment at Kodeljevo. At the same time Slava Saks invited her to move in with her at *Miklošičeva cesta No. 4*, opposite the Franciscan church. Ivanka found temporary lodgings for Vlado with one of my schoolmates Viktor Babnik who had a very nice villa on *Zalokarjeva cesta*. It is interesting that one room in that villa was, on paper at least, rented to Ljubo Sirc who was related to the Babniks but Vlado only met him when he was studying at Oxford in 1962, more than ten years later. Vlado was temporarily occupying a room rented to Ljubo Knop who was in Germany as an expert on the Yugoslav war reparations mission. The mission was picking up old machinery in Germany with which the Yugoslavs — or anyone else for that mater — could not do much. Germany on the other hand, with American assistance was able to replace that machinery with modern new equipment. When Knop returned from Germany Vlado camped in the hall next to his room. The Babniks did not like that at all and Ivanka finally found lodging for him with the widow of professor Karl Ozvald where he shared a room with her son Marjan. Ivanka and Breda found a small room in an apartment below Saks'. It was only two meters wide and four meters long, but they were able to squeeze a small upright piano into it so that Breda could continue her music studies.

Milena Exiled to Austria

During the war my wife's sister Milena married Franci Slapar, a lawyer and a reserve lieutenant junior grade in the *Royal Yugoslav Army*. Before the war broke out she was transferred to Kranj and remained there when the Germans occupied Upper Carniola as the occupation authorities were precluded by international law from dismissing her from the civil service. After the communist takeover on the day before the New Year 1946 around midnight a partisan woman brandishing a revolver came to Milena's apartment and ordered her to get ready at once to be taken away. Milena's one year old infant son Peter was, of course, asleep. Milena wanted to get her things and had to go to several rooms in order to gather them but once she would leave a room the woman would lock the door and take the key so that Milena could no longer return to that room. When she thus gathered a few things she woke the infant and dressed him but there was no time left to get something for herself. An autobus waited in front of the house. When it was full of the deportees it drove off in the direction of Jesenice. There all the prisoners were packed in a hall. Milena's child became ill for lack of proper nourishment. She asked a guard, Fonzi Roš by name, whom she happened to know, to telephone her mother-in-law in Ljubljana to come and get her grandson. When her mother-in-law came she was not allowed to enter the camp but had to wait at the fence. The guard did not allow Milena to bring the infant to the fence; he took him and handed him over to the grandmother. It was a very cold winter. After a few days communists took all the deportees over the frozen mountain roads to the border and told them to walk down the other side into the British occupation zone of Austria. Anyone who would attempt to return would be shot. Some people who had a lot of luggage simply had to leave it by the road. At daybreak after what seemed to be an interminable night of wandering in the snow they reached Podklošter where the British occupation authorities wanted to send them right back to Yugoslavia in accordance with the British policy of refusing asylum to refugees from the communist tyrannies. The deportees, however, told them that they would simply lie down and that they may shoot them if that was all that England could do for them. Surprisingly, the British relented and even offered them some food.

Author and friend Jože Majce on a visit to San Marino (1947).

In Austria Milena found an acquaintance who let her stay with her. She was very uncomfortable because she did not have enough suitable clothing for the severe winter cold and her shoes were all but worn out. Her husband Franci Slapar was with the refugee *Chetniks* under the command of General Prezelj in Eboli, Italy. As soon as Milena made a contact with him by mail he found an acquaintance who knew how to get across the closely guarded border between Italy and Austria and who agreed to take Milena across. The guide brought Milena to Riccione where she was received with great hospitality by the women in the camp and where she stayed until her guide could take her to Eboli. I took Milena to some kind of a market where we got her a pair of shoes. The kind women in the camp gave her some dresses so that she finally had something suitable to wear. Eventually she reached Eboli and remained with Franci until they departed to the United States. Their son Peter, however, remained with his grandmother in Yugoslavia and, after she died, with his aunt Anica.

In July 1946 I wrote to the Technical University in Vienna for a certified duplicate copy of my diploma which duly arrived within three weeks of my request. In August my niece Angela Schneider from Napa, California sent me her *Affidavit of Support*, a guarantee required by the American authorities for the issuance of an immigration visa, which I sent to the American consul in Genova. In February 1947 the American consulate notified me that I had been issued in August 1946 a Yugoslav quota number 354. At the end of April of 1947 I went to see off Milena and Franci at the railway station. They were leaving Eboli for Germany where the detachment of the *Royal Yugoslav Army* in exile to which Franci belonged was relocated.

Mirko Bitenc and the British Intelligence Service

One day professor Mirko Bitenc called on me and invited me to dinner at which he explained that he had been making clandestine trips to Slovenia and collecting information of interest to a foreign intelligence service, most likely the British. He would change his appearance and hide on a farm in a barn where a hiding place was prepared for him in a hayloft. He was brought food by the farmer. He proposed that I, too, accept such assignments and accompany him on such missions. I flatly refused because engaging in the espionage was altogether inappropriate and repugnant to me.[145] Having been turned down, Bitenc requested names and addresses of some acquaintances of mine in Yugoslavia who might be willing to work with him but I could provide him with none as it was impossible for me to know how various people aligned themselves under the new regime. In brief, Bitenc's mission to recruit me for the British espionage failed. Later we learned that the communists apprehended Bitenc and sentenced him to death. He was another victim unnecessarily sacrificed to the communists. I observe that nearly all attempts known to us to insert intelligence operatives in Yugoslavia with the assistance of the British met with failure. The communists apprehended all of those sent to Yugoslavia. Several years later it finally came to light that the chief of the British Secret Intelligence service for the Soviet Union Kim Philby was in fact a colonel in the *KGB*[146] who was systematically betraying all British operations to the communists.[147] In retrospect, it is not unreasonable to suspect that even our people had been betrayed by Philby and his operatives.

---

[145] Note by editor. Jerca Vodušek Starič, *Slovenski špijoni in SOE* (*Slovenian Spies* and *British Special Operations Executive*) Ljubljana, 2002, cites almost all Slovenian liberals as collaborators of the *British Intelligence Service*, but not the author who always had a low opinion of the British.

[146] *KGB — Komitet Gosudarstvenoy Bezopasnosty* — Soviet Political Police.

[147] See, for instance, Kim Philby: *My Silent War*, Grove Press, Inc., New York, 1968.

Much later, in 1950 in the United States, colonel Glušič filled me in on some background of professor Bitenc. During the war Bitenc was associated with *Hauptmann* Brandt, chief of the German *Bandebekampfung* (*Terrorism combat and control*) staff. As late as February 1944 Bitenc confided this liaison to Glušič and suggested that our clandestine organization become only semi-clandestine by accepting the support of Brandt and his organization. Glušič turned him down but behind his back Bitenc nevertheless arranged for a liaison between Lieutenant Janez Marn, commander of the *Chetnik* company in Mirna and Brandt. Marn was given German papers, enjoyed German support, and was allowed to move freely in the German controlled territory. In agreement with the German authorities Bitenc then organized the *National Guard* and *Chetniks* in Carniola (Gorenjska). People refused to support either of these groups because of their open collaboration with the Germans. Similar *Quisling* formations could never gain a foothold in Lower Styria and Prekmurje. Bitenc also called on colonel Glušič and demanded the removal of Zdravko Kalan from the Mihajlović organization in Novo Mesto; he told Glušič that Kalan must be removed in one way or another because, according to him, Kalan supported the communists in Novo Mesto. Glušič, naturally, would not go along with this as Kalan's loyalty to Yugoslavia was well established. Kalan was arrested by the Gestapo on June 22, 1944, along with the so-called *Vidovdan group*, which included Glušič.[148]

The Riccione Accord of Slovenian Democrats

In the second half of the nineteenth century three democratic political groups emerged in Slovenia: the liberals, the Catholics and the socialists. At around the turn of the century, in 1900, the Slovenian liberal intellectuals at the universities formed a group called *The National Radical Youth (Narodno radikalna mladina)* which also began to take part in the Slovenian political life. It was in contact with Serbian and Croatian students in Vienna and Prague. This group initiated a movement for the liberation of Slovenia from the Austro-Hungarian empire and for its union with the Serbs and the Croats in one state.

---

[148] In connection with colonel Glušič's characterization of Bitenc's activities professor Janez Grum communicated the following (The South Slav Journal, v. 18 No. 3-4, 1997, pp 152, 153):

"As a former lieutenant in the same *Chetnik* unit as then Majors Glušič and Bitenc from February 1944 on, I wish to correct the faulty statements expressed by Colonel Andrej Glušič to Ladislav Bevc (SSJ, Vo. 18, No. 1-2, pp. 95/96).

The statement "During the war" is too broad and misleading. During the Italian occupation Bitenc was a Yugoslav prisoner of war in a German camp in Northern Italy. He returned to Ljubljana after the Italian collapse. (Afterwards Ljubljana Province was under German

As promoters of the Yugoslav ideal in Slovenia these young liberals led by Gregor Žerjav were secretly making preparations for the liberation while one of them, Bogumil Vošnjak, was active abroad working closely with Thomas Garrigue Masaryk. Their activities had a decisive role on the creation of the State of Slovenians, Croats, and Serbs. Following liberation in 1918, this group assumed the leadership of the liberal party passed on to it by the elder generation whose chairman of many years had been Ivan Tavčar.

In the new nation state the Slovenian democrats built a strong political organization and were frequently represented on the cabinet of many governments. Their influence was particularly strong in the industry, banks, with the urban population and, to some extent, with the farmers. Because of their strong individualism and diverse views concerning the organization and conduct of the government, the liberal ranks often split into newly established groups and parties. The most important of such groups in the early twenties was the farmers' wing.

... occupation. In October 1943 Bitenc joined the newly formed Slovenian *Home Guards* [*National Guard*] (*Domobranci*) but in early 1944 was dismissed on German demand. In late February he joined the renewed *Chetnik* unit that was part of the Mihajlović movement in Slovenia (JVvD).

At that time Lieutenant Janez Marn of the *Chetnik* unit established his personal contact with German Hauptmann Brandt through a certain Slovenian journalist (*sit nomen odiosum*). I learned this during the summer and autumn (of 1944) when I witnessed two contacts between Brandt and the journalist on behalf of Marn *Chetnik* group. Therefore I am convinced that this particular journalist was already serving as intermediary between Marn and Brandt in February 1944 and that it was he, and not Bitenc, who provided German papers to Marn. When Bitenc learned about those papers, he "suggested" to Glušič "that our clandestine organization become only semi-clandestine by accepting the support of Brandtč"After the passage of so many years, let me add that the *Chetnik* Staff tolerated Marn's contact with Hauptmann Brandt.

Allow me also to correct another erroneous statement by Glušič, as reported by Ladislav Bevc, namely that Bitenc, in agreement with the German authorities then organized the "Home Guards" and "*Chetniks*" in Gorenjska (Carniola)." Both the:"Home Guards and these *Chetniks* are labeled by Glušič as "Quisling formations." I won't discuss the alleged "quisling" status of the Home Guards. But I must point out that the *Chetniks* in Gorenjska were also members of the Mihajlović movement in Slovenia. And let me mention that Bitenc was proud that he was personally promoted by the then Colonel Draža Mihajlović (Celje, 1940) to the rank of Lieutenant-colonel.

The *Chetniks* in Gorenjska were established on the Lipoglav (name of the hill, village and parish), southeast of Ljubljana in early July 1944 in accordance with the plans of the *Chetnik* Staff in the presence of a staff member.

Three months later this group crossed the German border near Dobrova (west of Ljubljana) to recruit new members in Gorenjska and stay there. Three days after their arrival, this *Chetnik* group was required, as comrades-in-arms, to come to the defense of the Home Guard post nearby, which was under attack by the Šlander Partisan brigade from the rear. The Partisans, caught between the Home Guards and *Chetniks*, were forced to withdraw and the Home Guard post at Lahovče was saved from annihilation.

Quite contrary to [colonel] Glušič's statement that "people refused to support the *Chetniks*," the anti-communists of Gorenjska (Upper Carniola) were in fact enthusiastic in their praise of the comradely *Chetnik* action on the night of October 4–5, 1944."

Because of its uncompromising stance which set the interests of the state above everything, the democratic politicians often had to confront difficulties. Their party chairman in the years prior to World War II was Albert Kramer. Kramer was always very favorably disposed toward me. I last saw him at the onset of the Axis occupation when I visited him at his home. He was staying with his niece who was married to Boris Puc,[149] son of the former mayor Dinko Puc. Boris Puc eventually took to the woods where he joined the partisans. Ljubljana wags observed in the papers that it was about time if he wanted to make a career. Kramer was seriously ill and confined to his apartment. After he died I was nominated to the directorate of the *Democratic party* consisting of seven members with Janko Mačkovšek as its chairman.

I knew Mačkovšek from the preparatory school (*Realka*) where he already had a reputation because of his work w on the national minorities problems. Following graduation from the University in Prague he was employed by the City of Ljubljana and was active in public life, particularly in the *Cyril and Methodius Association* where he succeeded Franja Tavčar[150] as leader of the Association. The Italians who occupied the southern part of Slovenia arrested and it took a long time before he was released. During the war we used to meet in the *Cyril and Methodius Association*'s office to discuss the situation. When Mačkovšek learned of the impending arrests of the liberals he requeste me to go to General Rupnik and ask him to prevent them. I would not go to Rupnik but I did ask Krener about it and he told me there was nothing to worry about which, unfortunately, proved incorrect. Anticipating arrest Mačkovšek invited me to his home and briefed me on the work he was carrying on and explained the cipher he used in sending his messages to London. Soon after that he was arrested and sent to the concentration camp in Dachau where, weakened by starvation and general deprivation, he died in an epidemic. When Mačkovšek was deported to the concentration camp in Dachau, he was succeeded by Ferlinc.[151] With the exception of a few individuals the entire older generation of liberals was wiped out during World War II either by the Nazis in Dachau, where Ivan Pucelj[152] and Janko Mačkovšek died, or by the communists.

The *Youth of the Yugoslav Nationalist Party* was initially formed within the fold of the *Yugoslav Nationalist Party* (YNP) before World War II broke out The

[149] Boris Puc (1907–1975), attorney in Ljubljana, joined the *Liberation Front* in 1941 and in 1943 joined the communist partisans where he was member of the communist judiciary, president of the military tribunal of the ninth corpus and president of the *People's High Court for the Slovenian Littoral and Trieste.*

[150] Franja Tavčar, wife of Ivan Tavčar, writer, lawyer, liberal politician and Mayor of Ljubljana.

[151] It is not clear from the manuscript if Dr. Ferlinc is the same person as professional engineer Bogdan Ferlinc whom the communists tried at one of their show trials after the war and imprisoned.

[152] Ivan Pucelj (1907–1945), politician and minister of the Yugoslav government, leader of the *Farmers' Party (Kmetska stranka)* in the 20's, later member of the *JNS*. In 1941 he was member of the *National Council* and the *Konzulta*. In 1944 he was arrested by the Nazis and interned at Dachau where he died.

recognized leaders of this party were Stojan Bajič and Andrej Uršič. Bajič had early attracted the attention of the *YNP* leadership which assisted him to attend a course in public speaking in Berlin while he was still a student at the university. Bajič was a born leader, bright, and an excellent speaker. He earned his doctorate in jurisprudence at the University in Ljubljana following which he was employed at the district court in Ljubljana where his brilliant performance soon brought an appointment as an adjunct professor (docent) on the university's law faculty. Andrej Uršič was equally bright, trustworthy and engaging.

At the beginning of World War II the *Youth of the Yugoslav Nationalist Party* was poised to assert itself in the party's leadership. The *Youth YNP* drafted a program on the goals and aspirations which it wanted the *YNP* leadership to adopt. At the time the *Youth YNP* requested a meeting on its program Dr. Kramer, the Chairman of the Slovenian *YNP,* was already seriously ill. No agreement, however, could be reached at the meeting. The principal spokesman of the *Youth YNP* was Bajič who at the conclusion of the meeting declared that the *Youth YNP* was leaving the *Yugoslav Nationalist Party.* Chairman of the *Youth YNP* at that time was Jože Rus[153] who was my personal friend and who often visited me in my office to fill me in on the new developments. I knew that Rus belonged to the less radical wing headed by Uršič but in view of the impending war he, as most of the members of the *Youth YNP,* was willing to go along with Bajič and his plans. Under Bajič's tutelage the *Youth YNP* decided to join the so-called *Liberation Front* but the association did not last and Rus was soon able to report that the *Youth YNP* had left the fold of the *Liberation Front.* Uršič attended the regular meetings of the *YNP* directorate as a delegate of the *Youth YNP.* The negotiation on re-establishing the affiliation in which principal spokesmen for the directorate and the *Youth YNP* were Otmar Pirkmajer and Andrej Uršič, respectively, were unfolding favorably.

Several attempts were made to form a single party of Slovenian democrats and on several occasions it did come to the fusion of stronger liberal parties under different names. In 1941 representatives of the *Youth of the Yugoslav Nationalist Party (Omladina Jugoslovenske nacionalne stranke — OJNS), Independent Democratic Party (Samostojna demokratska stranka)* and *Alliance of Young Men and Women Farmers (Zveza kmečkih fantov in deklet)* signed a working agreement in Ljubljana. During World War II Milan Lenarčič tried to achieve a fusion of several progressive parties but did not succeed. In exile, in the Italian seaside resort Riccione, on August 18, 1946 representatives of the *YNP, Young Yugoslavia, Independent Democratic Party,* and certain independent progressives signed a draft agreement.

In March 1946 representatives of all progressive democratic factions gathered in Riccione, where some of us were in the refugee camp. Most of the delgates came

[153] No relation to judge Rus, the grave digger of the *Sokol* organization.

from Trieste and Rome. Altogether about fifty persons participated. We were meeting for several days and resolved that all progrressive democratic factions join in a unified Slovenian Democratic Party; we adopted bylaws and regulations. The final agreement was signed on March 26, 1947. The Riccione agreement is the charter of Slovenian Democratic Party which is the successor party of all its constituent groups and their political organization. At the head of the Party are the *Principal Assembly, Principal Committee* and *Executive Committee.* By agreement the program of the *Slovenian Democratic Party* includes by incorporation the program of *YNP (JNS)* adopted at the assembly of *YNP for Slovenia* in December 1945. The *YNP* program is presented below.

PROGRAM
OF THE YUGOSLAV NATIONALIST PARTY

The following are our goals:

I. Fundamental:

1. Liberation, expansion and renewal of the Kingdom of Yugoslavia under the dynasty of Karadjordje.
2. Liberation and unification of the entire Slovenian nation in Yugoslavia including all its historic, ethnic, economic and defense territory.

II. Liberating:

1. Removal of any national, economic and social alienage from among us
2. Measures for efectively securing just and proper compensation of war damages sustained since 1914 onward.
3. Temporary government supervision and control of banking institutions and insurance companies, all mining operations and such other industries which because of their size or production serve the essential national interests, large merchant enterprises and large estates to ensure their rational utilization at a time of shortages caused by upheavals, until such time as normal economic conditions are restored.
4. Preventing evasion of these measures.

III. Moral:

1. Imposing penal sanctions on all foreigners and aliens who prior to World War II, during the war and during the time of the enemy occupation on our national territory commited acts contrary to state and national interests or any illegal act or act contrary to the legal instinct, against individuals or the whole.
2. Imposing penal sanctions on the traitors to the state and nation.
3. Imposing sanctions on all who share in the complicity of the misfortune that had befallen our nation and state because of active anti-national or anti-government alignment, incompetence, lack of responsibility or for profit.

4. Judicial review of all politically motivated trials and judgments issued after April 6, 1941 at the request of interested parties.

IV. Constitutional:

1. United Slovenia shall be a part of Yugoslavia having equal rights.
2. Constitutional restructioning of Yugoslavia should be carried out on a federal plan with the agreement and consent of all its constituent part and with equality of rights. The restructioning should reflect the nation's will without any limitations or reservations.
3. The tasks of the federalgovernment should be: foreign affairs and foreign relations, defense of state's independence and territorial integrity, administration of federal finance, determination of an economic plan based on the principle of economic unity of the state's territory, federal communications, general principles of social legislation and determination o general directives for a harmonious coexistence of all constituent parts of the state.
4. The political structure should be based on the principles of democracy as conceived in Western democratically constituted states.

V. Economic and Social:

1. In economic and social matters national and public interest should take precedence before private interests.
2. Private property recognized within the scope of constitutional provisions shall be inviolable.
3. The Party is opposed to class strugle and favors social concord which implies a harmony of social forces.
4. All citizens should be guaranteed at least sustenance, clothing and living quarters.
5. Private initiative in the economy should be encouraged.
6. Farmers, laborers, and intellectuals should be free from exploitation of their physical or intellectualefforts. Work both physical and intellectual should be the source and measure of rewards as a basic social value and a right and obligation of all. General social security should be provided as insurance for all contingencies. Women and minors should be protected against exploitation of their labor.
7. Administration in all fields should be simplified. Employees should be suitably compensated. Compensation of employees in the public and private sector should be the same for equivalent service.
8. Amassing of real estate in the hands of an individual should be limited.
9. Just compensation should be paid in case of possible expropriations.
10. Agrarian reform should distribute land so that farmers will be according to their size and production a basis of a sound farming family within the framework of the village and farming cooperative communities. Large grazing areas and forrests should be nationalized. Farm houses should be established the sale of which should not be allowed. Land ownership should be limited in case of persons who do not work the land themselves. Regulations should prescribe a just pricing schedule for

agricultural products which should correspond to the price structure of industrial products.

11. Nationalization or government regulation should be instituted for mining operations, energy resources, large means of production, processing and distribution and the insurance industry with the objective of reducing the cost of all basic goods that serve the general interest.
12. Planned development of economic activities for the national needs and full employment of the labor force should be instituted.
13. Integration of Yugoslavia into the social and economic order of the Balkans and Europe should be planned.
14. Corruption in all walks of public and private economy shall be supressed.
15. Cartels and trusts in the private sector will not be allowed.
16. Government should regulate large private economic enterprises.
17. The Party joins under its aegis the common man and the middle class. Its "objective is to apply its activities and influence to assit them in a permanent improvement of their living standard. The Party rules out any influence of the capital on its affairs which are in its totality dedicated to the common man and the middle class and to the interest of the entire nation.

VI. **Cultural:**

1. Fostering the desire for a community of all Yugoslavs.
2. A thorough transformation of education and development in the spirit of Christian ethics.
3. Freedom of conscience, religion and persuasion and cultural and religious tolerance.
4. Slovenian cultural creativity, education and enlightement must proceed from the nation's essence.
5. Education at all levels at the appropriate age should be available to everyone who is mentally qualified. Of particular interest should be a wider and deeper professional education in all fields of the economy to prevent hyperproduction of intelligentsia which tends to gravitate to the public sector. The party realizes that advancement of the economy and professional instruction in economics is a prerequisite for a favorable solution of the national and social questions.
6. Achievements of the sciences and arts should be enjoyed by all citizens so that they may become a universal treasure of the nation.
7. Sokol is the central noble national organization for physical and mental development based on the lofty ideals of the Sokol thought and honorable traditions of the nation's battle for democracy, physical fitness and intellectual freedom.
8. Guarantee of the four freedoms:
   a) freedom of speech and expression;
   b) freedom of religion and worship;
   c) freedom from fear;
   d) freedom from want.

II. International:

1. Establishemnt of a confederation of Balkan states
2. Close political, economic and cultural cooperation with states that are democratically constituted according to the model of Western democracies.
3. Cooperation with all states to secure world peace and prosperity.

Adopted at the assembly of the JNS for Slovenia

This 9th and 10th day of December 1945

J. N. S. Committee for Slovenia

The *Yugoslav Nationalist Party* undertook to support the proposal that representation in the *Slovenian National Committee* be structured so that if the progressive delegation in the *National Committee* should be limited to five members, the *JNS* would allow to the other parties to the agreement two places. However, if, the number of the progressive delegates should be even, the places would be evenly divided.

Many refugees, members of the various factions of the liberal party, found themselves in Trieste at the end of World War II in 1945. These people initiated a broadly conceived action for the unification of all progressive parties into a single party, the later Slovenian Democratic Party. Among these were Boris Sancin, Slavko Uršič and Bojan Ribnikar. Sancin and Ribnikar would often visit me in Riccione for consultations. A draft program of the unified party was prepared and an emissary took it to Riccione for me to review. The emissary was surprised at my immediate agreement to the program. Within a short time another delegate arrived bringing the final text of the program to which I likewise agreed. Following that, in August 1946, we held a founding convention of the Slovenian Democratic Party in Riccione. The convention was attended by some fifty delegates most of whom came from Trieste and Rome. Presiding at the convention, which lasted several days, was Dr. Pavle Pestotnik, the doyen of Slovenian liberals. It was agreed that all the progressive fractions would join into a united Slovenian Democratic Party; the governing documents and the by-laws of the party were also agreed upon at that time. The final form of the agreement was signed on March 26, 1947. That agreement was the charter and the political program of the Slovenian Democratic Party which was the successor of all former progressive parties and factions. The principles of the YNP committee for Slovenia adopted in December 1945 were incorporated in the Riccione agreement.

Representatives of political organizations who concluded the *Riccione Agreement* and members of the Interim Committee of the Slovenian Democratic Party

(SDS) were: Edvard Antosiewicz, Dr. Stojan Bajič[154], (Walter Gerstenmeyer), Ing. Ladislav Bevc, Dr. Vladimir Borštnik, Dr. Bogdan Drnovšek[155], Vlado Gorazd, Albert Herič, Dr. Franc Kolterer, ing. Milan Lenarčič, Dr. Boris Mihalič, Dr. Gustav Omahen, Dr. Janko Orožen, Dr. Edvard Pogačnik, Tugomer Prekoršek, Dr. Bojan Ribnikar[156], Ing. Boris Sancin[157], Dr. Avgust Sfiligoj, Rado Slanc, Miroslav Urbas, Andrej Uršič, Dr. Miloš Vauhnik, Živko Vedlin, Jože Verovšek, Dr. Bogumil Vošnjak, Ivan Vrančič, Dr. Marjan Zajec, Dr. Vinko Zorc, and Rudolf Žitnik.

Text of the Riccione agreement follows.

AGREEMENT BETWEEN REPRESENTATIVES OF SLOVENIAN PROGRESSIVE POLITICAL PARTIES ABROAD

I.

Intending to achieve a basis for a new political life of progressive Slovenians we conclude this agreement which should bring about the establishment of a mutual and unified progressive party. The undersigned who belong to progressive parties hitherto in existence consent to this agreement with proviso that on the return to the homeland their appropriate party forums will ratify it. They assume moral obligation to work within their party toward the implementation of this agreement.

Noting the great direct importance and and long term potential of this resolution and convinced that they are expressing the will of a large majority of progressive Slovenians they will henceforth conduct their political action in accord with the feelings and obligations stemming from mutual belonging and unity. With regular decisions of appropriate political forums issued from time to time the new party will become a full fledged and respectable successor of its constituent parts and will thus with a program

[154] Dr. Stojan Bajič — Walter Gerstenmayer (1901–1974), lawyer, judge and professor of law in Ljubljana, member of Omladina *JNS* (Youth of JNS) and *Nova Jugoslavija* (New Yugoslavia). Emigrated to Switzerland in 1944, after World War II became professor in the United States.

[155] Bogdan Drnovšek (1915–?), lawyer and journalist at the daily *Jutro*, member of *Omladina JNS*, associate editor of *Zarja Svobode*, publication of *Nova Jugoslavija (New Yugoslavia)*. Emigrated to Trieste in 1945 where he became member of *Slovenska Demokratska Zveza* (*Slovenian Democratic confederation*), later emigrated to the United States.

[156] Boyan Ribnikar (1914–1990) lawyer, journalist, businessman, owner of the daily *Večernik* in Maribor. Moved to Ljubljana to avoid German occupation, left for Trieste in 1945 where he was on the editorial board of Radio Trieste Station B [Slovenian], emigrated in 1947 first to Argentina then to the United States where he owned travel agency Andersen Travel in Berkeley, California. He was an active member of the *Slovenian Democratic Party* and editor of the party's paper *Domovina (Homeland)*.

[157] Boris Sancin (1912–1989) born in Trieste, lived in Ljubljana from 1923 to 1945, director of rice refineries in Ljubljana, Milano, and Trieste, leader of the Slovenian section of Radio Trieste, leading member of various Yugoslav nationalist organizations, active member in non-communist resistance movements during World War II, editor of various journals, political editor of the Trieste anti-communist weekly *Demokracija* (1947–1952), author of numerous political and economics articles, papers and lectures.

for the future as well as with reference to the past assume the shining tradition of the progressive camp in Slovenia. As such it will become the protagonist of the freedom loving idea in the cultural, political and social domain and a champion of the Yugoslav idea in the area of national politics.

II.

1) In emigration we shall work out a program and make all the technical preparations necessary for the establishment of the Slovenian Ddemocratic Party (SDS). At this we shall pay particular attention to the specific immediately needed work and act with full consideration of a carefully monitored internal and external political situation.

2) In formulating the program of the Slovenian Democratic Party we shall proceed from the program of the JNS (Jugoslovenska Nacionalna Stranka — Yugoslav Nationalist Party) adopted on December 12, 1945 in exile and with due consideration of »Political, social and economic principles« of OJNS (Omladina JNS — Youth of the Yugoslav Nationalist Party) adopted in 1940. Needed changes, reductions or amendments will be worked out by the Interim Committee at the earliest possible time with due consideration of programs of individual member parties which are merging into the Slovenian Democratic Party. In the program we shall declare ourselves in support of a democratic and federative government structure with the largest possible autonomy of individual federal units. In the federal state of Slovenia the government administration should be decentralized. We leave the decision concerning the administrative structure to the decision of the electorate.

3) The Party is governed by: the General Assembly, the Principal Committee and the Executive Committee

(a) The General Assembly will meet in the liberated homeland as soon as this will be possible. The Genaral Assembly will decide on the final formulation of the Party's program and will establish the Party's organization. Until the General Assembly meets the Interim Principal Committee and Interim Executive Committee have all the responsibilities and prerogatives of the Principal Committee and the Executive Committee respectively. All the relationships and proportions established by this agreement shall be in force only until the General Assembly convenes at which time it will decide on all questions of concern to the Party.

(b) The Principal Committee consists of up to 60 members. The amendment dated August 15, 1946 lists all the initial members of the Interim Principal Committee which has the right to co-opt other members. As members of the interim Principal Committee should be members of the Principal Committees of the constituent parties who have demonstrated with their work that they will contribute to the common cause, who did not betray the ideas of democracy, and who did not act against the national interests nor transgress against the interests of the common progressive political formation.

In the Interim Principal Committee we admit under the same condition also certain reputable progressive public personalities who never served on the Principal Committee of any progressive party or who belonged to no political party. — The

Principal Committee defends the purity of the program, sets the course of policy, resolves questions of Party politics, issues instructions to the Executive Committee, other agencies of the Party, and Party's representatives at various public or private institutions and forums. The Principal Committee sets the policy of the Party in accordance with the resolutions and decisions adopted by the General Assembly. Subordinate to it and responsible to it are all the agencies of the party and bearers of a madate. The only guidance for the Principal Committee which has the final authority in giving assignments to individual members within the Party or outside of it as the Party's representatives is the tactical and political effectiveness of individuals.

(c) The Interim Executive Committee consists of 10 to 15 members elected by the Interim Principal Committee when circumstances permit the Party to go public. The Interim Executive Committee will be constituted according to the priciples of fairness and with consideration of all parties that are now merging into the new party. Here special attention will be given to the principle of tactical and political effectiveness of individuals. Interim Executive Committee adopts its decisions with a two-thirds majority, if three of its members state that the subject on the agenda constitutes an important matter of principle, otherwise a simple majority is required. — The executive Committee conducts the political tactics of the Party according to the instructions and under the supervision of the Principal Committee.

(d) The Chairman of the Party is elected by the General Assembly. The chairman represents the Party everywhere. His is the initiative in all the activities of the Party and its agencies and he has oversight of all operations. He is assisted by vice-chairmen who have the same rights and obligations as the chairman when he is unable to act. The Chairman of the Party is also the chairman of the General Assembly, the Principal Committee and the Executive Committee.

(e) The Secretary General of the Party is elected by the Principal Committee for as long as he enjoys the trust of the Party. The Secretary General is at the same time the first secretary of the General Assembly and the Executive Committee of the Party. He manages the administration and provides for the organizational affairs of the Party. The Secretary General may neither be a candidate for nor hold public office nor is he allowed to be in private employment. His principal task is to ensure the continuity of the Party's administration and he reports to the Executive Committee of the Party. The service of the Secretary General must be properly acknowledged and rewarded. The same applies for compensation of other personnel of the party's main offices.

4) In assigning public offices the Party must consider all capable and worthy adherents and offer them an opportunity for political establishment. Creation of inaccessible narrow circles that might want to have exclusive control over the assignment of public offices must be prevented. Assignments in this sense should be controlled by the Principal Committee; tenure in a public office may be subject to a time limit if necessary. — All political and tactical questions must always be resolved within the framework of relevant party offices subordinated to the Party's democratic majority. The Party should prevent and censure dissidence and any breach of Party discipline.

III.

1) All preparatory work along the guidelines of this Agreement will be carried out by the Preparatory Committee, the members of which are listed in Amendment No. 2 of August 15, 1946. The Preparatory Committee appointed therein has the right to co-opt up to three members and the right to replace sitting members. The Preparatory Committee must work out primarily the draft program of the Party, its bylaws and formulate a position with respect to the [Slovenian] National Committee. The Preparatory Committee wil decide when the Interim Principal Committee will start operating. At that time the assignments and responsibilities of the Preparatory Committee shall terminate. Only in case of a political exigency will the Preliminary Committee already begin its work in exile. Such a decision, however, must be adopted unanimously. For this contingency all the signers agree that they will defend such decision at the last general assemblies of their parties in Slovenia which will be charged with the task of terminating the activities of their respective parties and their incorporation in the Slovenian Democratic Party.

2) Members of those progressive parties which have hitherto not been represented in the National Committee in principle support participation in the National Committee. Should a need arise for changing the present progressive delegation to the National Committee, the YNP will support the proposal that the question of representation on the National Committee be resolved in accordance with the principle of fairness. If the progressive delegation to the National Committee should be limited to five members, YNP will give up two places to other progressive parties that are entering in this Agreement; if, however, the number of places should be even they will be evenly divided. When the time for changing the progressive delegation in the aforementioned sense has arrived will be determined by the Preparatory Committee by unanimous vote. Unanimous must also be the decision as to the constitution of the newli formed delegation. Until the change is put in efect the current delegation of the JNS will report on the proceedings of the National Committee at the meetings of the Preparatory Committee. — The position of the newly created party with respect to the National Committee should be set forth by the Preparatory Committee at its first session in a separate declaration.

3) Until such time as the Interim Principal Committee begins to operate all resolutions and obligations of hitherto separate parties will bind only them except, of course, in such cases as they are unanimously reaffirmed by the Preparatory Committee.

4) This Agreement shall remain secret until the Prepartory or Interim Principal Committes decide otherwise. We shall, however, confidentially advise of its existence, without particulars, the Slovenian People's Party (SLS — Slovenska Ljudska Stranka) and the Yugoslav Socialist Party, Committee for Slovenia, as political partenrs. The notification shall include item III/3 of this Agreement. The notification will be made by the leadership of the JNS, Committee for Slovenia, following the signing of the Agreement by members of the Preparatory Committee.

IV.

Upon return to the homeland progressive press, progressive enterprises and institutions will be orgaanized so that they will become a living part of the new Party. All those institutions will have to be guided by the will of the party as expressed by the Principal Committee and carried out by the Executive Committee of the Party.

V.

This Agreement shall become effective when signed by members of the Preparatory Committee; it replaces in its totality the »Draft Agreement« dated August 15, 1946 and »Draft Agreement« dated September 15, 1946. Amendments to »Draft Agreement« of August 15, 1946 shall remain in force as amended by Sections 1a and 2a dated March 26, 1947.

Written and signed in two identical, equivalent copies, in exile, this 26th day of March 1947. (Riccione)

[SIGNED]
1. Ing. L. Bevc, 2. Dr. B. Drnovšek, 3. Ing. M. Lenarčič, 4. Dr. B. Ribnikar, 5. Ing. B. Sancin, 6. Dr. M. Zajec, 7. R. Žitnik

The purpose of the union was to organize all Slovenian émigrés of progressive persuasion wherever they might be to preserve and promote the traditional goals and ideals of the liberal party. Confrontations with the communist regime in the homeland were not on the agenda of the Democratic Party, we held the view that in time the Slovenian nation would regain its freedom as has been the case with all enslaved nations throughout the recorded history. In the meantime it was the task of the Democratic Party to prepare its members for the time when it could return to a liberated homeland and be of assistance in resuming its constructive activities for the benefit of the Slovenian people. Hence the first and foremost task of the SDS was to work for the liberation of the Slovenian people from communism. The SDS abroad represented many Slovenians in the homeland who were not allowed to speak and act according to their convictions. The Slovenian Democrats, as well as their émigré leadership were convinced that the joint and several interests of the Slovenians, Croats and Serbs, at that time as well as in the foreseeable future, of necessity required a common state. The *Slovenian Democratic Party* did not change its ideological viewpoint and it never wavered in its support of the unity and integrity of Yugoslavia whose future internal organization was expected to recognize the specific needs of its constituent nationalities.

Most of the leading members of the *SDP* were preparing for emigration to Argentina, accordingly it was agreed that the initial headquarters of the party would be in Buenos Aires. I was nominated president with the understanding that I would assume my duties when I got to the United States at which time the *Executive Committee* of the party would also be elected.[158]

Camp at Bologna

On June 21, 1947 we were moved to the *Displaced Persons Center at Viale Enrico Panzacchi 14 in Bologna.* The camp commander was an American, Humphrey by name. He had an assistant from Malta who served as his interpreter translating Humphrey's directives to Italian. The commander took no particular interest in the running of the camp, but the food, although not abundant, was quite satisfactory. Here we first noticed some new arrivals from Yugoslavia who were received as refugees and employed mainly as truck drivers. Some of them were known as communist sympathizers by the other refugees; we could do nothing about it except exercise caution in dealing with them. American Slovenians sent a large shipment of clothing, shoes, and underwear to Bologna which was transported from Rome by trucks. Enroute the dishonest drivers distributed the better items to their contacts and fences so that only leftovers reached Bologna. The refugees were not particularly interested in those leftovers. A camp employee urged me to go and select something for my use. I was reluctant to do this because there were others in the camp whose need was greater than mine and so I simply looked at what was available without selecting up anything. My friend Novak Marcikić and I used to go to the market where there was plenty of fruit. Once we went to the cathedral where the famous preacher Lombardo propounded the case against divorce in the presence of the cardinal and an enormous crowd of listeners. The preacher was giving reasons against divorce while another priest, standing next to him in the pulpit, quoted the arguments in favor of divorce which the preacher invariably refuted.

Novak Marcikić (1960)

[158] Delaying the announcement of author's nomination turned out to be a wise precaution because the communist intrigue which the United States Department of State was always willing to accommodate could well have prevented his immigration to the United States if it were known that he intended to be active in the émigré politics. An example of the United States political assistance to Tito's regime is the statement of George Vest, Assistant Secretary of State for European Affairs, made on June 23, 1997 to the delegation of the Assembly of Bosnia-Herzegovina. In connection with the most recent activity of extremist émigrés, George Vest assured the visiting communists, that the United States government would do everything possible to remove obstacles to the development of the U.S. — SFRY relations. According to the U.S. Department of State, *extremist émigrés* were all who opposed Tito's regime.

We were soon joined by Pestotnik who followed us from Riccione. His wife, née Geršak, owned a business which traded in agricultural products. The communists seized all her property on grounds that, during the time of the Italian occupation, she had been doing business with clients in Italy and sentenced her to several years of imprisonment. Terminally ill she was confined to her home when the communists decreed that she had to move to some garret on *Congres Plaza (Kongresni trg).* She died in a truck with which the communists were moving her to her new quarters. We concealed this tragic news from Pestotnik because of his heart condition.

Pestotnik had an acquaintance by name of Jelenič, a native of Suhor near Metlika who worked for a relative somewhere on the Zaloška Road in Ljubljana as a meat cutter and who regularly attended the gymnastics at *Sokol I at Tabor.* Jelenič emigrated to San Francisco and there opened a grocery store on Vermont Street not far from the *Slovenian Hall.* At that time most of that neighborhood was Slovenian. He became a representative of Mrs. Pestotnik's business which was shipping mushrooms to the United States. The sales of the mushrooms went well and Jelenič was regularly sending the proceeds to the Pestotniks in Ljubljana. When the war broke out, the trade was interrupted and Jelenič, who could no longer send money from the most recent sales to Yugoslavia occupied by the Axis powers, deposited the residue of Pestotnik's funds in a bank account. As the sum was substantial I suggested to Pestotnik that he withdraw the money and leave for Switzerland where he would be able to live independently but he did not take my advice.

In the meantime Beatrice Hunter, an adventurer apparently of Slovenian origin, found out that Jelenič was holding Pestotnik's funds in trust and pestered him so persistently that he loaned her some money out of those funds. The woman met Tito's delegates at the *United Nations Organization* which was being formed in San Francisco and was able to join them on their flight back to Yugoslavia. There were no regularly scheduled flights to and from the New World at that time. The refugees were told that all transportation was still needed by the military. Of course there never was a lack of ships for transporting the refugees from Russia to Stalin's mass execution orgies on the coast of the Black Sea. In Ljubljana Hunter, representing herself as the president of the import-export firm *Inter-Allied*, called on Mrs. Pestotnik and pressured her to "lend" her the money kept by Jelenič but Mrs. Pestotnik would only refer her to her husband in Bologna. Hunter readily left for Bologna where she invited Pestotnik to a conference in her ostentatious quarters at Hotel Imperiale. It was quite clear that the woman's purpose was fraud and embezzlement as she was representing to Pestotnik that she was trying to get him a higher return on his funds out of compassion she felt for his children. I again urged Pestotnik to have nothing to do with that woman and instead withdraw the money from the bank and leave for Switzerland as his condition was making it inadvisable for him to remain in the camp. At my request Hunter saw me, too, and promised to visit my sister in Napa, California, forty miles north of San Francisco, to give her a report on my situation. Of course, she never visited my sister. Pestotnik authorized Jelenič to "lend" his money to Hunter who gave him an unsecured promissory note written on a standard form available at any bank and told him that the note was guaranteed by the bank that furnished the form. The note stated that money would be repaid with 5 percent interest, which was considered to be high at that time. As I expected, Pestotnik never saw any proceeds of that loan and only part of the capital was later recovered in the United States through extensive litigation. Such were the circumstances to which we were exposed in exile.

In the course of my public service and as a lecturer at the university I met and maintained friendly relations with many Russian émigrés who were either my students or my colleagues in the civil service. To the camp in Bologna also came Grigori Stanislavski who left Russia at the time of the revolution and settled in Ljubljana where he owned a grocery shop on Resljeva Road. His customers were mainly military officers with whom he was maintaining social contacts. When Stanislavski came from Yugoslavia he could bring along only so much as he could pack into three trunks.

Stanislavski was a friend of the industrialist and wholesale merchant Milutin Jelačin and acted as his advisor after the communists seized all his property. As already mentioned, in addition to losing all his property, Jelačin had a hard time avoiding long imprisonment; he avoided it only by showing that he was a substantial supporter of the *Liberation Front.* Stanislavski told us how the new Soviet men behaved in Yugoslavia. Into his friend Jelačin's villa, across the street from his cork factory, moved the Soviet Mission headed by Colonel Bogomolov. Bogomolov's soldiers and his partisan guard of honor wasted no time in stripping Jelačin's villa of furnishings and everything that was not nailed down including even the electrical and plumbing fixtures. Jelačin sent Stanislavski to intercede with his compatriot Bogomolov who, after listening to Stanislavski, only shook his head and said: "Oh, these fellows really are a wild bunch!" Of course it was Bogomolov himself who had ordered his men to haul everything away to the Soviet Union.[159]

Stanislavski, who never quite learned to speak Slovenian fluently, had many other stories to tell about the events that took place during the Russian revolution; he boasted that he had seen the Czarina when she came to visit the wounded soldiers in a hospital. His stories had the ring of fabrications, or perhaps expressions of unfulfilled aspirations and desires. Later on, in Camp Senigallia he was arrested because of some intrigues and sent to the dreadful internment at Lipari from where he returned embittered and justly offended.

[159] An interesting encounter with Bogomolov's men in Jelačin's villa is described by Ljubo Sirc: *Nesmisel in smisel* (*Nonsense and Sense*), Pika Print, Enfield, England 1968, p. 7-8

Camp at Regio Emilia

On September 11, 1947 orders were issued transferring all refugees who were not accompanied by their families or relatives to the *13th Displaced Persons Center at Reggio Nell' Emilia*. That camp used to be an enormous barracks with sprawling horse stables. In these stables the camp administration installed two-tiered bunks on which were packed hordes of refugees from the Near East whom the Germans had recruited as laborers and who were at that time returning home. In the midst of those masses were shoved Pestotnik, Franc Kolterer, Novak Marcikić and I. Although the camp had not yet been fully organized, the food was no worse than everywhere else but the sanitary facilities were terribly overcrowded so that I had to get up very early in the morning if I wanted to use them. Friends who had remained in Rome succeeded in securing the good offices of Miha Krek who had some influence with the *International Refugee Organization* to have us transferred within a month to the *9th Displaced Persons Center in Senigallia* where the majority of Slovenian refugees lived. The Maltese aide of the Bologna camp commander arrived to pick us up and took us to Bologna from where we were sent to Senigallia the following day. Soon after that the aide disappeared because he reportedly embezzled one million lire.

Camp at Senigallia

Everything in Senigallia was ready for our arrival. The camp was actually a camp for Slovenians who also actively participated in its administration. We were received by the commander's aide Padar, a Slovenian who had a good command of English. The commander was an Englishman who was able to communicate with us only through his interpreter.

Padar placed Kolterer and me in a separate room where some ten persons presumably meriting special consideration were billeted. Among them was Vladimir Borštnik, at one time the secretary of the Russian consulate in Trieste in the era before World War I, who was fluent in the Russian language. I had known him since 1902 when we were in the group of twelve students from the provinces to whom Franja Tavčar provided a daily lunch in her home at Breg in Ljubljana. I remained in touch with Borštnik also after we came to the United States. Unfortunately he was already terminally ill when he came to America. He went to a Serbian orthodox institution near Chicago where he soon died. Other members of our group, among them Pestotnik and Novak Marcikić were placed in other rooms.

The Senigallia camp organization was exemplary. The food, prepared by Slovenian cooks, was good and the same for everybody. The only exception were the employees of the camp administration who had a separate mess hall. It was gratifying to see that the Slovenians maintained their own secondary school which provided an opportunity for uninterrupted schooling of the refugees' children.

Novak Marcikić and I often took a fifteen minute walk to the market in Senigallia where we could buy fruit, mainly apples. I was not interested in a job within the camp but when an opening for a secretary in the camp infirmary appeared Padar called me in for an interview with the camp commander who appointed me to that post. Usually there were forty to fifty patients in the infirmary. As a secretary I, of course, had nothing to do with the treatment of the patients; my task was to maintain the records, keep track of the patients who were sent to the Italian hospital, and distribute the food supplements provided by the welfare administrator. Akulov, a Russian émigré from Yugoslavia, who was in charge of the hospital, was very considerate toward me.

To everybody's consternation Akulov was suddenly arrested and imprisoned at Lipari. The reason for his arrest was unknown but it was rumored that as a dentist he was not authorized to act as a general practitioner. Akulov's replacement was Nuredini, an Albanian from Yugoslavia, who was graduated at the Medical School in Belgrade. Nuredini's spoke impeccable Serbian and was always correct in his demeanor. He considered himself to be a Yugoslav rather than an Albanian, probably because he disliked the annexation of Albania to the Italian empire during Mussolini's reign. Nuredini gave me a letter of recommendation when we were moved to Bagnoli although I did not ask him for it.

Gustav Omahen and his fifteen year old daughter Nikica also were in Senigallia. Omahen set up a dental clinic for the refugees in which Nikica worked as his assistant. Omahen's wife and mine were friends and colleagues since the time when they both taught elementary school in Beltinci. Toward the end of World War II Omahen came to Ljubljana from Metlika where he was practicing general medicine because Metlika was no longer safe: the Germans used General Vlasov's troops in their operations against the partisans. General Vlasov founded the *Russian Army of Liberation* and fought on the German side against the bolsheviks. Vlasov's troops included many Kalmuks, non-Russians from whom no woman was safe. Omahen was thus already a refugee when he was in Ljubljana. As the camp's dentists he and his daughter were drawing salaries in addition to having free room and board like the rest of us. They were able to buy a knitting machine with all the necessary accessories which they later took with them to Argentina where Nikica could make a living by knitting. There were many other acquaintances in Senigallia and associating with them to a large extent mitigated the feeling of isolation we felt in other refugee camps.

Refugees in Senigallia were interviewed by a commission sent by the *International Refugee Organization* to determine if they qualified for special assistance of *UNRRA* (*United Nations Rehabilitation and Relief Action*) All members of the commission were Jews who were not interested in being of any help; they were obviously influenced by the communist propaganda which maintained that the refugees were persons who had collaborated with the Germans.[160] As expected, the outcome of the commission's recommendation was that *UNRRA* denied any kind of assistance or support to the refugees; instead it was sending trainloads of its supplies to the Yugoslav communist regime which was selling them to the population. The activities of *UNRRA* in general soon lost all vestiges of a humanitarian operation and became simply a prop for various dubious regimes throughout the world. When I was already in California, I once received from the international organization *CARE* a request for contribution to its program of sending food packages to Yugoslavia. To my inquiry whether I could designate the recipient of the package bought with my contribution *CARE* responded that it was sending packages only to the Yugoslav authorities which alone determined who would benefit from the aid. Naturally, I would not help support an organization like that; I found other ways to assist my family.

I wrote to the *Foundation for Assitance to Students (Fond de Secours aux Etudiants)* in Geneva inquiring if it could send me some technical literature. The organization responded promptly and sent me a copy of the *General Engineering Handbook* published by McGraw-Hill and a series of books on mathematical methods for scientists and engineers by Baule.[161] A few weeks after the shipment of the books arrived, one of the welfare officers casually mentioned to me that there were some books for me in the office. I sent Baule's tomes to my son who was studying advanced calculus on his own and would make good use of them.

On January 20, 1948 the International Red Cross provided me with an identity card which would serve as my passport for the immigration to the United States. I had already been registered on the waiting list at the American consulate in Genoa for some two years when I was notified that my application would be handled by the American consulate in Naples where I sent all the required documents on July 20, 1948.

The camp administration announced that we would have to vacate the camp at Senigallia to make room for the Jews who were due to arrive from Germany enroute to Palestine. The preparations for the move were dragging on for a long time.

[160] In the early eighties many Soviet Jews got in trouble with the communist regime because they applied for emigration to Israel which was considered to be a disloyal act. At that time, of course, Jews were no longer uncritically accepting the communist line.

[161] Bernhard Baule: *Die Mathematik des Naturforschers und Ingenieurs*, in seven volumes, 3rd edition, Von S. Hirzel, Zürich, 1948.

Slovenians, most of whom applied for emigration to Argentina, were sent to the refugee camp in Bagnoli near Naples and only those of us who had some kind of employment, like Marcikić and I, were allowed to stay until the arrival of the Jews.

Once a refugee who was guarding the entrance to the camp called me and told me I had a visitor. To my great surprise the visitor turned out to be Quinto Zecchini accompanied by a retinue of his engineering and administrative staff. Zecchini's chauffeur, who used to be with him in Ljubljana, had spotted me on the street in Senigallia and reported this, for him rather sensational, information to his chief who came to visit me the next time he was passing through Senigallia. When the Italians occupied the southern part of Slovenia, Zecchini, a civil engineer, was put in charge of the Department of Roads and Bridges at the former regional administration (*banovina*) in Ljubljana. Under the international law the Italians were required to leave in place all Yugoslav government employees and they observed the law but nevertheless brought in their own officials and engineers and set up a parallel organization, *Azienda Autonoma Stradate Statali* (*Autonomous Agency for State Highways*) which had the task of carrying out our former functions. When Zecchini arrived to our offices I was assigned to turn our affairs over to him because I was one of the senior engineers and spoke Italian. Zecchini did not require much assistance; he would only need an occasional clarification but he never expected us to work. During one of our first contacts, when it must have appeared to him that I was very reserved, he told me that he was from the Abruzzi and a relative of Benedetto Croce, a renowned Italian philosopher and, at least initially, a well-known opponent of fascism. I then invited him and his family for a visit at my home. Zecchini's anti-fascist background probably stood him in good stead in getting appointed the Director of the Italian Department of Roads after the war. He showed kind interest and concern for the fate which has befallen me and for the well-being of my family at home. In appreciation of his thoughtfulness I used to send him holiday greetings from California every year but when he stopped responding I lost contact with him; he might have retired or even died, I never found out what happened.

Finally the group of Jews from Germany whom we have been expecting arrived. The group's leader was a Dr. Katz, apparently a medical doctor, who used to announce his demands in a peremptory and overbearing manner. The camp commander ordered, as part of the standard operating procedure, a medical examination of all new arrivals. Dr. Nuredini directed me to maintain, as customary, a record of the examinations in my capacity as the secretary of the clinic but this brought hysterical objections from Katz so that I felt no small relief when Dr. Nuredini accommodated his wishes.

The camp employee in charge of housing placed the new arrivals in the various apartments which several of them had to share. When he later inspected those units he found unsanitary conditions which caused a considerable consternation when he sent his report to the camp commander. Instead of using the common kitchen the new tenants were cooking in their rooms which was specifically prohibited. The situation with respect to the use of sanitary facilities was likewise unacceptable. The Jews were receiving food and support from an association of charitable organizations which joined in this effort and were commonly referred to as "Joint." Groups of these new arrivals carrying geese and other poultry to the camp where the birds were slaughtered and prepared according to the kosher usage became a common sight on the road to Senigallia. The Jews kept to themselves and associated with no one else.

It soon became clear that Senigallia was only a staging station for the emigration to Palestine: Katz's charges were waiting for a ship that would take them to Palestine which at that time was still under the British administration and where unlimited immigration was not allowed. People arriving without permits were refused admission and those who were discovered to be in the country illegally were deported notwithstanding the imminent advent of the new state of Israel which would accept unlimited numbers of Jewish immigrants on whom Israeli citizenship would be conferred at the instant of their landing. Katz was obviously planning to get his people to Palestine even before the proclamation of the state of Israel; presumably there would be some advantage in having a larger Jewish population in the emerging Jewish state. Not all of Katz's charges, however, were looking forward to risking an illegal entry to Palestine and working on a *kibbutz*:[162] thus, unbeknownst to Katz, a young couple managed to obtain immigration visas for Peru and one fine morning they slipped unobserved away to the railway station from where they departed to the country of their choice.

Camp Bagnoli near Naples

On August 6, 1948, the remaining Slovenians departed for the *No. 2 Displaced Persons Center at Bagnoli near Naples*. Bagnoli was an enormous transit camp for all refugees who wanted to emigrate overseas. Each receiving country set up its own office there which screened the prospective immigrants with respect to their medical fitness, issued visas and arranged for group transportation to Australia, Canada, South America and elsewhere. When we arrived, those who opted for emigration to Argentina had already left. Dr. Nuredini again employed me as a secretary at the clinic.

The camp at Bagnoli was rather well laid out; it used to be a former fascist institution for orphaned youth. We were billeted in a large hall divided into several

162 A collective farm in Israel.

rooms by temporary wooden partitions where Marcikić and I could have a room to ourselves. Marcikić became the chief of the workshop and later persuaded me to leave my job as a secretary at the clinic and join him as a technical expert in his shop. When the two of us were appointed to those jobs we got a larger room with a separate entrance from the dormitories. The American camp commander took care that our food was good and adequate. He also organized soccer and baseball teams so that the youth was provided with recreational opportunities.

A light rail service to Naples with trains running every fifteen minutes made Naples easily accessible from Bagnoli. I was naturally interested in the large museum which I frequently visited. Its collections contained many rare objects found at the excavations in Pompei. Another special event was a trip to Pompei which Marcikić, Časlav and Marija Gaković and I made together.

Gaković, Marcikić and I once bought a large tin of condensed milk in the marketplace which we planned to consume at dinner at Gaković's home. When we opened the can we found that it contained some foul smelling paste. The vendor had obviously filled the container with that paste and carefully soldered the cover back on it. Gaković and I took the tin back to the vendor and demanded our money back which the vendor refused. There ensued an argument which attracted a large crowd of curious onlookers. Gaković finally cut the argument short by placing the can on the vendor's bench saying that he was reporting the matter to the police. On hearing that the vendor promptly refunded the money.

On another occasion, Marcikić was shopping for new shoes in the same marketplace. As he was comparing the various items a crowd of loitering youths gathered around him. Fortunately Marcikić could not decide on a purchase. The same day a camp resident reported that he bought a pair of shoes in that market and that those ubiquitous vagabonds obligingly assisted him in wrapping the shoes up; when he came home, however, he found that the package contained a pair of battered old shoes which the hoodlums switched on him when they were wrapping up his purchase.

One day when I was changing a few dollar bills at a money changer's kiosk in Naples, a young man appeared and offered to buy dollars from me at an unusually high rate of exchange. To my inquiry as to how he could offer such high price he said that he was buying cigarettes from American sailors for which he had to pay with dollars but that he was making good money with the sales of American cigarettes. I fended him off by saying that I had already changed all my dollars and had nothing left. Later on the same day I read a report in the paper that someone bought Italian lire with American dollars from a young man who neatly wrapped the purchased notes in a newspaper before he handed them over; when the victim came home and opened the package he found that the package contained only pieces of ordinary paper.

A Visit to Rome

On October 1, 1948 I went to Rome to have my diploma and professional credentials verified by a commission of the *International Refugee Organization.* A member of the Commission happened to be a graduate of my alma mater, the Technical University in Vienna,[163] and we could talk about the professors we both knew. The *IRO* commission issued to me a document stating my professional qualifications although, as I already mentioned, the University in Vienna had sent me a certified copy of my diploma promptly and free of charge when I requested it soon upon my arrival to Italy.[164]

On that occasion I visited Lieutenant-Colonel Karlo Novak whom I had met during the war. At that time Novak was in charge of a resistance group[165] and had invited me to a meeting at which he proposed that the *Sokol Legion* join his group. The *Sokol Legion* did not accept his proposal because the *Sokol War Council* had been acting in accordance with the *Sokol* principles and ideals and was operating within the fold of the *Royal Yugoslav Army in the Homeland* under the command of General Dragoljub Mihajlović.[166] Novak, however, established contact with Pavle Vošnar who was at that time commander of the *Sokol Legion.* Under Novak's influence Vošnar once proposed at a meeting of the *Sokol War Council* that we join the communist *Liberation Front*[167] which we naturally declined. Induced by Novak, Vošnar subsequently joined the *Village Guards,*[168] and had to leave the *Sokol Legion* because of that. When the *Village Guards* were discontinued Vošnar came back to

163 *Technische Hochschule in Wien.*

164 To my surprise nobody in the United States where I sought employment was in the least interested in examining my diplomas.

165 Lieutenant Colonel of the General Staff Karlo I. Novak was since November 1, 1941 to the time of the collapse of Italy in September 1943 commander of the Slovenian troops of the Royal Yugoslav Army in the Homeland. That position was originally held by Colonel Jaka Avšič during whose tenure Novak was his chief of staff. When Avšič defected and went over to the communist partisans, General Dragoljub Mihajlović appointed Novak commander of his troops in Slovenia.

166 Author's account appears to differ somewhat from Novak's on this point. [Karlo I. Novak, *Moj vojno-politički rad, Sokolski Vestnik,* **v. 23,** No.1, 23-30; No.2, 15-25; No.3-4, 33-55; Milwaukee, 1987.] Novak states: "As the first relatively well organized group our forces were joined by the *Sokol Legion* with about one thousand men and women headed by the *Sokol War Council.* The *Sokol Legion* consisted of those Sokols who remained faithful to their oath of allegiance and did not want to join the so-called *Liberation Front* because it was dominated by the communists. Unfortunately its leadership (Dr Boris Mihalič, Ing Ladislav Bevc, and others) was not sufficiently aggressive and had long resisted the demands of the command in Slovenia for the establishment of operating units on the terrain and commencement of active military resistance..." *Ibid.* v. 23, No. 1, p.26.

167 The *Liberation Front* (*Osvobodilna Fronta*): communist partisans who were activated when the Germans attacked the Soviet Union in the Summer of 1941 and who carried out the communist revolution in Slovenia during the time of the Axis occupation.

168 The *Village Guards* (*Vaške straže — Milizia volontaria anti-communista*): volunteers armed and authorized by the Italian occupation authorities to combat communist terrorism.

the *Sokol Legion.* After the war Novak sought me out in the refugee camp with a request that I sign a document concerning the welfare of the refugees sponsored by Queen Mary and, I believe, Lady Astor. I acceded to his request because it was of benefit to all refugees.

Novak took me to see the sights of Rome and then invited me to his palatial apartment in which there was an impressive library. The apartment, which used to belong to a fascist hierarch, had been commandeered by the *Allied Military Government* and put at the disposal of Reverend Long, an American catholic priest working for the American intelligence who, having no need for such sumptuous quarters, in turn placed them at the disposal of Major Novak and his Serbian consort Jovanka. Novak gave me the telephone number and address of an Irish monsignor and diplomat in Vatican through whom I could get in touch with him.[169] It was apparent that Novak worked for the American intelligence service. Jovanka prepared an excellent dinner for the three of us. When, at the end of 1949, I finally obtained my visa for the United States, Major Novak had already been here for some time, but was on an assignment in South America. Under the immigration law in effect at that time persons with Novak's accomplishments could be brought to the United States without having to run the obstacle course the ordinary mortals were put through.[170] Upon my arrival to the United States Jovanka wrote me several very complimentary letters in which she assured me that I was certain to assume an important position in Slovenian politics although I never told her that I was interested in politics.

Farewell to Italy

My sister was sending me large packages of food and clothing which I was able to put to good use. I carried on an extensive correspondence with my wife and children, my niece Majda Della Santa in Lucca, whose husband Pino once came to visit me when he was in the Neapolitan area, and other friends and

[169] Monsignor Hugh O'Flaherty, organizer of the rescue operation that saved several thousand allied prisoners and other persecuted persons from the German captivity during World War II. O'Flaherty's exploits are described in J. P. Gallagher's novel *The Scarlet Pimpernel of the Vatican* and in a somewhat more romanticized motion picture *The Scarlet and the Black* directed by Jerry London, CBS Fox Video.

[170] Colonel Andrej Glušič was offered a similar arrangement when the American intelligence service decided to terminate his clandestine operations against Tito's regime. Glušič wrote to the author that he already had been poised to take action against the *Central Committee of the communist party* in Slovenia but his American sponsors decided such action would be undesirable. Glušič succeeded in obtaining an immigration visa on his own which he used in coming to the United States so that he was not dependent on the irresolute temper of the American intelligence establishment.

acquaintances in other camps, in Argentina and throughout the free world, and, of course, with my sister and her children in California.

Vida Mihalič, wife of judge Boris Mihalič and daughter of Ljudevit Pivko,[171] was employed as a gymnastics teacher in a secondary school in Maribor. At the 1945 elections Vida had the courage to cast her vote in the box provided for voters who disapproved of the regime. That box, naturally, had no slate of candidates. It was simply there to show how democratic and tolerant of dissent the communists were. The ball she cast into the voting box hit the bottom with a loud noise of which the monitors, all of them communists, duly took note and the next day Vida found herself without a job. She, however, had sufficiently good connections with which she managed to obtain a passport for Trieste to visit her husband Boris from where she did not come back.[172] Vida helped me in sending packages to my family in Yugoslavia.

In January 1949 arrived the American commission charged with the resettlement of the refugees who would qualify for immigration to the United States under the newly adopted legislation providing for the admission of 300,000 refugees over and above the quotas determined by the national origin formula. The commission, lackadaisical as most government agencies, was in no hurry to begin its work. Although Congress passed the law on the immigration of the refugees, President Eisenhower, under the tutelage of Henry Cabot Lodge, did all he could to frustrate the implementation of that legislation. To that end he appointed as Commissioner of Immigration one Scott McLeod who was notorious for his opposition to the immigration of the Slavs. McLeod was so effective in his task that only some two thousand refugees were admitted under the new legislation in the first two years during which it was in effect. The Washington Post carried a cartoon by the famous cartoonist Herbert Block – Herblock – depicting McLeod as a walrus weeping over the fate of the oysters he was about to eat. This is how immigrants who knew the English language and could be easily assimilated were treated at that time. Today, however, American military transport planes are bringing in masses of Asian immigrants with virtually no obstacles, not to speak of the Mexican immigrants flooding over the border.

---

[171] Ljudevit Pivko, professor of philosophy and languages at the teacher's college in Maribor, crossed over with his unit from Austria to Italy at Carzano in South Tyrol during World War I, in September 1917, and organized the Yugoslav Legionnaires who fought on the side of the Allies for the destruction of Austria. See: Ljudevit Pivko: *Proti Avstriji*, Klub Dobrovoljcev v Mariboru, Maribor, 1925.

[172] Boris Mihalič and his wife Vida returned to Yugoslavia in 1960, some time after the Allies and Tito agreed that Trieste should be given to Italy. Dr Mihalič apparently relied on the assurances of the Yugoslav authorities that returning émigrés would not be persecuted. As soon as Mihalič crossed the border he was arrested and taken to the prison in Rajhenburg for a thorough brainwashing where he had to endure the humiliations and deprivations that are the hallmark of communism. The author has never been in touch with him since he left the free world, all the efforts to establish contact with him or his family were in vain.

The American immigration commission, with its deliberately slow pace, first selected its offices and then placed an order for the furniture to be built by our workshop. After a wait of several months it was finally announced that the applicants may come forth. I received a relatively low number but, to my surprise, the commission began to call in first those who registered later than I. On July 13, 1949 my turn came and I was interviewed by a young refugee. He demanded the name of the hotel in which I stayed when I was in Lienz. Experiencing a momentary memory block I could not recall the name of Hotel Sonne, which, of course, should have been of no consequence but when to my annoyance the youth continued to press me in his pompous manner I ostensively got up, showed the chair away, and left. After that incident I simply had to wait for my regular quota number to come up. Later, when I was already in California, I was informed that I did qualify for a visa under the Refugee Resettlement Act, too, but of course at that time I no longer needed it.

On August 18, 1949 the welfare officer got around to telling me that a week ago the American consulate in Naples telephoned that I should present myself at the consulate. I promptly went there and was sent to the international clinic in Via Tasso for a medical examination. At that time there were incredibly stringent medical standards and barriers in place for the prospective immigrants. The slightest suspicion of tuberculosis, for example, could be grounds for the refusal of an immigration visa. Today, however, even persons infected with loathsome and contagious venereal diseases may freely come to the United States where they continue spreading their ailments through their aberrant lifestyles. On September 1, 1949 I went to the consulate again only to be told that a Jewish family by name of Levy from Shanghai who was somehow allocated to the Yugoslav quota showed up and claimed that it had a lower quota number than I. The week's delay caused by the welfare officer had pushed me behind the Levys. I was told to call again in about a week to see if something had turned up.

In the meantime my sister sent me $460 for my travel expenses provided by my nephew Warren Mufich who also had prepared an Affidavit of Support for me, that is, he guaranteed that I would not become a financial burden to America. Fortunately for me, a Slovenian woman who was ahead of me on the waiting list temporarily waived her visa because she wanted to wait until her husband's number came up so that the two could travel to the United States together and I was allowed to use her quota number. In this way I finally received the visa in my Red Cross passport. The next day I went to the office of the Home Lines transoceanic company where I bought a passage on the liner Italia scheduled to sail for New York on September 29, 1949. I notified my wife and my friends of my departure. I decided to take leave from my niece Majda in Lucca and from my friend Milan Lenarčič in Preganziol near Treviso before I should sail for the United States.

Milan Lenarčič Dies

My itinerary was to take me to Lenarčič by September 19, 1949. When I arrived on the early morning train from Lucca to Venice and boarded the bus for Treviso I was puzzled by the compassionate look the conductor gave me when I asked him to let me off at Villa Pace in Preganziol. On entering the courtyard of the mansion I saw there my former neighbor from Ljubljana, Hrovat, who said: "You are just in time for the funeral." He explained that two days ago, when Lenarčič was returning home from Treviso with his bicycle, a speeding car from Venice hit him from behind and dragged him and the bicycle for at least sixty feet until it finally came to a stop. Lenarčič was taken home unconscious; all the medical help was of no avail, and he died very soon after the accident. While my neighbor was telling me this the funeral services inside were concluded and the coffin was sealed. I was allowed to go to the room where my devoted friend's body was lying so that I could take leave of him.

Milan Lenarčič (1945)

Many Slovenians from Trieste came to the funeral, among them Dr. Angeleto, a well-known political leader, and there was a huge attendance from the local population by whom the deceased was well liked. The funeral cortege stopped in front of the cathedral where nine masses were offered at the same time: one at the principal altar and the other eight in the apse. The choir from the diocesan seminary sung at the services. When the rites in the church were concluded the procession went on to the cemetery where the casket was placed in a mausoleum. The widow pointed out to me that the fact that she and the deceased were Serbian Orthodox religion did not seem to stop the Catholic clergy from officiating at the funeral.

My friend's widow asked me to stay for a few days to help sort out the papers in his engineering office. Lenarčič and Hrovat had been working to patent a process for the lumber industry. That was the reason for Lenarčič's numerous bicycle trips to Treviso. Lenarčič was a mechanical engineer and a forestry expert. While I was working in the office his widow came to ask if I had any idea where her late husband's will might be. As I did not know, she went to check the coat her husband wore when he was hit by the car and, indeed, she found the will in it. When I read it I saw that it was still the same will as the one my friend had shown me in Ljubljana and which he always carried with him. To the question what should be done with the will I replied that it should be entrusted to Dr. Agneleto for safekeeping.

A notice of the accident also appeared in the Ljubljana papers where the family of my deceased friend published an obituary. My wife to whom I had written that I was going to Preganziol was terribly frightened when she saw the notice — she thought that I, too, might have been involved in the accident. My son went to Vrhnika to ask Lenarčič's relatives about the details of the accident. The communists had confiscated all the property of Lenarčič's relatives and evicted them from their house; my son found them somewhere else. At first they were distrustful which was only natural because they did not know my son but eventually he managed to explain to Lenarčič's mother what was the purpose of his visit. She sent for Lenarčič's brother who showed him a telegram which, of course, did not mention me at all. My family spent a few days in uncertainty until my post-card arrived from Treviso with reporting the tragedy. Today everything could be settled with a simple telephone call but in those days a telephone call to Italy from Ljubljana would be enormously difficult to put through, if it were allowed at all.

Some, like for instance Dušan Lajovic, the Slovenian industrialist in Australia, conjectured that the automobile accident was only a camouflage for murder organized by *OZNA*. Lenarčič had been in close contact with Colonel Glušič and supported his publication *The Voice of King Mathias (Matjažev Glas).*[173] On September 29, 1949, after having spent more than four years in refugee camps I sailed for the United States, a new world, and a new life.

[173] According to an ancient Slovenian legend the once mighty King Matjaž (Mathias) is sleeping with his army under the Peca Mountain and will some day come out to free Slovenia.

America

Arrival to America

On the day set for my departure, September 29, 1949, I got ready to embark on the ocean liner Italia which should, as agreed, send a car to pick me up. The ship's car was late and I asked the camp commander to give me a lift in the camp administration's car. He was glad to oblige but when his car arrived the ship's shuttle showed up too. Of course I took the latter. The ship sailed at 7 PM. I shared a stateroom with an American, owner of a drugstore. He said that there was not a thing that you could not buy in an American drugstore, including bicycles. The next morning we landed at Palermo and on October 2 we were sailing through choppy sea through the straits of Gibraltar. We sailed through the straits at night. I went on deck to see the rock, one of the most important bases of the British Empire. It was dark and there was nothing to be seen except a few lights in the distance, the wake of the ship was rushing by with a characteristic noise. I recalled an almost forgotten dream I had a few months ago in which I lived through an identical experience. On October 3 at 2:30 PM, with the sea still rather rough, we docked at Lisbon. Passengers had the opportunity to go sight seeing but I was not interested. A few locals came on board offering to post letters and other services, expecting a tip, of course. But they seemed to be such unsavory characters that I would not be surprised if they stole even the postage stamp from the envelope. Two days later, on October 5, we reached Azores but the sea did not quiet down until October 9.

It was on the eve of Columbus Day, October 11, at 9 PM that I disembarked in New York where I was admitted as permanent resident of the United States. The ship's personnel took our luggage to Pier 54 where it was arranged in alphabetical order. In the customs shed at letter B was already waiting the faithful Bogumil Vošnjak with whom we exchanged friendly greetings. Vošnjak directed the porter to take my luggage to a taxi which we took to his hotel. At 11 PM we were in Hotel Woodrow where he was staying and where he had reserved a room for me. That evening we stayed up talking for a long time.

First of all Vošnjak gave me a clear picture of the *Free Europe* organization. The spokesman of that organization, Ambassador Grew, said in a statement to the press, that all political leaders from the countries behind the iron curtain would be receiving subsidies. Later, however, that commitment was changed to the effect that subsidies will go only to the established national committees and not to individuals. Thus the Yugoslavs and Poles were excluded from the program because we did not yet have our own national committees. Individuals did receive subsidies from *Free Europe*[174] in the form of paid research assignments given under condition that they refrain from political activity. If a national committee were established — in this way Dimitrov got his and so did the Albanians — it did receive a substantial sum as subsidy, the expenditure of which was under rather strict American control. It was thus inappropriate to accuse Krek of failing to support Slovenian émigrés because Krek was receiving no funds from *Free Europe* which he could distribute among his adherents. All that he was receiving was his monthly salary which was probably much higher that Vošnjak's honorarium under his three month contract for writing a political history paper. Krek's area of activity was in the press service of the Library of Congress in Washington and it was expected that he would be put in charge of the press service of the *Free Europe*. Other so called political leaders, were receiving a monthly stipend amounting to $300. If we succeeded in forming a national committee of our own, the subsidy would be big enough to pay for the sustenance of our people, for the travel expenses of Žitnik and Zajec, and even to send some funds to our people in Europe. Hence, for this reason reason, it would have been very important to establish a national committee for the entire Yugoslavia. Former Yugoslav ambassador Konstantin Fotić advised that such committee would receive approval by the State Department. According to him it would be possible to organize such a committee in the foreseeable future only if the committee consisted of former deputies in the Yugoslav parliament. The subsidies, however, would not be limited only to former deputies. If it ever came to that, Fotić would be among the decision makers and would make sure that the Slovenians would not be overlooked. Rumors that even Pavelić would benefit from the program were nonsense as the first point of our platform was to exclude adherents of totalitarian methods among which are included members of *Straža (Watch)*.

Vošnjak did not agree that an assembly of deputies should be a political representation. Parliamentary unions are only preparatory steps for a political representation or a path leading to it. It would be proper to consider the proposal of made by Zajec that the gathering at Tabor should be given parliamentary status and that in any case a delegation of political parties and the Executive Secretary of the *Slovenian National Committee* should have such status. His plan was to include also members of the *Yugoslav Committee* from the time of World War I who lived in the United

174 *National Committee for Free Europe:* An organization established for the purposes of propaganda which gathered and to some extent supported émigré politicians from Eastern Europe if such support was in agreement with the foreign policy of the United States. Ostensibly the Free Europe was funded by wealthy individuals but in reality it was under covert control of the Department of State. The leading administrative positions in the organization were, according to Dr. Vošnjak, held by former second rate American diplomats.

States, primarily Ivan Meštrović. He mentioned the problem concerning admission to the *London Committee* in 1946. At that time the main reservation to our joining was antagonizing the Croats who would in any case not join a *Yugoslav Committee.* At the time of that decision it was not known what the actual tenets of the politics of Vlatko Maček were. The situation in 1949 was considerably different, according to Vošnjak. Krek had formally stated that his deputies would join the *Parliamentary Union*. Vošnjak's liberal parliamentary colleagues were of the opinion that Krek had committed himself by that statement. They told the same to Maček. Of course Krek could not be forced to go through with it, if he did not want to. He would however, make a bad impression if he reneged on his word. Krek should have agreed at the very least to include Vošnjak in the *Parliamentary Union* since that would not have endangered the unity of the *Slovenian National Committee.*

In 1948 Maček became quite different from what he was in 1946. His relationship with King Peter II as well as with the Serbs became quite cordial. His public statements, however, had to be properly interpreted. In view of the fact that Krek, that is, the *Slovenian People's Party (Slovenska ljudska stranka — SLS),* had eleven

Author's sister and her family who made it possible for him to come to the United States at his sister's golden wedding anniversary in 1956 — From left to the right: Lillian, married Lang; Florence Winifred, married Cunny; Josephine (Dolly) married Moore; June, married Crokrite; Doris, married Lloyd; sister Josephine, brother-in-law Martin Mufich, Silvester Peter, Albin Anthony, Martin Joseph, Rudolph George, Warren Ladislav. Warren had on request of his mother acted as author's sponsor to secure his visa and paid for his travel to the United States, absent is Kay (Catherine) who was deceased

deputies while the liberals had only one it was agreed that there would be no majority rule. If Maček entered the *Yugoslav Committee* or *Parliamentary Union* he would with his five deputies be given an equally strong voice as the Serbs who would have had tens of deputies. Likewise Vošnjak intended that each party would have only one voice in each national committee regardless of the number of deputies. His Serbian friends had agreed to that. King Peter had been in Washington on political business at least eight times where he was received by Secretary of State Dean Acheson but even the most prominent Serbs could not tell what went on.

Next day, October 12, was Columbus day, a holiday. Nevertheless Vošnjak was able to arrange for a meeting with Yarrow, former State Attorney General of New York and chairman of the campaign for the election of John Foster Dulles as senator from New York. Yarrow received us with great kindness. He is Russian by national origin and had come to the United States after World War I. He knew Tito's regime and Yugoslavia well and held governor Šubašić in high regard. He was particularly interested in the fate of those Slovenians from the Littoral to whom the *International Refugee Organization* denied the right of immigrating to the United States. He recommended that Vošnjak write a letter to Dulles describing the situation and insisting that the senator take action to straighten out that scandalous situation. After an extended conversation we parted expressing a mutual desire to continue our discussion at an appropriate time. Yarrow made a very favorable impression on both of us. Vošnjak told me that when I was still in Italy Dulles had cabled the American consul in Rome that he was interested in my case.

On returning to the hotel we found a telegram from Miha Krek, chairman of the *Slovenian People's Party (Slovenska Ljudska Stranka — SLS)* and minister of the former Yugoslav government in exile in London during World War II, who was at that time chairman of the *Slovenian National Committee*. Krek advised that he would be in the Times Square Hotel on Thursday and Friday. The following day Vošnjak telephoned Krek and told him that I was in New York. Surprised Krek exclaimed: "How did **he** get here?" His surprise was rather strange if one considers that a few months before Krek wrote to me that my relatives had sent the required guarantees for my immigration to the refugee organization at Barett Avenue in New York where Krek had some influence. It is still unexplained where is the first affidavit which my niece Angela Schneider at Krek's request sent via Barett Avenue. Angela could have sent it directly to Washington as my nephew Warren did with his affidavit. At first Krek wanted to speak with Vošnjak alone; only at Vošnjak's specific request did he agree that I might come, too.

In the morning we went to Krek's hotel where we were received by his wife and Mr. Godina, who was part of his entourage. Eventually Krek arrived. I cannot say that he received us very cordially, I would rather say superciliously. It did not occur to him to describe the political situation to us. The outline of what he was talking about was vague, scattered, and not well focused. Everything was pervaded with Krek's innate streak of fatalism and, as Vošnjak put it, lack of initiative. We wondered what impression he was making on the Americans when he made such ineffectual statements.

Vošnjak initiated a businesslike discussion by pointing to Kuhar's *Klic Triglava*[175] which was attacking the *Slovenian National Committee*, claiming that it was not active enough. Krek was making excuses that he had to devote all his efforts to the refugee question. That work was nearing the end and Krek would again be able to give his full attention to the *Slovenian National Committee*. I pointed out the statement of Ciril Žebot in the New York Times. All three of us noted that Žebot's declaration which demanded some kind of "New Economic Policy"[176] for Yugoslavia as well as milder treatment of the Catholic Church was some kind of overture toward currying Tito's favor which indicated that Žebot had also gone over to the "nudists." We recalled that Žebot's father was also a great political opportunist.

Vošnjak expressed surprise that nowhere in the American press appeared news of mass arrests in Slovenia and forcible collectivization of land. The Czech were another matter altogether: *The New York Times* was publishing articles on arrests in Czechoslovakia. Krek apologetically pointed out that the Czechs had succeeded in establishing their own press agency but we had not. Vošnjak remarked that the Czechs received money from *Free Europe* with which they could conduct their campaign whereupon Krek retorted that we, the Yugoslavs, were in a completely different situation than the Czech, Rumanians and Hungarians. Vošnjak contradicted that saying that according to his inquiries the *Free Europe* wanted to help the Yugoslavs too, but under certain conditions. Krek was pessimistic about the possibility of establishing some kind of Yugoslav representation but did claim that although he could set up a committee for Yugoslavia similar to that of Rumania's there would be no sense in doing so. He said he was instrumental in establishing the *Free Europe*.

I specifically asked Krek what efforts were being made for all the members of the *Slovenian National Committee* to eventually gather in the United States. He responded that it would be nearly impossible or at least very difficult to help individual émigrés in Argentina to come to the United States. He declined to do anything for Franc Kolterer under the new immigration law for displaced persons. In response to my question if the *Slovenian Catholic Unit* in Joliet could be persuaded

[175] *Klic Triglava* (*Call of the Triglav Mountain*), émigré political journal published in England.

[176] New Economic Policy — NEP: policy under which Lenin in view of the total collapse of communist economy temporarily permitted a small measure of private initiative so that the Soviet economy might recover. Naturally the communists soon terminated the project, suppressed private enterprises and, according to their established practice, confiscated them.

to approve loans for economic rehabilitation of émigrés in Argentina, specifically Žitnik, Krek ruled out that possibility saying that no American lending institution would approve loans outside the United States. Although both Vošnjak and I assured him that there is no risk in case of Žitnik, Krek would not hear of it.

Finally we agreed that Krek would prepare a declaration which would emphasize that the intention of the *Slovenian National Committee* was to secure to the greatest extent possible our ethnic and human rights in the homeland, that a memorandum to the United Nations concerning Tito's recent abuses would be prepared and, finally, that a memorandum of our meeting in New York would be prepared for the *Svobodna Slovenia* (*Free Slovenia*) in Buenos Aires. Vošnjak asked me to write a letter about that from California.

We both objected to Krek's assertion that initiative in certain matters must be left to the Americans who, according to him, knew what was to be done. We emphasized that we are responsible for our future. Krek and Maček must, of course, remain in the good graces of the Department of State. Vošnjak pointed to the extensive publication activities of Bogdan Radica which were carried out more or less under the aegis of *Free Europe*. Krek agreed to deliver the memorandum of the *Slovenian National Committee* on UNRRA, prepared by Vošnjak, to the office of John Foster Dulles.

I was under the impression that Krek was too far from the center of action and that he was not *au currant*. Moreover he was leaning toward pessimism and dependent on the people around him. Barett Avenue, although it could, certainly would not act in the sense of the motto:"Open your heart, open your hands and dry up your brother's tears." Moreover it is probable that any favorable inclination it might have had toward the progressive poor was restrained by fanatical new arrivals who would not forget the feuds of the past. Krek, of course, did not invite us to lunch.

In the evening we were visited at the hotel by A. Jerič, a Croat from peninsula Pelješac, one time economic attaché and a very active public personality who was highly critical of Maček's and Krek's "grandism". We talked for several hours and Vošnjak and I had an opportunity to find out how the Dalmatian Croats think about the relations between the Serbs and Croats. The Croats were rather hopeful that the Slovenians would be impartial mediators but nothing had been done by Krek along those lines thus far. We all agreed that it was both in the Slovenian and in Croatian interest to repel any attempt at introducing totalitarian methods and that the Third Yugoslavia must address a decisive democratization of the civil administration and the judiciary. Jerič was very satisfied with what we two had said on the subject and later presented Vošnjak with a rather fantastic plan for the representation of our émigrés which would operate under the name of *Provisional*

South Slav Confederation. He stated, however, that he would withdraw his plan if Vošnjak so desired.

A long and very friendly discussion we had with Žujević, secretary of the *Serbian National Defense (Srpska narodna odbrana)* was very interesting. Žujević was grateful that Vošnjak maintained constant contact with him although he had often told him that he did not support the activities of *Serbian National Defense*. He stressed that the need to have a national representation was urgent. He, too, mentioned that Slovenians could play a historically important role in helping establish a representation. To Vošnjak's question if Krek had done anything along those lines Žujević could only answer in the negative.

On Friday Vošnjak and I discussed at length what was to be done to reactivate the role of Slovenians and arrived at a completely unanimous conclusion. Dr. Vošnjak was going out of his way and spent all his time with me during my stay in New York. He was at my side until I left for California on October 14 at 11 AM. In this way we had enough time for discussions and exchange of opinions. I felt uneasy at having caused him so much expense particularly because I realized that his income was none too secure. At the University of California at Berkeley his contract was terminated after one semester with the excuse that because of his age — he was born on September 4, 1882 — he could not remain on the faculty if he had not already been employed for five years. It is entirely possible that there were political overtones to his dismissal as some influential members of the academic senate had great sympathy for the communists and Vošnjak, of course, made no secret of his political views. The president of the university, Robert Gordon Sproul, who ruled the university with an iron hand, was so high handed that he refused to see Vošnjak and talk the matter over with him.

Vošnjak was well acquainted with professor Joseph Kerner[177] at the University of California. Many years later we found in Kerner's files part of Vošnjak's correspondence among which was his historical Corfu diary subsequently published by the South Slav Journal.[178] It was news to me when Vošnjak told me that most of the university professors in the United States do not have permanent appointments so that they are always uncertain whether their contract would be renewed for the following year. Only those who achieve permanent tenure can afford to speak freely and take part in political action. When I was with him in New York he had a three month contract with *Free Europe* to write a paper on political history. Later he told me that he, the author of several books published in England

---

177 Professor Kerner was the expert on Slavic countries attached to the American delegation at the Paris peace conference after World War I. Kerner and colonel House were favorably disposed toward the Slovenians and their just demands that Carinthia be incorporated into Yugoslavia.

178 V. Bevc, *The Corfu Diary of Dr. Bogumil Vošnjak*, The South Slav Journal, v. 9, No. 3-4. London, 1986, pp. 25-32, *supra*.

prior and during World War II as well as numerous papers in the field of political science, had to endure counseling from one of the second rate former diplomats on how to write.

My relatives had sent me money for an airline ticket from New York to San Francisco but I thought that I would see more of the New World if I traveled by bus. Thus I took my leave from Vošnjak, who saw me off at the depot, and boarded a Greyhound bus for California. The bus went directly to California, no changing was required. At the none too frequent stops we could get refreshments and freshen up. Although I had been studying English I still had difficulty in communicating with the personnel of the line. If, for example, I ordered orange juice they served me tomato juice. Every so often the driver was changed and each new driver introduced himself and told us how far he would take us. I noticed that the roads must have been built rather early because one could see the effects of their growing pains. Thus, for example, the exit ramps were curved far too sharply for the exit speeds of modern automobiles. I was able to draw comparisons with the German *Autobahn* which I had the opportunity to inspect prior to World War II. A particularly strong impression was the lavish lighting everywhere. Lights are characteristic of America. When I went to Europe again after several years all European cities seemed to be rather on the dark side, including the City of Light — Paris itself. In fact, my wife and I went to *Sacre Coeur* one evening to savor the nightly panorama but we were disappointed. All the city lighting seemed very faint and only in the distance was there a better lit island — the *Place de la Concorde* and its fountains. As it happened, the bell tower clock just struck ten and at that instant they turned off even those lights.

After three days the road to California was gradually becoming monotonous but it became more interesting when we begin to climb toward the Rocky Mountains. Those mountains, however, are not similar to the Alps at all which rise sharply from the plains but are more like a series of plateaus above which rather low summits rise from the traveler's altitude.

In this way I arrived to Napa in the early afternoon where I telephoned my sister from the depot. My brother-in-law Martin arrived almost instantly, gave me a brotherly greeting, and took me home to their large Victorian house which is today restored and owned by my nephew Rudy. I met my sister Josephine again after forty five years. Immediately at my arrival Martin presented me with a nice tie and showed me the large room which they had newly furnished for me.

The telephone began to ring and the drawing room began to fill with nieces, nephews, and their families. Niece Florence[179] and nephew Albin were still living at home. Nephews Rudy and Martin, the latter with his wife Lucille and daughter

[179] Florence Winifred, married Cunny, was for many years the County Clerk in Napa, which is the highest administrative officer of the county elected by the voters.

Judy, lived nearby and came right away. So did niece Kay (Catherine) Walrawen with her husband and son Billy Arata from her earlier marriage, niece Dolly (Josephine) Sherman with her husband Al, son Joel, daughter Sharon and the twins son Ricky and daughter Vicki, nephew Warren "*Buddy*", veteran of Iwo Jia, who was my sponsor and had sent me money for the trip, nephew Sylvester with his wife Thelma, niece Lillian Lang with her husband Martin. From somewhat more distant ranch under Mt. Veeder came niece Angela Beryl with her husband Ewald Schneider, son Gary and daughter Marvel. Niece Doris with her husband Richard Lloyd and daughters Jerry and Jackie drove from Berkeley. There was a great deal of greeting with nieces and nephews whom I knew only from the photographs which my sister was diligently sending to us before the war.[180] Everybody was very friendly and attentive.

The local paper *The Napa Register*, published the news of my arrival with a headline that my sister, who was citizen of Napa for many years, and I had not seen each other for forty-five years,[181] and mentioned that a festive dinner would be held on the date of publication to celebrate my arrival.

Niece Angela introduced me to Archer Taylor, professor of German, and his wife Hasseltine both of whom were on the faculty of the University of California at Berkeley. The Taylors lived in the hills not far from the Christian Brothers Winery at Mt. LaSalle near Napa on a sprawling estate with a big house where they kept a large professional library. They undoubtedly also had a residence in Berkeley because at that time, when there was as yet no freeway, an automobile trip to Berkeley took at least an hour and a half. Professor Taylor immediately offered to introduce me to Professor Harmer Davis, director of the Institute of Transportation in the Department of Civil Engineering at the University of California at Berkeley, where I might inquire about employment. A few days later she personally drove me to Berkeley where I talked to the director but the department had no interest in me. The procedure for recruiting faculty in the United States is quite different than that in use at European universities prior to World War II. The Taylors also recommended me to the Department of City Planning for the City and County of San Francisco, where their friend Francis Violich held a key position but nothing could be developed there either. They made excuses that one would have to be resident of San Francisco for at least one year to qualify for employment.

[180] At my first encounter with the *Gestapo* in 1941 the agents made off with all my American correspondence but a few pictures, which were in the family albums, nevertheless survived that raid.

[181] After the end of World War II both I, from Italy, and my wife, from Yugoslavia, tried to re-establish contact with my sister. She had in the meantime moved from the East Bay to Napa but it required the assistance of the Red Cross to find that out. The Red Cross notified both of us that it located my sister and that it would deliver to her any message of up to 25 words. This is considerably more, I imagine, than what the Red Cross would be willing to do today.

In my search for employment I found a mentor, Jaroslav Polivka, a Czech in Berkeley, who had already been in the United States prior to World War II. Polivka referred me to various enterprises where suitable employment for me might be found and he also wrote to his friends in such companies, for example to Kaiser Engineers, to Bechtel, whose owner is a Czech, and the like. I also applied at the *Engineering Societies Personnel Service* in San Francisco, an organization of professional engineering societies — civil, mechanical, electrical, mining, and shipbuilding — which was some kind of an agency for engineering positions. There they were not in the least like professional colleagues, the main thing they wanted to know was who had referred me to them. As is the practice here, they treat professional engineers simply as labor force in a competitive market. If a position were found through that agency, a fee amounting to two percent of the annual salary payable immediately on receipt of the first month's salary would have to be paid to it. At that time relatively good engineering salaries amounted to approximately $3,000 a year; that amount was subject to a rather high income tax which was even higher for persons who were not American citizens.

Jaroslav Polivka

In time it became clear that it would be very difficult if not impossible to overcome the prejudices against employing somewhat older persons. When I was still in the refugee camp in Italy my niece Angela turned to the *National Catholic Welfare Committee* with **a** request to assist me in obtaining a visa for immigrating to the united States. When I was already in Napa she brought me the reply of the *NCWC* in which that organization was warning her that she should not be trying to help me because I was already 55 years old and could well become a burden for her. At Angela's request I wrote a reply to this humanitarian organization stating dryly that subject of their concern had already arrived in the United States.

Vošnjak naturally wanted me to come to the East Coast where we could work together on Yugoslav political émigré politics. To this end I applied for a research position at the Library of Congress which turned me down. Krek suggested that I might try to get employment at *Free Europe* in New York and gave me the addresses of DeWitt C. Poole, president, Frederic R. Dolbeare, vice president, and Edgar P. Dean, executive secretary. I wrote to those three gentlemen, whom Vošnjak considered to be diplomats promoted upstairs — in the attic — who readily responded that the activities of *Free Europe* did not include civil engineering — as if I sought a technical employment with them — and that they accordingly had no suitable position for me. In addition, employment with *Free Europe* on the East coast would be in a completely unfamiliar place. A further problem problem with that organization was that although it was posing as a "private" organization it was in fact an agency of the State Department with personnel which the government might need if it succeeded in replacing the governments in the various countries of eastern Europe and from which it always expected support in carrying on its propaganda. Prohibition of engaging in political activities of one's own choice imposed on the employees of *Free Europe* would, moreover, negate my aspirations concerning Slovenian political émigré politics. Because Vošnjak acted according to his own convictions — he did not come to the free world to be told what to do as he could have that at home — he was in a very short time dismissed from his employment at *Free Europe* in the process of which Krek, according to Vošnjak, played a rather sad part.

Eventually Polivka wrote to A. G. Streblow, owner and president of the *Basalt Rock Company* in Napa, manufacturer of prestressed concrete beams and other masonry products, where Polivka had been consultant and perhaps also a shareholder for many years that it would be to the advantage of the company if it acquired a professional with my qualifications. Streblow then hired me, at first for a very modest job which, however, improved in time. Streblow had begun with a quarry and made quite a large hole in a small hill to the south of Napa. He then expanded to produce concrete building materials and ultimately to prestressed concrete beams. He paid his first employees and associates with stock certificates of his company who then became directors and key administrators of the company. He was primarily a businessman and had no particular understanding of what engineering involved. Hence when I obtained California registration as a professional engineer, instead of giving me a raise because with my registration removed the need for his company to pay expensive outside consultants for approving the company's designs, Streblow sent me a letter congratulating me on passing the necessary examinations — in English.

At Basalt I soon became friends with managers Don McCall and Harold Price who were very friendly and favorably disposed toward me. From time to time we went together to attend some professional talks in San Francisco after which we would have dinner in one of the better restaurants in a skyscraper with a breathtaking view of the Bay Area. Other coworkers in the company's engineering department were Sam Hedges, who was fond of making an occasional philosophical observation, Maurice French, Ed Godsil and Woody Lewis. I got along very well with all of them. Woody arranged for some consulting work for me which produced a little extra income, Ed Godsil recommended me to a surgeon, Dr. Hegler when I had to have an operation for a hyperactive thyroid gland which was successful. The company used to give annual parties at Christmastime, usually at the Silverado Country Club near the Silverado Trail of Robert Louis Stevenson fame to the north of the town where there was a great deal of drinking and dancing to loud music. Returning from one of those parties Don McCall once had a serious automobile accident and had to be confined to a wheel chair for a long time. Fortunately he eventually made a complete recovery.

Napa also had a junior college which provided evening classes in English for immigrants and courses on American history and American institutions required at that time for prospective citizens. The English courses were at various levels ranging from elementary to intermediate and advanced and since I had already studied the language in the refugee camp I could enroll in an advanced class. The teachers were very interested in the students and tried to establish friendly social relations with us. Such friendly relations are characteristic for the American system of education. Teachers are sincerely interested in the success of their students and do everything possible to help them achieve without ever compromising the grading of their examinations. At these courses I was at times something of a help to the teachers who would first try new material on me to see if I understood it which in turn encouraged the other students.

Lester Reukema

At these English lessons I met two teachers Leatha Stockmon and Ruth Smith Pritchard who became good friends. Mrs. Pritchard knew Lester Reukema, professor of electronics at the University of California at Berkeley, who had been working on the on the calutron, part of the Manhattan Project,[182] during World War II. When my son Vlado came to the United States she recommended him to professor Reukema who was in position to get Vlado admitted to the College of Engineering.[183]

Another friendly family that we met were the Von Uhlits. Von Uhlit was member of the Hanover family, that is, a relative of the British royalty but he always maintained that he did not care for his nobility and he was devoting all his attention to his sizeable *ranch* and the factory of Levis in Napa which he owned. He was usually wearing blue jeans. It initially seemed rather odd to me that a person of his

182 Project Manhattan was name of the program for the development of the atomic bomb, Calutron was a cyclotron converted into a mass spectrograph used to separate various isotopes of uranium and plutonium.

183 Vlado and I went to Berkeley to see professor Reukema who examined Vlado's academic record from the University in Ljubljana then took us to the dean's offices. Miss Vivien Lane, secretary of the College of Engineering and *de facto* acting dean — Dean Morrough O'Brien being more of a public relations personage — thought that because of the overflowing enrollment Vlado could not be admitted. Professor Reukema, who as we learned later was one of the most influential professors, however, said that he had reviewed Vlado's record and that it showed only excellent and very good marks and that he simply wanted to have Vlado in the department. That settled the matter. Miss Lane then immediately called an assistant and instructed him to do all the formalities necessary for Vlado's enrollment in the College of Engineering so that by February of 1956 he was already attending the University as a senior, was graduated in electrical engineering with *highest honors* in 1957, and received his doctorate in 1961.

stature and husband of a college professor would be so informal in appearance but we soon got used to that. Von Uhlits were fond of holding picnics at the swimming pool on their ranch to which they often invited us.

Clairvoyance and Telepathy

During some of my unusual experiences I had the impression as if something were unfolding in front of me that I had already seen. I wondered if it was possible to sense events either at a distance or in advance in some way that is not yet known to science. To this question we shall probably not have an answer for some time although American psychologist J. B. Rhine at the Institute of Extrasensory Perception has for a long time been trying to shed some light on it by systematic analysis of events that may be evidence that telepathy and clairvoyance exist. Moreover, in recent years studies based on quantum mechanics indicate that the existence of parallel universes and time travel from the future to present would not violate the laws of physics. I certainly experienced a few events which I cannot interpret as purely a replay of recorded memory.

Miloš Stare

For example: In the night when on a trip to Germany with Dědek we almost plunged into the Adige River my wife, at home in Ljubljana, could not sleep all night which seemed unusual to her.

Another instance happened in 1944 when I was invited to the founding session of the *Slovenian National Committee* where the three major Slovenian political parties decided to establish the *National Committee*. At that meeting I met certain people whom I had never seen before. A few days before that I dreamed that I met on a wide field Miha Krek who was member of the Yugoslav Government in Exile in London. He was surrounded by a large retinue. I walked up to him and spoke to him. Next to him stood one of his associates who immediately pulled a notebook from his pocket and wrote down what I was saying. When I spoke at the meeting of the *National Committee,* Miloš Stare, who would later own and publish the paper *Free Slovenia (Svobodna Slovenija)* in Buenos Aires, was taking notes of what I said at which I realized that the man in my dreams had been taking notes exactly as Stare did in real life. I had that dream when I did not yet know whom would I see at the meeting the following day. Then, when I looked at Stare who was taking notes of what I said I saw that he bore an extraordinary resemblance to the associate of Krek's in my dream. He had exactly the same face, stature and demeanor — as the man in my dream. I was wondering at the similarity between the dream and its realization a few days later.

As already recounted above, in1945 I was escaping from the partisans and the British across the Austrian border into Italy at Silian in Tyrol. Although he was not alerted in advance of our arrival the local rector gave us shelter and promised to provide our small party with guides who would take us safely across the difficult snow covered mountain trail and the well guarded border. On the night prior to our crossing I dreamed that four young men aged between 15 to 18 years arrived to be our guides. One of them looked like my thirteen year old son who had remained in Yugoslavia. In the dream I immediately observed that my guide was taller than my son. Next morning when we gathered at the parish the rector introduced our guides. To my surprise they were the young men I had already seen in my dream and whom I had never seen before. The rector assigned me to the boy who looked like my son without me having expressed any preference.

In July of 1949, after almost four years of waiting in the refugee camps in Italy, I dreamed that I was on board a ship that was sailing through the Straits of Gibraltar. I had never read about the area around Gibraltar and had only a vague idea of what it may be like from the photographs that appeared from time to time in the papers. In my dream I was disappointed by what I could see and thought that the view was certainly overrated in the various reports I had read. The following September I finally got my visa and embarked on the ocean liner *Italia* which sailed from Naples via Gibraltar to New York. I had all but forgotten my dream of July. The ship sailed through the Straits of Gibraltar between nine and ten o'clock at night. I wanted to see the famed fortification and went to the deck to take in that unique view. I was almost alone on the deck. Suddenly, as the ship was passing the coast, I heard the strong noise of its wake and the waves from the bow. It was dark and there were hardly any lights to be seen so that I was rather disappointed. At that moment I recalled my dream and realized that the reality was just as I had seen it in the dream.

It seems that in all of those instances the circumstances were too precisely defined for the experience to have been a mere happenstance.

Saving my Family Held Hostage

When I saved enough I began sending small four pound packages with some staple contents to my family. At first those packages were duty free but later, around 1951, the communists imposed stiff customs duties on them. Ivanka had to return the last of those packages because the customs duty exceeded the value of its contents. I also used to send a large air mail envelope a pair of nylon stockings which cost one dollar here and could be sold in Yugoslavia for 2,000 dinars. Ivanka's salary at her modest job amounted to 10,000 Dinars so that the sales of nylon stockings were of considerable help to her.

Money could not be sent through regular channels, and sending currency in letters was not safe because all mail was being opened. I was sending mail to my wife with the name of my sister in the return address on the envelope. The communists were, of course, aware of it. Once Ivanka wrote that she found in my envelope a letter of a man totally unknown to her who was writing to his girl friend in Ljubljana that he had steady but physically very demanding employment in the United States and was in a position to pay for her passage which could be arranged if she could somehow come to Italy and there wait for a visa to the United States. When Ivanka mailed the letter back to me I saw that my letter had been opened along with the letter of someone else but the police censor after "inspecting" them in his indolence placed each letter in the other's envelope. I realized that the young woman was in danger and that the communists would forcibly prevent her from emigrating. I published a notice to the unknown man — I only knew that his name was Filip — in a Slovenian paper in the United States requesting that he get in touch with me but I received no response.

Eventually it became impractical to send packages and I began sending money. Arrangements could be made so that someone who needed foreign currency gave Ivanka the equivalent value in dinars at the black market rate and I would send him the dollars to an address outside Yugoslavia. In the beginning of 1954 a travel agency was offering one thousand dinar banknotes at a very good price. I bought some and was sending them to Ivanka in letters. Everything went well but the customs opened and confiscated the last shipment of 5,000 dinars saying that the law prohibited sending Yugoslav currency from abroad. Because that law did not prohibit sending United States dollars I was sending dollars in my letters from then on. A customs official whom Ivanka knew told her that the customs would not confiscate dollars but would merely register them.

When I came to the United States Ivanka began to apply for a passport so that my family could join me. The applications which required the filling of the interminable forms that can only be thought up by the warped minds of communist bureaucracy were routinely rejected by a form letter stating that "there is no reason for you to travel abroad because your husband can return home". At the request of Vošnjak's Senator Leroy Johnson, with whom Vošnjak was well acquainted, expressed his interest in our case to the Secretary of State. The American Embassy in Belgrade summoned my family to the embassy and initiated the procedure for issuing their immigration visas although they did not have passports. They went to Belgrade without telling anyone, Ivanka sent word to her employer that she was ill.

American consul Crook in Belgrade spoke Italian with my wife and English with the children and he said that the embassy would write to the Yugoslav government. After some time the embassy wrote that the Yugoslav government ignored its letter but that visas were available for them if they could somehow get out of Tito's paradise. My friend Colonel Glušič knew that the American ambassador Allen had enough influence with Tito's regime to get anyone out of Yugoslavia if he wanted to. In our case he obviously did not want to. Hence the trip to Belgrade was in vain. Today the Embassy would not do even as much as it did then because it would not want to annoy Tito.

I also turned to the *International Red Cross* in Geneva requesting its help to get my family out of Yugoslavia. After a few weeks I received an inane reply that my wife should apply for a passport with the police authorities. As I recall the Red Cross even provided the address of the police as if there were any difficulty in finding that out.[184]

In July 1952 I presented a memorandum concerning the detainment of my family to the *Human Rights Commission* of the *United Nations Organization.* In it I documented how the communist regime in Yugoslavia systematically violated Articles 12, 13, 14, 15, 16, and 17 of the *Universal Declaration of Human Rights* of which Tito's Yugoslavia was a signatory. The Commission responded that it would circulate the memorandum to member states and wanted to know if it might divulge my name. I naturally consented. There was no response to my memorandum. The Executive Secretary of one of the offices of the United Nations, Protić, told me that it was not likely that anything would be accomplished at the United Nations because, as he put it, one crow never pecks another crow into the eye. I never heard of any political émigré from the communist countries who had been helped by the United Nations.

It was obvious that it was necessary to exert political pressure on the Yugoslav government to release the families of political émigrés which were held as hostages or just because of the innate malice of the communists. My friend Vošnjak, who knew how to reach important American papers, went into action. He succeeded in publishing a long article in *The New York Times* which elucidated the plight of the families of Yugoslav émigrés from the viewpoint of history of international law concerning émigrés and unmasked the base hypocrisy of Tito's regime which evinced a systematic disregard and contempt for human rights.[185] At the same time

184 Author's spouse Ivanka Bevc had after her arrival to the United States performed countless hours of volunteer work as a Grey Lady for the *American Red Cross.* Her work consisted of serving as interpreter for foreign languages at the State Hospital [for mentally ill] at Imola near Napa when the medical personnel was unable to communicate with patients whom they sometimes found derelict and without documents in the streets; she also was translating letters for the patients. But the *Red Cross* did not want to help her.

185 Dr. Vošnjak's article: *To the Commission on Human and Civic Rights of the UNO*" was reprinted in The South Slav Journal, v. 9, No. 3–4, London 1986, pp. 70–74.

he was applying pressure at the *UN Committee on Human Rights*, in the United Sates Senate where he had friends among influential senators, and also at the Department of State which for unfathomable reasons was very fond of Tito.

Because of pressures of this kind and the unfavorable publicity Aleksandar Ranković made a public statement that the communists would allow families of political émigrés to leave Yugoslavia and that they would be issued passports and exit permits. Naturally my family immediately applied for passports but their applications were denied again. My sister wrote to Ranković directly but he did not deign to reply. My daughter Breda and her friend Anda Glušič, daughter of my friend Glušič, decided that they would go to Belgrade and ask Ranković to make good on his word. As it was not advisable to write openly Glušič did not know why the two girls were going to Belgrade. He thought they were going to the American embassy where, as he knew from experience, the naive vice consuls would not be discrete and everything would soon become known to the consul's secretary, Žukova, about whom everyone was saying that she was a communist informer. Glušič, with whom I have been in constant communication, was concerned and wrote me twice because he thought that the visit in Belgrade would only make the situation worse. He was also afraid of intrigue because he thought that the advice to request an audience from Ranković might have come from the communist bigwig, Jože Vilfan, who was a distant relative of the Glušičs. He thought that Vilfan might possibly have retained a streak of humanity but he was sure that his wife was a dyed in the wool communist.

On December 18 1953 the two girls went to Belgrade. Breda stayed with the Gaberšček family at Kosančičev venac and Anda with friends of her own. Anda was received immediately by herself and Breda had to go to see Ranković all by herself which she did not expect. When she arrived at Ranković's headquarters she found a huge crowd of different people in the antechamber, among them persons who had come to see the minister over disputes involving common use of the kitchen, bathrooms and similar problems that are the inevitable side effects of a communist society. It looked hopeless but the receptionist to whom she reported said that she would be received first because it was the ancient Serbian custom to receive first those who had come from far away. She was soon ushered before an aide who asked what was the nature of her request. Breda insisted on seeing Ranković and the aide explained that the minister would see her, but that it was his job to get first all the information about her request and her background. Breda told him that she was going to ask for passports and an exit visa which the aide took note of and told her that the minister would see her in a few hours. After a couple of hours waiting someone came for her and escorted her up a staircase into the presence of the mighty Ranković. Ranković was very reserved at first and questioned Breda curtly in the police interrogation style. Breda told him that the family had heard of his statement and that all she wanted was permission to leave Yugoslavia because the family had nothing there as everything was taken from them and being marked as class enemies had no future either. Ranković then said that I and my friends had collaborated with the occupying forces. Breda denied that emphatically as a false accusation whose absurdity she substantiated by the fact that I had been imprisoned by the *Gestapo*. Apparently in those two hours during which Ranković's office was making inquiries they were not advised of that — in fact the communists in Ljubljana were quite capable of maintaining that my imprisonment was just for the sake of appearances. The minister's demeanor then visibly changed: he asked when and where and Breda told him. She then told him that my wife was at her wits end and that she could not carry on in that state any longer. Ranković comforted her and told her not to be upset and that he would make sure that things would be straightened out. He had given similar assurances to Anda but Glušič was skeptical and did not expect much from him. However, two weeks later two postcards came from the *Administration of Internal Affairs* directing my wife and daughter to bring two photographs of each and a fee of nine thousand dinars to the passport office so that they could be issued passports. No forms or anything else was required. American immigration visas were issued to them relatively soon and when they had them I asked my friend Urbanc, who had a travel agency in Toronto, to arrange for their travel to the United States. Vlado, however, had to remain behind because he was subject to military conscription.

In the meantime Anica Slapar, who cared for her nephew Peter, son of Franci and Milena Slapar whom Milena left with Anica's late mother before she was exiled to Austria, succeeded in obtaining a passport for the boy. For many years the communists had been denying him an exit permit. While Peter was being detained in Yugoslavia there was a great deal of publicity about a trial in Germany in which the communists demanded return of a ten year old boy whom as an infant the Germans took to a concentration camp where he was separated from his mother. A kind German couple rescued the boy and adopted him so that his mother only found out that he was still alive after many years. His adoptive parents refused to give up the child. The American military government in Germany, by then aware of the true nature of the communist regime, refused to return the child to his natural mother because that would consign him to communist captivity.[186] The Yugoslav regime appealed and the appeals court ruled that the child was to be returned to his

186 American authorities are no longer so sensitive to the problem of forced repatriation of refugee children. The Clinton administration forcibly sent a refugee boy named Elian whose mother died on the raft with which a group of Cuban refugees tried to reach the United States back to Cuba and communist captivity. At that operation Attorney General Janet Reno sent a detachment of submachine gun toting special forces to forcibly remove the boy from his relatives and foster parents in Miami.

birth mother. In announcing the decision of the court the presiding judge, who had opposed the repatriation, pointedly asked Makso Šnuderl,[187], representative of the Yugoslav government: "When do you wish to receive your prisoner?" Peter Slapar, prisoner of the communist regime, however, did not have access to an impartial court that would return him to his parents in the United States.

When Peter's exit permit was granted he naturally could not travel by himself across two continents and an ocean. Anica, who was fluent in English and a few other languages, wanted to take him to the United States and personally deliver him to his parents. To do so she requested a visitor visa at the American embassy when she brought her nephew to the embassy in order that his immigration visa could be processed. The unsympathetic American vice consul, Coppih by name if I recall correctly, was indifferent to the plight of a child who had been separated from his parents for ten years and found a contemptible way to refuse Anica a visitor's visa. He manipulated her into saying that she would be quite glad to stay in the United States whereupon he told her that this was grounds for denying her a visa and left the room saying that his time was precious. It is true that persons who intend to stay in the United States are not entitled to a visitor's visa because they are considered prospective immigrants and have to obtain an immigration visa. Since only 933 immigration visas per year were issued for the entire Yugoslavia this obviously meant a wait of several years and, in effect, exclusion from the United States in the meantime. The fact that Anica was fluent in English was apparently also held against her. On the other hand more than 933 persons per hour were illegally crossing the Mexican border and nothing was done about it. The irony was that Anica did not have the slightest intention to remain permanently in the United States because she was well off in Yugoslavia and had close family ties there. But since she had heard that Americans liked flattery she agreed with what the consul was suggesting, thinking that it would impress him favorably. Anica then never got to see her brother Franci again because when in 1989 she did get to visit America, Franci had long been dead. He had a heart attack in 1979 and died instantly.

With Ivanka finally allowed to go to America another opportunity for Peter to travel opened up and Ivanka offered to take the boy with her to California. After briefly thinking the matter over Anica consented and they agreed that she would

---

[187] Although Makso Šnuderl was professor of law in Ljubljana he had to retain local attorneys because the American courts there would not be impressed by speeches public prosecutors make in Yugoslavia. At home, however, Šnuderl was brilliant in presiding at the politically inspired show trial for expulsion of Janez Merčun from the university because the latter had the temerity of criticizing Marxism in a seminar and irreverently called Marx's *piece de resistance* "Das Kaput-all" almost a capital offense in Yugoslavia. In addition Merčun had resigned his membership in the *People's Youth of Slovenia*. Merčun was, of course, expelled, barred from studying at any university in Yugoslavia and subsequently arrested when he tried to leave for the free world where he would not be denied academic education. Eventually, however, he made it to Switzerland and is now free.

accompany Peter as far as Paris. There was no way Vošnjak or any of my friends in Washington could get Anica a visa.

Urbanc arranged for travel on a brand new Greek ocean liner *Olympia* but when the time for sailing to America came the travel agency in Paris with which Urbanc was working telegraphed that there would be a delay because the ship was still in the dry dock. When this happened again Ivanka got tired of waiting. She wired the travel agency she was coming to Paris and that they should get her passage on another ship rather than haggle with the Balkan line. She was concerned because in a communist country it was always quite possible that Ranković could be replaced by some Stalinist like Mitja Ribičič who would revoke all exit permits.[188] It was difficult to leave Vlado behind. He accompanied them as far as Postojna and on return to Ljubljana moved into their "apartment."

In Paris the *Transmundial Travel Agency* found passage for Ivanka on the famed French liner *Liberté* as nothing seemed to be moving with the *Olympia. Liberté* was the pride of the steamship company *Compagnie Generale Transatlantique* which also operated the liners *Ile de France* and *Flandre*, the latter a converted and renamed German ship.

In the bustle of preparations and because of all the delays Ivanka forgot to notify me of the sailing date. At the Basalt Rock Company, where I was employed, everyone was in pleasant anticipation of Ivanka's and Breda's arrival. Harold Price took me to a room and in confidence showed me a large trunk in which the employees were placing their housewarming gifts: kitchen utensils and everything needed in a household. When I rented an apartment in a duplex on Clay street not far from my sister's house they brought in much of the needed furnishing including a refrigerator, washing machine and the like. My sister gave me the furniture in my room, which her children had furnished at my arrival.

It was decided that Barney Dawe, a company employee, would take me to the San Francisco airport. Days went by but there was no word from Ivanka. Harold Price, the head of my office, began to make calls through the maritime operator trying to reach the liner *Liberté*, the ship's name the travel agency in Toronto provided. It turned out that there were several ships of that name at sea but none was the liner *Liberté*. Everyone was in despair. Finally I called Boyan Ribnikar in New York, whose address I had given to Ivanka, to find out if he knew anything about her; Ribnikar answered "She's here." When Ivanka came to the telephone I explained the confusion about her arrival. She had simply forgotten to send me the details about the ship's sailing and arrival, but she did notify my friend, who made

[188] Many years later Ranković had been in fact removed from office because he introduced too intrusive surveillance of Tito himself. The official reason given, however, was that he was illegally trading on commodity markets with government owned ships and their cargoes as if they were his personal property.

all the arrangements for the travel to Napa. Ivanka also asked me if I agreed that they travel to California by air. We decided that they would fly on April 9, 1954.

At Basalt there was general excitement, I wired to my sister-in-law Milena to come and get Peter whom they had not seen in nine years. Barney Dawe welcomed Ivanka and and Breda and presented them with a bouquet of roses. We wired Vlado from the airport about the happy arrival which happened to be exactly on his birthday. My sister and brother-in-law received Ivanka and Breda with the same kindness as they had received me. Visits and greetings of relatives followed and there was a great deal of excitement. We stayed at my sister's for another two weeks and then we moved to the new apartment where the company had already delivered the trunk with presents Harold Price had shown me and some additional furniture to supplement that which my sister gave us. A surprise for Breda was that there was a piano in the apartment which the landlady, Mrs. Stoddard, did not need so that it was available for her use.

Through available black market channels we were sending money to Vlado every month so that he was taken care of financially. He was learning English and went twice to the British consulate in Zagreb to take examinations in English language administered throughout the world by the *Local Examinations Syndicate* of the University of Cambridge on the same date every year. He was then well prepared for the examination in English required of all students as part of the entrance examination at the University of California. He was in no hurry to complete his studies at the university because his conscription would be deferred so long as he was a student and he felt great revulsion toward the *Yugoslav People's Army.* He eked out some additional funds from honoraria received for translating motion pictures from English into Slovenian at *Triglav Film* and later as a teaching assistant at the Faculty of Electrical Engineering. The communists protested that appointment but they did not sway the faculty. In any case Vlado got a preview of what prospects he had in the socialist Yugoslavia for an academic career or any other professional career for that matter.

Vlado was hiking in the mountains most of the time, exploring the terrain for a possible escape across the border. Unfortunately the possibilities of a successful crossing were slim because the border guards were well concealed and it was difficult to remain too long in the border areas to learn the time when their patrols were active. He once came from the Kamnik Pass (*Kamniško Sedlo*) to the alpine lodge *Frischauer Lodge (Frischauerjev dom)* at Okrešelj where the keeper, a police confidant, immediately asked him if he had seen a young couple in the lodge at the Kamnik Pass the previous night. She said, they had information that they intended to escape to Austria via Belska Kočna. Vlado had seen no such couple in the lodge at the Kamnik Pass but, as he wished with all his heart that the two young people succeeded in escaping from the hated communist tyranny, he immediately told the informer that he remembered well seeing that couple at the Kamnik Pass lodge. As he was descending from Okrešelj to Logarska Valley he was overtaken by a border guard who was returning from the direction of Belska Kočna and who probably had abandoned the pursuit of that couple when he was informed that they had been seen in an entirely different place. He probably had to forgo making two more notches on his rifle butt for the time being.

Vlado's friend Milan Dietz, however, did not succeed in escaping across the Stol mountain, where Tito's bloodthirsty headhunters had shot hundreds of people. Milan was so sure that he would be able to reach Austria that he imprudently wrote in the guest book at the lodge under the summit the following ditty:

Adieu, Tito's paradise
I say forever my goodbyes!

Unfortunately the informers at the lodge noticed that entry and alerted the border guards who wasted no time in apprehending Milan. He then spent six months in jail. In fact, he was lucky that they did not shoot him as the border guards enjoyed doing that, particularly at Stol. Eventually Milan succeeded in escaping and is now free in Germany.

Vlado took a two week course in sailing at Pirano hoping he might be able to find a way to Trieste. All the boats were, however, under constant surveillance of the maritime police. The border guards had a lookout equipped with telescopes in the ruins of the castle on top of the hill above Pirano and constantly observed all boats. No boat was allowed to leave the port at night, everybody had to return by six o'clock. If they did not they had to do a lot of explaining and risked imprisonment. Two weeks after Vlado returned to Ljubljana he learned that two young men from Serbia escaped to Trieste on the boat they were using for sailing instruction. The Italians returned the boat but, fortunately, not the two young men.

An acquaintance of Vlado's went swimming late at night at San Nicolo in the Yugoslav zone of occupation of the Free Territory of Trieste, swam far out to avoid underwater netting and similar obstacles, came at dawn out of the sea on the other side of the line of demarcation, and was free. Vlado then went to see professor of ichthyology, Miroslav Zei, director of the oceanographic institute at Rovinj, with whom my wife was acquainted, and questioned him about the habits of sharks in the Gulf of Trieste. Professor Zei very kindly and didactically explained everything to him and told him that sharks are not active at night, a point of greatest interest to Vlado. On parting, however, professor Zei said:" I think that you should consider a safer way to get where you want to go," without Vlado ever even remotely hinting at his intentions.

One day in the fall of 1954, in front of Daj Dam where he would sometimes lunch, Vlado met Lote Marcikić, wife of my good friend Novak Marcikić who was with me in the refugee camps. Lote had, like Ivanka, received a passport for herself and her daughter Anna but not for her son Savo who was also subject to military conscription. But Lote did not want to leave Savo behind and was trying in every way possible to obtain an exit permit for him, too. When she saw Vlado she told him that Savo obtained a statement from the military district that the *Yugoslav People's Army* had no objection to Savo's going to the United States for good. She said that there was a newly appointed major in charge of the district who thought that if someone did not want to live in Yugoslavia the *People's Army* did not want him and that he should go wherever he wanted. Mrs. Marcikić did not quote the Serbian expressions used by that major and so what his exact words were has to be left to the imagination of the reader. Naturally, Vlado made a beeline to the military district where he filed his application for release with a surly receptionist who assured him that the application would be denied. Three days later when he came to pick up the decision the sourpuss clerk handed him the approval of his request. Two months later he received a passport restricted for travel to the United States only so that Vlado could not simply board a train and leave for, say, Trieste, which is what he would have done if he could. The pressure from abroad on the Yugoslav government to respect human rights and allow hostages to leave must have been really strong. It is also possible that the communists had found out that Vlado, who had become twenty-one years old, had lost the preferential status of a minor child of permanent residents of the United States and could not qualify for an immigration visa for the foreseeable future. Although both I and my wife were permanent residents in California and were as such paying taxes which supported the University of California, Vlado could not qualify for a student visa because he, of course, intended to stay here permanently.[189] Vošnjak immediately swung into action to correct this situation. He promptly convinced Hubert Scudder, the representative in Congress from the Napa district, to sponsor a private bill which would allow Vlado to come immediately. Similar legislation had been sponsored in the case of Anda Glušič by Senator Kilgore, a friend of Glušič, because Anda had also reached her twenty first birthday. In Anda's case the casual handling by the American embassy in Belgrade was particularly objectionable because Anda had full fourteen days until her 21st birthday. If the Embassy had issued her a visa immediately, as it should, she would have been able to fly to the United States and be admitted as a permanent resident before she reached the magic age of 21. After that age she would no longer be considered her parents' child under the American immigration law.

[189] Today, however, persons who are in the United States illegally are not only allowed to enroll at the University of California but are even exempt from paying the nonresident fee of several thousand dollars which American citizens from other states have to pay.Note by Editor.

Because I was not a citizen of the United States at that time Scudder required recommendations from three citizen who were voters in his district. These were readily provided by my fiends in Napa one of whom was Mayor George Warner, owner of a Chrysler dealership. Scudder ordered the FBI to investigate and to gather all relevant information. Many years later when Vlado secured access to his *dossier* in the office of the Immigration and Naturalization Service he found, among other things, that the investigation discovered that I had never been unemployed and was never on relief. The private law Scudder introduced read as follows:

> *Private Law 351 - 84th Congress, Chapter 605 — 1st Session, H.R. 4289*
>
> *Be it enacted by the Senate and House of Representatives of the United States of America in Congress Assembled, That, for the purposes of Section 203(a) (3) and 205 of the Immigration and Nationality Act, Vladislav Bevc be held and considered to be the minor child of his parents, Ladislav Bevc and Ivanka Sket Bevc, lawful permanent residents of the United States.*

Through the efforts of the indefatigable Vošnjak who knocked on many doors and sought out many officials who could expedite the legislative process the private bill was indeed adopted by Congress and sent to the President for approval. At that junction Vošnjak had to intervene again to prevent a pocket veto by the president as his office showed no intention of approving the act. Eisenhower, under influence of Henry Cabot Lodge, was very much opposed to any exceptions from the unreasonable strictures of immigration laws. But Scudder was a Republican and a Republican president was reluctant to affront a congressman of his own party. The legislation was finally signed on August 5, 1955. To secure a preferential visa for Vlado I had to file a petition with the Immigration Office. I went to San Francisco where that office is located and was directed to an administrative law judge. When he glanced at the data presented on the form he said that the private law was inapplicable, or rather that it was flawed, not valid because it referred to me as "permanent resident" but I had in the meanwhile become a citizen the previous July. The private law had been in process since February and it was impossible to change the wording once the process began. When I expressed bewilderment over the unexpected position of the so-called administrative law judge who, if he had any legal training at all, should be able to see that a citizen is a permanent resident *a fortiori*, he relented a little and said that if I wanted to risk a fee of $20 he would accept the petition and forward it to a higher level official in Los Angeles with a negative recommendation. Apparently he was one of those persons who because of their limited professional horizon do not dare do anything that would require a decision based on a logical application of the law to facts. Naturally, I paid the required fee and immediately requested Vošnjak to file an appeal with the appellate level of the Immigration Office in Los Angeles. Vošnjak persuaded congressman Scudder to do so and the appeals office in Los Angeles in reversing the decision of the official in San Francisco amended the petition by determining that Vlado was entitled to a nonquota visa which was to be issued immediately without charging the quota of the 933 Yugoslavs who were allowed to come to the United States every year.

I recall that prior to World War I traveling to the United States was just a matter of buying a ticket from a steamship company. The only documentation required was a worker's booklet issued by the local mayor. Presidents Roosevelt and Truman, however, already had to contend with a Republican Congress which imposed all kinds of restrictions on immigration to which all of us who immigrated to the United States up to, say, 1965 had to conform. In 1948 President Truman sent Congress proposed legislation which would admit 300,000 refugees above the existing national origin quotas. Congress amended the proposed legislation with numerous restrictions and limitations before it was adopted and sent to Truman for signature. The press reported that Truman signed the law but expressed a protest about the congressional amendments. I have already written about how that law was administered and implemented. From 1965 on, however, Congress, urged by Presidents Kennedy and Johnson, eliminated all discriminatory provisions and now married children of immigrants and even their nephews are eligible for immigration visas.[190]

As soon as Vlado received his American visa and collected the required transit visas — the Swiss visa was the hardest to get — he left Belgrade for Ljubljana in great haste. He was in such a hurry that he hired a taxi to chase the airport shuttle down the Terazije until he had flagged it down, boarded it, and arrived at the Zemun airport in time to get on a DC-3 plane for Ljubljana. On the recommendation of professor Vurnik and without any effort on Vlado's part he had just been appointed assistant at the *Institute for Electrical Economics* presided over by Milan Vidmar, but had not yet been notified of the appointment. The engineering faculty also wanted him to continue his appointment in the electric measurement laboratory under the direction of professor Aleš Strojnik[191] When Vlado was already home in California Vurnik wrote that he wanted to notify him of the appointment but that he had al-

---

190 Immigration laws seem to be a very attractive area for demagogue politicasters who are again pushing the subject to the forefront of political debate where the current economic recession is attributed to the immigrants. (Note by Editor: Written in 1969.)

191 Many years later Strojnik wrote Vlado from Australia where he was a visiting professor inquiring about possibilities of employment in the United States. Strojnik, encouraged by Mitja Ribičič of all persons, had written an article in the communist journal *Nova Obzorja* entitled: "Who is a Scientist?" That article made him universally resented in the Slovenian academic circles because it exposed the rotten core of the Slovenian academia. It was advisable for Strojnik to look for employment abroad. Vlado happened to have on hand a circular from the National Academy of Sciences which listed several opportunities of academic employment and he sent that to Strojnik. The result was a visiting appointment for Strojnik at Cornell University in electron microscopy. Later Strojnik became full professor at the Arizona State University in Tempe. Vlado stayed with him one summer making electron optics calculations for the scanning electron

ready "flown the coop." Loving grandmother Katarina Sket accompanied Vlado on the Orient Express as far as the border town of Sežana where they parted, forever as it turned out. In Trieste Vlado was met at the railway terminal by Boris Mihalič and his wife Vida whom he had known from their days in Ljubljana during World War II. In Paris he called on the Transmundial Travel Agency that had handled my wife's travel but they knew nothing about any travel arrangements for him. Urbanc's agency had given the instructions to another agency named Patratour. The confusion was readily clarified and in a few days that agency found him a place on the liner *Flandre*. In the meantime he had an opportunity to see the sights of Paris on the ten dollars he was allowed to take out of Yugoslavia. The agency of course, settled all the hotel and restaurant bills with the funds we sent. Knowing the language from his kindergarten times and some extracurricular learning he had no difficulty in communicating with the French. At embarkation in Le Havre a worry-wart line official maintained that passports of travelers to the United States must be valid for at least six months after their visa expires but Vlado's passport had only about one month until expiration. When Vlado showed the man's superior a copy of his private law the man muttered something to his subordinate to the effect that this was a thing the Americans could sort out between them and the official relented. Vlado telephoned me from the ship that there might ber a problem in New York. I again asked Vošnjak to be prepared for such a contingency. The affable immigration officer in New York who examined Vlado's travel documents, however, readily stamped his papers and there was no problem at all.

Dr. Bogumil Vošnjak and Vlado Bevc visit the Capitol in Washington in 1955

Vošnjak also sent a representative of an immigrant organization who came to the pier kindly offering her assistance. She was under the impression that I had been a former Yugoslav ambassador to the United States and that that was the

...

microscope he was developing. Strojnik rekindled Vlado's interest in aviation and took him flying on a few occasions. When they next met Vlado already was a pilot in his own right. Strojnik wrote three books on the design of ultralight glider with power assist and built his own plane which was housed at a CIA facility in Arizona.

reason why the private legislation was passed. Also at the pier were Tone Osovnik and Božo Drnovšek who as very gracious and friendly hosts accompanied Vlado to see the sights of New York, the Hayden Planetarium, Broadway shows, and so on. From New York he went to Washington where he could express his personal thanks to Vošnjak and call at the office of Hubert Scudder. The congressman was not in but his secretary who had done all the work was present and they gave her a bouquet of roses as a small token of our gratitude. That secretary or legislative counselor had been steering Vlado's private bill through the labyrinth of legislative procedures all the way to a successful conclusion. Such people are the true driving forces of the American Congress who invisibly and often without acknowledgment perform important and demanding tasks.

In the company of Vošnjak and Colonel Glušič Vlado saw the sights of Washington, the citadel of democracy and the world's last best hope for liberty. Glušič made exquisite color photographs from vantage points only a Washington resident

Author's family and relatives in later years at the author's 85th birthday — first row from the left: Rajka (daughter of Dr. Vlado Bevc), Ellen, Marie (daughters of Breda Bevc, married George), author's wife Ivanka, Annie (daughter of Breda Bevc), Breda Bevc (married George), Milena Slapar (sister of Ivanka Bevc), Franci Slapar (author's brother-in-law); second row from the left: Nada Bevc (née Jarošin, wife of Dr. Vlado Bevc), Vlado Bevc, Ladislav Bevc, Richard (son of Breda Bevc), Dr. Richard George (author's son-in-law)

would know. At the suggestion of Tone Osovnik, Vlado also visited the offices of Voice of America where he was able to give his comments concerning the Slovenian language broadcasts to one of the directors. It was soon time to leave for California where he arrived toward the evening after an uneventful flight despite a thunderstorm around the Chicago area. Breda had come from Monterey where she was employed as instructor at the *Defense Language Institute* and worked on the preparation of a Slovenian language textbook. Nephew Martin drove us all to the airport where Vlado eventually stepped off a different plane from that on which he was initially scheduled.

Thus ended my arduous and prolonged efforts to save my family from the communist captivity in which Ivanka and Breda had spent nine years and Vlado ten and one-half years.

When Ivanka and Breda came to California they immediately enrolled in English courses at Napa Junior College where I had good friends among the faculty. Breda also attended some preparatory courses for her future studies at the world renowned University of California at Berkeley. Ivanka, on completing her coursework in English, found that by taking a few more additional courses she could qualify for the degree of Associate in Arts which she duly achieved in 1958 just when our first granddaughter Marie was born.

Through the acquaintances and connections made at the college Ivanka began to gather a group interested in folk dancing. The group was called *Ivanka's Dancers* and they ever more often appeared at various local events. This took a great deal of time and effort but she was rewarded by the appreciation shown by the interested public and complimentary articles in the local papers.

Pestotnik's Lawsuit

In camp Bagnoli near Naples which was the last staging area prior to departure for overseas countries we had to undergo a medical examination to determine if we were fit to immigrate to certain countries. Pestotnik was found to be unfit because of advanced sclerosis of the arteries and was not acceptable to any country. He was sent to another camp nearby where the facilities and medical care were better. After I had already been in America for some time certain western countries agreed to accept certain numbers of incurably ill as permanent residents. Under that program Pestotnik could come to the United States soon after me. He was sent to an assisted living facility in Denver ran by nuns called T*he Little Sisters of the Poor* where all the necessities of life were provided. Of course it was unthinkable that he could get some kind of employment. Once established in Denver he began to inquire with Beatrice Hunter — the woman to whom he loaned the money his San Francisco business partner Jelenič was keeping for him — if she would return the loan or to at least pay regular interest. Corporate "president" Hunter, however, was evasive and found many excuses. She claimed that her "enterprise" was doing very well but she kept evading making any payments. All Pestotnik had in his hands was some kind of a form from a bank on which was written a guarantee that Hunter would repay the loan at an early and suitable time. That "guarantee," which was not worth the paper on which it was written, was guarded by Pestotnik as his most precious possession. He was even thinking about using that money to leave the institution and live independently on his own. He asked a friend in Chicago to find out from the bank in San Francisco whether he could cash in that "bank guarantee." The bank, of course, knew nothing about any guarantee. Such guarantees which used to be given in Yugoslavia where Pestotnik was vice-president of Ljubljana *Credit and Loan Bank (Ljubljanska kreditna banka)* on the strength of his wife's investment were unknown in the United States. He also asked me to find Hunter in San Francisco and find out what is the matter. I went to the city where Jelenič gave me Hunter's business address. The woman had her office in Hotel St. Francis on Union Square in the center of the city. In the lobby was a liveried receptionist whom I told I wanted to see Hunter. He announced me and then said that Hunter could not see me as she was too busy.

Pestotnik finally asked me to find him an attorney who would file a suit against Hunter to recover his money. I found Mark Stanton in Napa who was very helpful and agreed to accept the case on a contingency basis: Pestotnik would be required to pay his fees only if something could be recovered from Hunter. Stanton engaged the Butters law firm in San Jose with whom they jointly prosecuted the case. In numerous conferences I explained to them the whole affair and prepared English translations of the extensive correspondence between Pestotnik and Hunter.

When the lawyers filed suit they were authorized by Pestotnik to name me as plaintiff. American law allows plaintiffs in such cases to state that they bought the claim for one dollar. As soon as the suit was filed Hunter disappeared. The attorneys then decided to file a suit against Jelenič who was liable because he betrayed Pestotnik's trust in giving to Hunter without Pestotnik's approval a large sum of money which was entrusted to his safe keeping. When the hearing was scheduled the attorneys advised that Pestotnik should also be present. I invited him to come to Napa a month ahead of time as a guest of my family. He indeed came and Ivanka took good care of him with her cooking and everything else. Pestotnik complained that his home for the aged served leftovers from restaurants and fast food places that did not agree with him at all. His memory was already failing him. In long conversations we went over the entire story. The attorneys too had talked to him.

We thought it might be a good idea if the court retained an official interpreter for the Slovenian language.

At the court hearing in Redwood City the presiding judge immediately noticed that Pestotnik's memory was failing him. At the beginning of the hearing the judge called the interpreter, a descendant of Croatian immigrants from Croatia who did not know the literary language, and told him to interpret. The interpreter then showed to Pestotnik his basic document, the so-called "bank guarantee" or promissory note from Hunter and inquired in his Dalmatian dialect, "Have you seen this book?" In the interpreter's dialect *book* means *letter*. Pestotnik, a doctor of philosophy in Slavic languages, angrily replied, "I do not know what you are saying. The money is mine." The judge then took over the examination and, to everyone's surprise, Pestotnik responded sensibly to him. Thereupon the judge dismissed the interpreter. But Pestotnik stubbornly insisted that he did not recognize the guarantee. By gestures we somehow succeeded to get him recognize the document. The judge admonished us not to signal to the witness.

After such dramatic events at the hearing the judge determined that Jelenič betrayed Pestotnik's trust because he gave to Hunter the greater part of his holdings. The judge ordered Jelenič to pay Pestotnik the entire sum with interest which had by that time accrued to a substantial sum. Jelenič was not satisfied with the judgment and advised that he would appeal. If he appealed and the appellate court ordered a new trial Pestotnik, after all the delays, would no longer be able to appear at a new trial. Therefore the lawyers agreed to settle with Jelenič for half of the contested amount, which was, for eight thousand dollars. I persuaded Pestotnik to accept the settlement particularly because the concurrent suit against Hunter was still active and if it were resolved favorably, she would have to pay the entire sum regardless of Jelenič's settlement.

The lawyers retained a private investigator to track down Hunter and get the address of her hiding place. The investigator found her in Alaska where she used the name Duval because she had in the meantime remarried. When she was summoned to the court in Alaska she told the court that she was a resident of San Francisco. The court demanded her address in San Francisco where the suit against her was then filed. Duval-Hunter did not appear at the hearing where she was represented by her counsel. All that he could offer by way of defense was the allegation that Pestotnik was a Quisling to which Pestotnik's lawyers objected strenuously and were sustained by the court. The court commented that the money was Pestotnik's property regardless of his political persuasion and that, of course, the opposing counsel did not offer any probative evidence in support of his allegation. The court sentenced Duval-Hunter to repay the entire sum as claimed. The next day, however, Duval filed for bankruptcy and her petition was granted as a matter of course by the federal judge to whom she applied. The lawyers told me that she had in this way been lawfully released from any obligations she had. We could perhaps go after her if she started a new gainful enterprise, but all trace of her had vanished. One must realize that the laws of California had been tailored to the needs of the Golden West where adventurers such as Duval-Hunter ensured through their political cronies that they would be beyond the reach of the law.

Slovenian Democratic Party

The *Slovenian Democratic Party* is the successor of the national progressive party which in the time before World War I was called the *Liberal Party.* Liberals had from as far as it can be remembered been the protagonists of the idea of Yugoslav unity. The secretary and the driving force of the club of Slovenian deputies in the Austrian Parliament was Gregor Žerjav who was one of the publishers of the Zagreb journal *Voice of the Slovenians, Croats and Serbs* (*Glas Slovenaca, Hrvata i Srba*). The journal was promoting the demands of the Declaration of May among all Yugoslavs in the Austro-Hungarian monarchy. Žerjav succeeded Ivan Tavčar as chairman of the *Liberal Party* and he was in turn succeeded by Albert Kramer. In Yugoslavia there was of course a different group of parties therefore the *Slovenian National Progressive Party* united with others in the *Yugoslav Nationalist Party (Juoslovenska nacionalistična stranka — JNS)* and as such became established in the Yugoslav political life. It had long ago been an advocate for Yugoslav unity and free thinking, a liberal party since its foundation that is to say from much earlier than 1908 when Austria introduced a general voting franchise.

Even before world War II *Youth of the Yugoslav Nationalist Party* (*Omladina JNS*) was established within the YNP. Best known among its organizers were Stojan Bajič (Walter Gerstenmayer) and Andrej Uršič. Stojan Bajič came to the attention of the Party as a university student and the Party arranged for him to attend a course in public speaking in Berlin. He was bright, an excellent speaker, and a born leader. He received a doctorate in law at the University in Ljubljana and became an adjunct professor at the law faculty. Andrej Uršič was equally bright and created a feeling of trust and confidence in personal relations. Initially the *Youth of the YNP* intended to become more involved in the Party's leadership. Hence they prepared a program which they expected the Party to adopt. The *Youth of the YNP* requested a meeting with the party chairman to be held at his residence because he was ill. No agreement was reached at that meeting where the principal speaker was Bajič who at the conclusion announced that the *Youth* was leaving the party. Jože Rus,[192] a personal

[192] Jože Rus (1904–1992), professional engineer, Chairman of the *Youth of JNS* in *New Yugoslavia (Nova Yugoslavia).* The communists arrested him in 1945 and sentenced him to ten year imprisonment.

friend of mine, belonged to Uršič's wing. He often visited me in my office and kept me informed of all further developments.

Because of the war psychosis the *Youth of the YNP* put their faith in the plans of Bajič, their chairman. Influenced by Bajič the *Youth of the YNP* decided to join the so-called *Liberation Front* but the association did not last long and Rus would soon report that the *Youth* left the *Liberation Front.* When Kramer died during World War II he was succeeded by a directorate of seven members; I was one of them.

Prior to World War II Kramer said at a Party convention that I was to have a leading position in the Party in the future. In other words, I was to succeed him. Accordingly Marjan Zajec, *Regional Executive Secretary (banovinski tajnik)* of the *YNP,* required my approval of all Party decisions, such as for instance agreement with the *Slovenian People's Party (Slovenska ljudska stranka — SLS)* for the establishment of the *National Committee for Slovenia.*

Following the Italian capitulation in 1943 Bajič went abroad to escape the Germans who occupied the Province of Ljubljana which the Italians contrary to international law had annexed to Italy. I learned that Bajič was for a short time in Switzerland from the refugees who had been there. Switzerland did not want to accept refugees for permanent residence and was pressuring them to leave. Hence Bajič was among the first to come to the United States where he changed his name to Walter Gerstenmeyer. He secured a position at a university, in the library at first and then as a lecturer and a professor. I got in touch with him and invited him to join the *Slovenian Democratic Party.* He turned me down citing reasons, the main of which was bitterness. He declined my repeated requests with explanations of matters that must have been unknown to him and remained embittered. We kept in touch with kind wishes at holidays. He died at an early age.

Meeting of Slovenian democrats in New York in 1953. From the left: author, Dr. Boyan Ribnikar, Dr. Bogumil Vošnjak and Dr. Bogdan Drnovšek

Uršič used to come to the directorate of the Party to attend regular meetings as a delegate of the *Youth of the YNP.* The directorate discussed with him the conditions under which the *Youth* would join the Party again. The leading members at these discussions were Otmar Pirkmajer for the directorate and, of course, Andrej Uršič for the *Youth.* The talks were evolving favorably.

In anticipation of the end of war in Europe the *Slovenian People's Party* and the *Yugoslav Nationalist Party* concluded an agreement on cooperation and established the *National Committee for Slovenia.* Later *Social Democrats* represented by Celestin Jelenc were added to the group of cooperating parties. An outcome of that agreement was the convening of the provisional *Slovenian Parliament* under the leadership of the *National Committee* on May 3, 1945 as narrated earlier in these memoirs.

At the end of the war in Europe many refugees, members of the various factions in the progressive party, assembled in Trieste. They initiated action for merging all progressive parties into one, the present *Slovenian Democratic Party.* Among them were Boris Sancin, Slavko Uršič, and Boyan Ribnikar. Sancin and Ribnikar would visit me in the refugee camp at Riccione. They prepared a program for a unified party and sent a delegate to Riccione who submitted the program to me for review and approval. The delegate was surprised when I immediately agreed to the proposed program. After some time another delegate came bringing the final text of the agreement, which I also approved. Next a founding assembly of the *Slovenian Democratic Party* was convened in Riccione chaired by Pavel Pestotnik as the most senior member of the party's committee. The participation was very good — people came from Rome, Trieste, and various refugee camps. The consultations lasted three days. All decisions were unanimous which is understandable because the purpose of the assembly was to join all the progressive fractions into one *Slovenian Democratic Party.* Following adoption of the bylaws the Committees of the *Slovenian Democratic Party.* The Party is led by the *Principal Committee* from whose members the *Executive Committee* is elected. The *Executive Committee* carries out the business of the Party under direction of the Chairman and Executive Secretary. All of the members of the Party constitute the *General Committee* from whom the *Principal Committee* is elected. In 1946 the *Principal Committee* consisted of thirty five members. It was agreed that I would be elected the chairman of the Party when I would have arrived to the United States and was able to actively participate in politics. It was also agreed that the *Slovenian Democratic Party* would become member of the *National Committee for Slovenia.*

Until October 1949 when I arrived in the United States the in Argentina was conducting the party's business. The most active members of the committee were Rudolf Žitnik, Tugomer Prekoršek, Boyan Ribnikar, Gustav Omahen, and Alfonz Pipan. At the first meeting of the *Principal Committee* held in Buenos Aires on September 10, 1950 I was declared active chairman of the *Slovenian Democratic*

Party as the first constitutional assembly had agreed. At that time the *Executive Committee* was also formed whose membership was gradually increased as needed by persons who could show that they had continuously belonged to and supported a progressive party.

I wanted to immediately elect a vice-president and a treasurer but encountered resistance from Žitnik who objected to nomination of Zorc as vice-president and Ribnikar as treasurer. Despite Žitnik's opposition Zorc was duly elected vice-president and Mačus as treasurer instead of Ribnikar. The elections were unanimous which shows that the nominating process was indeed based on the directives of the founding assembly. When Zorc and Mačus died no one came forward to assume their responsibilities. Ribnikar soon found an opportunity to resign from the Executive Committee.

Peter Urbanc

Eventually all the work fell on the shoulders of Peter Urbanc, the party's executive secretary, and on mine. In his capacity as the owner of a travel agency Urbanc did a great deal of traveling and had the opportunity to personally discuss matters with members of the party. All correspondence, however, went through my hands at my expense. As long as Zajec remained as executive secretary some kind of party headquarters existed in Buenos Aires. After he left there were no other volunteers helping me and Urbanc.

In 1953, when Bogdan Drnovšek was still executive secretary, we convened a meeting of the *Principal Committee* in New York. The meeting had an extensive and important agenda. As a rule members would send me authorizations to represent them. In attendance were Vošnjak as vice-president, Božo Drnovšek as executive secretary, and Boyan Ribnikar as a member of the *Executive Committee.* Edi Antosiewicz, the last leader of the wartime *Sokol War Council*, also came but as he was seriously ill and his memory was failing him he could not contribute much to the meeting. At that meeting we adopted numerous action programs one of which was to invite *Slovenska Pravda*, the publisher of the monthly journal *Call of the Triglav Mountain (Klic Triglava)*, which was at that time mounting sharp opposition against the *National Committee*, and offer it to become a member of the *National Committee.*[193] After the conclusion of our meeting a delegation consisting of Vošnjak, Drnovšek and me went to Krek who agreed with our proposal to invite *Slovenska Pravda* to the *National Committee. Slovenska Pravda*, however, did not respond to our invitation. Shortly before our meeting I

[193] *Slovenian Democratic Party* had six members on the *National Committee* and the *Slovenian People's Party* seven members. The *Slovenian Democratic Party* gave one of its places to the *Social Democrats* represented by Celestin Jelenc.

received a communication from Buenos Aires demanding that we defer action on numerous organizational proposals but those present unanimously turned down that demand. The headquarters strongly objected to it and its opposition caused a crisis in the Party and alienated a number of active members.

Soon after there was another blow. Drnovšek resigned from the *Executive Committee* and Ribnikar followed suit. The reason was that they both acquired American citizenship and interpreted their oath of allegiance by which a new citizen renounces allegiance to any foreign entity as barring them from engaging in the politics of a foreign country.[194] Prior to resigning Drnovšek prepared the minutes of the meeting and delivered them to Urbanc who took over his responsibilities. Only sixteen members remained on the *Executive Committee* and, with rare exceptions, all party work had to be shouldered by Urbanc and me. It was an odd situation. I do not recall any instance of a member of the party coming forth with an initiative or a proposal of any kind. Urbanc and I were just not physically able to carry out the extensive program of the *Principal Committee.* We concentrated our efforts on the recruitment of new, younger members but our efforts were not met with success. We could find no one who would take up the work. In 1983 the *Slovenian Democratic Party* nominated Danica Kanalc Petriček[195] to the *Slovenian National Committee.* Petriček brought a dynamic personality to the party and revitalized to an extent its operations by working closely with Peter Urbanc and the author.

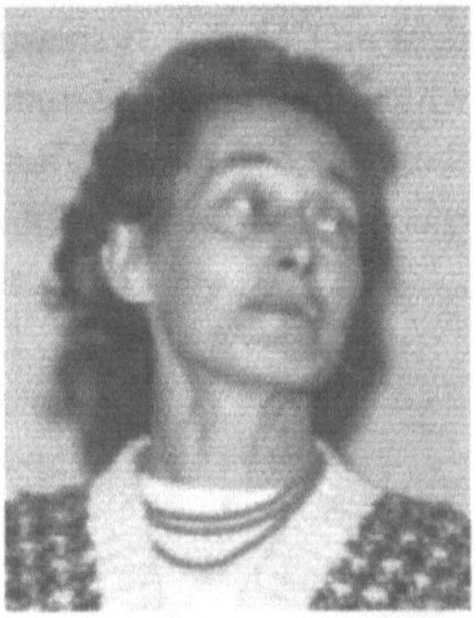
Danica Kanalc Petriček

In the first number of its paper *Domovina* (*Homeland*) which came out on April 6, 1951, on the twentieth anniversary of the German attack on Yugoslavia the *Slovenian Democratic Party* declared that it supported a democratic federal Yugoslavia and therefore desired political representatives of all three parties come to an agreement on the political representation abroad. To this end the *Principal Committee*

194 The matter of dual citizenship always represented a bone of contention in the United States. If naturalized American citizens exercised the rights, claimed the benefits in a foreign citizenship or accepted employment of a foreign country their American citizenship was taken away from them. Even prolonged residence abroad could be grounds for losing citizenship. There was much litigation on the subject and finally the U.S. Supreme Court in *Afroyim vs. Rusk*, 387 U.S. 456 (1967) decided that an American citizen cannot be deprived of citizenship unless he himself voluntarily gives it up. The concerns of Drnovšek and Ribnikar were thus unfounded.

195 Danica Kanalc Petriček (1911–1990), in Carinthia, graduate of Business Academy in Ljubljana, active in nationality affairs in Carinthia, returned to Yugoslavia after *Anschluss*, married Robert Petriček in 1940. Member of Mihajlović's resistence, imprisoned by communists, emigrated to Argentina in 1955. Member of the editorial board of *Free Slovenia* in Buenos Aires, *Slovenian Democratic party* and *Slovenian National Committee*, moved to the United States in early eighties, died in Trieste in 1990

had already in 1950 hailed the initiative of London's *Poruka* (*Report*) and was always favorably disposed toward any action that would unite Yugoslav democratic forces on a common foundation. Democratization of public life not only in Slovenia but the whole of a federal Yugoslavia was in the opinion of the Party the basis for the future of a democratic Yugoslavia.

In coalition with the *Slovenian People's Party* and the *Socialist Party*, the *Slovenian Democratic Party* was represented on the *Slovenian National Committee* in accordance with the *National Declaration* of October 29, 1944. That declaration called for integration of all Slovenian territory in a United Slovenia in a Federal Democratic Yugoslavia.

The task of the Party's leadership in exile is to gather political friends who had to leave the homeland, to reinforce among them the commitment to the Party's ideals, seek the means by which those ideals can be realized, and to prepare for the time when the homeland and the nation are free.

Realizing that we can achieve and retain liberty only with the participation of the entire nation, the *Slovenian Democratic Party* cooperated in the formation of the *National Committee for Slovenia* of which it was a full partner from its inception. In this common effort we see a guarantee that the Slovenian national

Slovenian democrats in Buenos Aires (around 1950); sitting in the middle of the first row is Dr. Miloš Stare, on his right Dr. Boyan Ribnikar, on his left Rudolf Žitnik; in the second row between Ribnikar and Stare is probably Dr. Franjo Mačus

leadership will unwaveringly support the basic principle that Slovenians belong to the Yugoslav nation and that there is no life for us outside of Yugoslavia.

This is true not only for the Slovenians who are one of the constituent nations of Yugoslavia but also for those Slovenians who exist as national minorities in foreign countries. In one's national state it is not necessary to fight for his rights, the citizen has those rights and can exercise them at will without impediment. In a foreign country, however, there is an incessant struggle not only to secure his rights, but also to preserve and protect those rights that have bee won, and to regain those that have been taken away. As long as part of the nation is compelled to fight for restoration of those rights that have been taken away by force and violence it has not attained the status of a constituent nation. The provisions of a treaty such as, for instance the *Austrian State Treaty*, is merely an empty form without content or relevance to reality.

The significance of the *National Committee for the Slovenians* is enormous. If we did not have it we would have to create it. Without the *National Committee* we would have no common representative and no leadership in national politics which must be unified throughout the time in which we live. From this realization proceed the demands and recommendations to attract and include all who recognize the significance and purpose of the policy of common representation. The *Principal Committee* of the *Slovenian Democratic Party* at its second meeting expressed the will of its entire membership in a resolution which proposed a process for restructuring the *National Committee*. So that the Party would be a regular participant in the proceedings of the *National Committee* it was decided to revive the activities of the presidency as provided in the basic agreement of 1944 and to establish a headquarters in a given place. To this end the *Principal Committee* decided to fill the two vacant positions in the *Slovenian Democratic Party* with two members who live in the vicinity.

The *Principal Committee* proposed further that in addition to the *National Committee* a broader national representation be established which would make possible participation of all national entities without limitation to the three political parties. The proposal for having the presidency in a given place is organizational in nature as the presidency had foreseen from the beginning. The *Principal Committee* of the *Slovenian Democratic Party* has by electing its representatives made the only possible move toward a true revival of the activities of the presidency of the *National Committee*. The second part of the proposal addresses the need to enable all national forces to directly influence the activities of the national representation or, in other words, to make their participation possible in accordance with democratic principles with the motto: All and everything for the common cause. It is our desire that every participant consider the proposals outlined from the point of view of the need to strengthen the Slovenian national representation.

Free Sokol

Initiative for Restoring the Sokol Organization

The *Sokol* organization in Yugoslavia was destroyed during World War II. There was no possibility of ressurecting it after war because *Sokol* was in the way of the new rulers who seized power in the Slav countries of Central and Eastern Europe. Modern democracies were instrumental in establishing in that part of Europe a tyranny unparalleled in the history of mankind. Countries like Yugoslavia, Czechoslovakia, Bulgaria and Poland that once were free fell under the rule of Soviet communism after the communists killed everyone who might oppose them.

The *Sokol* organization which served as a model for the largest and most progressive nations of the world where similar gymnast organizations were established is not allowed to exist in the above Slav countries but the *Sokol* idea remains alive. It lives in the hearts of all progressive Slavs and will flourish again when the tyrannies in the Slav countries are gone. It will rise again reborn because its resurrection is in the interest of the Slav nations and their mutual solidarity. The work of the *Sokol* is not easy to advance because its activities were barely in the first phase of its program.

These facts demand that *Sokols* living as émigrés throughout the free world, organize and establish close ties. Our youth too must be brought up in accordance with the *Sokol* principles and engaged in *Sokol* activities.

Hence, émigré *Sokol* leaders, members of *Sokol Association of Buenos Aires*, at their regular meeting on November 17, 1957, proposed to initiate action for reestablishment of *Sokol* activities in all larger émigré centers where our people live. The *Sokol Association of Buenos Aires* adopted their proposal and established a committee of Sokol leaders who had signed the proposal. It was called *Central Committee for Reestablishment of the Yugoslav Sokol Organization.*

The *Committee* was chaired by Janko Jazbec[196], deputy elder (*starosta*) of the *Ljubljana Sokol District* (*župa*) and elder of the *Ljubljana Sokol Association* in Ljubljana, who initiated this action. Elder Mate Veža of the *Sokol Association of*

196 Janko Jazbec (1890–1965) Professional watchmaker and jeweler, leader of the *Sokol Association of Radovljica*, founder of the *Sokol Association of Lovran*, district coach, veteran defender of Slovenia's northern frontier, deputy district coach of the *Sokol Association of Rijeka*, presiding elder of the *Sokol Association of Ljubljana*. During World War II member of the clandestine *Sokol Legion*. Emigrated to Argentina following World War II, President of the *Committee for Restoration of Yugoslav Sokol Organization*.

Buenos Aires was also a member of the committee. Other committee members were Šime Lovrič, coach (*načelnik*) of the *Sokol Association Zagreb-Kapitol*, Miloš Prelog, deputy regional coach of *Sokol Association of Celje*, Herman Kralj, secretary of the *Sokol Association of Maribor*, Peter Capuder, coach of *Sokol I Association at Tabor in Ljubljana* and manager of the journal *Vestnik* (*Herald*), Ivo Vučković, coach of the *Sokol Association of Supetar*, Petar Josipović, deputy secretary, Abadjić Stojadin, and Franjo Markov.

It is noteworthy that the *Sokol Association of Ljubljana,* where Jazbec presided as elder, and a few other *Sokol a*ssociations rather ostensibly admitted the communists whom we at *Sokol I at Tabor* had expelled. However, during the war Jazbec realized the error of his ways and wound up among us in exile. In 1958 Jazbec sent me from Cordoba, Argentina, where he had a watch making and jewelry shop, several letters proposing that we cooperate in our efforts to reestablish the Sokol organization. I had in fact as early as 1952 written to all the *Sokols* living abroad whom I knew, but the response I got was rather meager. There is a time in every emigration when apathy seems to set in. Naturally I was pleased to join the effort carried on by the *Central Committee for Reestablishment of the Yugoslav Sokol Organization* as a member of the *Action Subcommittee.*

The task of the *Committee* was to initiate action as outlined above as thoroughly as possible. Accordingly it began to seek cooperation of *Sokols* living in countries throughout the free world. Circulars were sent to all known émigré *Sokols.* Also cooperating in this effort were Vinko Zorc, Rudolf Žitnik[197] and Franjo Mačus[198] who took over as Chairman of the *Committee* after Jazbec died in 1965.

The need for decentralizing the organization soon became apparent. The *Action Committee* was accordingly divided into several regional committees coordinated through the *Central Committee* at the *Sokol Association in Buenos Aires.* The regional committees were to be located as follows: in South America Buenos Aires in the United States, Canada and Australia — a city to be selected by each country's Sokols, Trieste for South Europe, Munich for Central Europe, and in West Europe London or Paris.

---

[197] Rudolf Žitnik (1904–1974), sheet metal contractor, coach of the Sokol Association Štepanja Vas, during World War II organizer of the *Yugoslav Army in the Homeland*, member of the *Principal Committee of the Slovenian Democratic Party* in Exile.

[198] Franjo Mačus, (1905–1974). Received doctorate at the business economics faculty of the Univerzity in Zagreb. Coach of the *Maribor Sokol District* (župa), *Sokol* ideologue, organizer and publicist. Member of the *State (banovina) Committee* of the *Yugoslav Nationalist Party (JNS).* During World War II member of the *Sokol War Council*, interned by the Italians. When Germany occupied Ljubljana in 1943 he moved to Italy and became member of the *Yugoslav Committee* as representative of the *Yugoslav National Party (JNS)*, member of the *General Commitee* of the *Slovenian Democratic Party* in Exile and its *Executive Committee*, emigrated to Argentina where he continued working in the *Slovenian Democratic Party*, founded *Sokol Association* in Argentina. In 1965 he succeeded Janko Jazbec as chairman of the *Committee for the Reestablishment of the Yugoslav Sokol Organization*, member of the *Slovenian National Committee.*

American postal service honored the centennial of Sokol by issuing a commemorative stamp

The exchange of ideas was reinforced through the periodical *The Voice of Sokols (Sokolski Glas)* published in Buenos Aires. The importance of the *Sokol* press cannot be overemphasized. The *Committee* called upon all *Sokols* to contribute articles to their periodical and to promote it wherever possible.

In the United States were Yugoslav *Sokol* associations which were established before World War II, such as, for instance, the *Sokol Association of Cleveland*. The later owed its renewed vitality to the organizational talents of Valentin Benedik.[199] In San Francisco, near my home in Napa, some kind of *Sokol Hall* existed but when I learned that it was closely connected with the Yugoslav consul I did not feel like approaching that association. It did not seem to have a true *Sokol* program there and there were no public activities. The hall was used principally for dances and parties which do not constitute the principal activity of a *Sokol* association. The members were so-called "progressives" and *Social Democrats* of the old immigrant generation who for the most part were fellow-travelers. It would not have been proper for me to associate with them: I would not be able to change their views and association with them could only compromise me.

Ivanka and I soon found a *Czechoslovak Sokol Association* with its own hall in San Mateo, south of San Francisco, where Jiři Španek was the presiding elder. Through that association we were able to establish ties between the Czechoslovak and Yugoslav *Sokol* organizations.

Božo Drnovšek once wrote me that he attended an authentic Czechoslovak *Sokol* rally in New York where everything was organized in accordance with the *Sokol* traditions as they used to be in Europe. He spoke with the leading members

[199] Valentin Benedik (1898–1978) received doctorate in law from the University in Ljubljana, 1925, member of *Sokol I Association at Tabor*, judge at Bled, Kranj and Ljubljana. During World War II member of the Supreme Command of the *Yugoslav Army in the Homeland*, after the WWII emigrated to the United States, presiding elder of the *Sokol Association of Cleveland*.

of the organization who promised to get in touch with me. However, nothing came of it — it was up to me to take the initiative.

The Czechs founded the first *Sokol* organization in the United States, *American Sokol Community (Američka Obec Sokolska)*, in St. Louis, Missouri in 1865. This is the oldest American *Sokol* organization and it had already held thirteen all-*Sokol* rallies. In 1965, it organized the largest of its rallies in Chicago to celebrate the hundreth anniversary of *American Sokol*. Many American *Sokol* associations participated in this centennial rally including the *Yugoslav Sokol in the Free World*. On that occasion it presented to the *American Sokol Community* a memorial scroll sent by the *Sokols* in Argentina. Thousands of *Sokols* marched in a majestic parade on Cermak road from Cicero to the stadium in Berwyn. The United States government was represented by a detachment of the armed forces. American and Polish veterans were in the parade and the US Postal Service issued a special stamp commemorating the event.

The *American Sokol Community (Američka Obec Sokolska)* consists of six *Sokol Districts (župa)*: Eastern, Northeastern, Central, Western, Southern and Pacific. These districts consist of more than seventy *Sokol* associations with a total of eight hundred adult members and eighteen thousand youths and children.

After the founding of the first *Sokol* association in the New World, the *Sokol* idea spread incredibly fast to other cities where Czech settlers lived. The idea spread also among other Slav nationalities: Slovaks, Russians and Croatians, Slovenians and Serbs who after their liberation from Austro-Hungarian domination united in the Yugoslav Sokol Federation. *American Sokol* has a supreme council whose members include: *American Sokol Organization* — the Czech *Sokol* organization; the *Czech Catholic Union Sokol;* the *Federation of D. A. Sokol in America;* the *Polish Falcons of America;* the *Slovak Gymnastic Union Sokol of the U. S. A.*; and the *Slovak Catholic Sokol.*

The *Yugoslav Sokol of America* (*Jugoslovanski Sokol v Ameriki*) has also joined. these *Sokol* organizations united in the *Sokol* movement. *Sokols* in America number one quarter of a million adult members, with youth and children they number over one-half million. In the United States *Sokols* have built hundreds of *Sokol* halls, summer camps, playing fields and libraries.

The *Central Committee for the Reestablishment of the Yugoslav Sokol Organization* and its associated action committees succeeded in establishing the primacy of the *Yugoslav Sokol in the Free World.* A major problem arose when Radmilo Grdjić, former secretary general of the *Sokol Federation of the Kingdom of Yugoslavia* (*Sokol Kraljevine Jugoslavije*), objected to the presence on the *Central Committee for the Reestablishment* of certain *Sokols* who were at the same time members of the Ljotić organization. The matter was eventually settled by acknowledging that *Sokol* is a democratic organization whose members may also belong to other movements or political parties provided that these are not contrary to the *Sokol* idea.

At the meeting held on October 1, 1974, Radmilo Grdjić[200] was elected presiding elder and mentor of the *Yugoslav Sokol in the Free World.* Elected as deputy elders were Ivan Drčar, Jerko Geršković, and myself. Other members of the leadership included secretary general France Vadnal, deputy secretary general Dušan Svetlič, coaches Nikola Dašković and my wife Ivanka Bevc, deputy coach Savo Barać, deputy mentor Jovan P. Nikolić, treasurer Dimitrije Kralj, deputy treasurer Lojze Galič, registrar and archivist Milena Svetlič. Those elected to the *Supervisory Committee* were Novak Marcikič, Spasoje Filipovič and Radivoj Rigler; and to the Court of Honor: Rudi Perhinek, Miloš Vauhnik and Dr. Valentin Benedik. In December 1974 however, Radmilo Grdjić suddenly died and I assumed the duties of the presiding elder of the *Yugoslav Sokol in the Free World.* The same year Rudolf Žitnik, another good friend of mine and a political collaborator in the *Slovenian Democratic Party*, died in Argentina.

Rally of the Czechoslovak Sokol in Zürich, July 1976

Despite the great distances separating us in the free world a respectable number of Yugoslav Sokols took part in the Zürich rally. This was our first gathering following the establishment of the Representation of the *Yugoslav Sokol in the Free World*, the successor of the *Sokol Federation of the Kingdom of Yugoslavia* destroyed by World War II.

The elders of the *Yugoslav Sokol in the Free World* selected a rally committee with Jovan Nikolić, whose home is in Switzerland, as chairman. Participating members of the *Representation of the Yugoslav Sokol in the Free World* were Ivo Mavko from Canada, Anton Komotar and Miloš Vauhnik from Germany, France Vadnal from France, Alojz Povhe from Switzerland, and my wife and I from the United States.

At this rally for the first time one of our Sokols heading the thirty persons strong Yugoslav *Sokol* delegation appeared in a resplendent classical *Sokol* attire. He was Savo Barać, standard bearer of the Representation of the *Yugoslav Sokol in the Free World.* I had the honor of leading the delegation.

[200] Radmilo Grdjić (1887–1974) Graduated in Law at the University of Zagreb, became journalist and publicist, editor of the Sarajevo *Narod* 1922, secretary general of the *Sokol of the Kingdom of Yugoslavia* (*Sokol Kraljevine Jugoslavije*) 1939, active member of the resistance ovement of Dragoljub Mihajlović, emigrated after World War II, founder of journals *Preporod* and *The Yugoslav Observer*, collaborator of numerous émigré papers and journals such as *Glas Kanadskih Srba*, *Ravnogorski borac*, *Amerikanski Srbobran* and *Sokolski Vestnik*. Presiding elder of the *Yugoslav Sokol in the Free World.*

The events of the rally took place at different locations and it was indeed fortunate for us to have the indefatigable and resourceful company of Jovan Nikolić. He solved the transportation problem by making special arrangements with the local rail system. In turn we tried to make his work easier by following him as a cohesive group.

Representatives of the Yugoslav Sokol in the Free World at the Fourth Rally of the Czechoslovak Sokol in Zurich in 1976. Third from the left is Miloš Vauhnik, fourth Ladislav Bevc, and the fifth France Vadnal

When Ivanka and I arrived in Zürich on July 2, Nikolić immediately came to greet us at our hotel. He secured a guide for us the very first day which made things much easier. When there was no program going on Nikolić either accompanied us himself or assigned someone else to us. When we took our leave at the end of the rally I expressed my thanks for his excellent management of our participation with the heartfelt remark," What would we have done without brother Nikolić."

On the evening of July 2 there was a get acquainted party in the hall at Kasino Zürichhorn. There was no structured program, the attendees who filled the hall had an opportunity to find old friends and make new ones while a live band was playing in the background. Nikolić finally saw France Vadnal who arrived from Paris overloaded with luggage for the Yugoslav *Sokol* exhibit at the rally. Vadnal had an unfortunate accident with that luggage hurting his hand and leg. But he carried on and was at Nikolić's side all the time, two inseparable sterling characters. The first evening all of our Yugoslav delegates met and we remained together until the conclusion of the rally.

The flag of the Representation of the *Yugoslav Sokol in the Free World* was solemnly unfurled in Zürich's Döltschill gymnasium. We had initially expected that our flag would be unfurled at the presentation of *Sokol* delegations at the Letzigrund stadium. The representative of the host *Sokol Association* in Zürich, however, was concerned that the unveiling of the flag of the *Yugoslav Sokol in the Free World* would be accompanied by a speech which would run contrary to the famous Swiss neutrality. In other words, the Swiss did not want to annoy Tito's regime. Hence we had our event for Yugoslavs and representatives of the *Czechoslovak Sokol* separately. Vadnal and Nikolić made arrangements with Jan Waldauf, the leader of the *Czechoslovak Sokol Center* (*Ustředi Československého Sokolstva*), to use the space in the large Döltschill gymnasium for the unfurling of our flag on July 3 at 11 A.M. When we showed up at the appointed time Waldauf called for an intermission in the competitions that were going on at that time.

Dušan Svetlič

Gathered around the flag were my wife Ivanka, Balta, Barač, Djokić, Djuraković, Komotar, his son, Lakonić, Mavko, Nikolić, Pavlović, Povhe, Vadnal, Vauhnik and I. Throughout the event Vadnal was playing softly a recording of the *Sokol March*. I greeted all present, mentioned the historic importance of the event at which we were displaying the new flag of the Representation of the *Yugoslav Sokol in the Free World* as the successor of the suppressed *Sokol Federation of the Kingdom of Yugoslavia*, and requested that each of them describe his association with *Sokol* and his or her work in the organization

The flag was made and embroidered by my wife, coach of the *Yugoslav Sokol in the Free World*. The flag is made of satin, with the Yugoslav colors and a grey falcon on one side; the other side is white, embroidered with Slovenian, Serbian, Croatian, Czech and Bulgarian national motifs; a large *Sokol* emblem and inscription "*Yugoslav Sokol in the Free World*" both in Latin and in Cyrillic script and the dates 1863–1974. It was in 1863 that the first *Sokol* association in the Slav South was established and in 1974 that the representation of the *Yugoslav Sokol in the Free World* was formed. The flag was designed by Dušan Svetlič and the ornaments added by my wife. At the unfurling Ivanka attached to the flag a memorial streamer, made according to her own design and gave a short speech in Czech which she learned when she attended a course in gymnastics in Prague. She reminisced how as a girl of nine years she marched with the *Sokols* to a small outing near Gorizia where the Austro-Hungarian authorities only allowed them to unfold their Slovenian flag when they were outside the city limits, so that the sensitivities of Austria's cherished Italian subjects would not be offended. When she concluded her greeting to the flag I handed it to our standard-bearer, Savo Barać, charging him to guard and protect it and to proudly unfurl it in front of the multitudes of his *Sokol* brethren and sisters. The strapping Savo Barać, dressed in a resplendent traditional *Sokol* uniform, took the flag into his care.

The audience responded to the speeches with vigorous applause. Immediately after the event concluded elder Hřebik ran out of the auditorium and heartily embraced Ivanka. The coaches Jan and Vlasta Waldauf stepped out and also expressed their congratulations. Afterwards Nikolić took us all to a nearby restaurant where we had a festive dinner in pleasant surroundings.

Sokols perform on the athletic field at the 1976 Rally in Zürich

On the evening of July 3 a *Sokol* cultural performance sponsored by the *Czechoslovak Rally Committee* was held in the *Kongresshaus* hall. The extensive program included recitals and dances of Czechoslovak and Swiss artists. The performance was at the highest artistic level and was enthusiastically received by the public which filled the large hall. We hope to see a professional evaluation of the performance published in the Czechoslovak reports on the rally.

On the following day, July 4, after a festive parade, the Yugoslav standard-bearers greeted the rally in front of the main stadium stand at Letzigrund. Ivo Mavko from Canada carried the flag of the North American group, Trifko Vrtikapa from Vienna the flag of the West European group, Vladimir Balta from Canada the flag of the *American Yugoslav Sokol Association* in Cleveland, Ohio, Savo Barać the flag of the *Representation of the Yugoslav Sokol in the Free World*, and a representative of the Yugoslav folk dancing group *Oro* from Osnabrück carried the group's own flag.

The same day the main performance of the rally took place at the Letzigrund stadium. The rehearsals of all groups competing were in the morning followed in the afternoon by the rally's climax — all attending groups performing calisthenics.

Our group arrived in the morning at the location of the Yugoslav *Sokol* exhibition. The exhibit was dominated by four Yugoslav *Sokol* flags including the new flag of our Representation and the flag of the Yugoslav folk dancing group *Oro*. We were pleasantly surprised by the exhibits presented with incredible resourcefulness by France Vadnal and Jovan Nikolić.

France Vadnal

There was always a large crowd of visitors at our exhibition, among them the presiding elder of the *Czechoslovak Sokol Center* (*Ustředi Československého Sokolstva*) Antonin Hřebik and his wife. A large number of photographers were taking pictures of our flag, among them was professional photographer, Spasoje Laconić, from Paris. Many exhibits were contributed by Dušan Svetlič and France Vadnal. I hoped that some day Svetlič and Vadnal would prepare a catalog of the exhibits.

Prior to the calisthenics performance every *Sokol* flag of those participating was formally decorated with memorial streamers which were handed out by the presiding elder of the *Czechoslovak Sokol Center* (*Ustředi Československého Sokolstva*), Antonin Hřebik, and secretary general, Jarmila Uhlifova. Then followed a parade of all groups with their flags including the entire Yugoslav delegation which walked with our new flag in front of us.

Following the calisthenics were exercises on gymnastic equipment, light athletics competition, and a performance by Swiss athletes on the trampoline. In addition the Yugoslav folk dancing group *Oro* of Krsto Todorović captivated the audience by its performance which I thought should have been included with the cultural events of the previous day.

The Czechoslovak *Sokol* organization demonstrated with its public performance that it has preserved the tradition of the pre-war All-*Sokol* rallies for this performance ranks among the unforgettable calisthenics performances with its organization, execution, and marvelous coordination which have been the hallmark of an equally harmonious *Sokol* spirit pervading those attending.

A reporter for the Radio Free Europe in Munich interviewed the representatives of the *Yugoslav Sokol in the Free World*, Trifko Vrtikapa, Jovan Nikolić and myself.

My short statement in the Serbo-Croatian language was translated by Vrtikapa as I spoke. I said:

> *Greetings to Sokol, in the Free World, true to its founder and mentor Tyrš!*
>
> *Yugoslav Sokols in the Free World are uniting under the leadership of the Representation of the Yugoslav Sokols in the Free World, whose task it is to continue with the Sokol activity interrupted during 1941–1945, in their homeland by brute force and violence.*
>
> *I call on all Yugoslavs in the Free World, who remained faithful to the Sokol ideals, to join us. With this it will become possible for us to continue the illustrious Sokol tradition with firm confidence in the power and triumph of the Sokol thought and Sokol ideals.*

It is interesting that the Voice of America, whose Slovenian section was at that time managed by son of the presiding elder of the *Sokol Associationa at Vič* near Ljubljana, simply ignored this assembly of freedom-loving *Sokols*. Slovenian language transmissions of Voice of America were more concerned with describing how the various Puerto Rican groups in New York adapt themselves to the American way of life — a topic which is of interest to Yugoslavs, at least in my opinion, as much as the proverbial snow of yesteryear.[201] The Swiss press was careful to say nothing about a rally that brought thousands of people to Zűrich lest it offend the communist rulers behind the Iron Curtain. So much for the free press of Western Europe.

On the evening of July 4 a party with dancing to the tunes played by a live band was given in the *Kongresshaus* hall. The large hall was overflowing with people exuberant because of the success of the rally. The Yugoslav folk dancing group gave a performance and an encore to satisfy the enthusiastic crowd. I again maintain that the performance of this group should have been included in the cultural program of the previous day because of high quality of its performances.

The *Sokol* rally at Zürich was proof — if proof was needed — to all the free world as well as to our people in the homeland that the Yugoslav *Sokol* is still alive and well and, more importantly, that it is still active and progressing both quantitatively as well as qualitatively, from one rally to another, decisive and unwavering, because of the strong commitment of its membership and in spite all the difficulties, obstacles, and distance that separate us.

Ivo Mavko from Canada noticed that at the rally, grouped behind the Australian *Sokols*, were *Sokols of Montreal* with the flag of Tito's Yugoslavia. It appears

---

[201] Note by Editor: Much later, in a publication focused on the tragic destruction of a detachment of the *Yugoslav Army in the Homeland* at Grčarice in 1943, the same Voice of America manager deprecatingly dismissed the activities of *Sokols* in the free world as "personal correspondence of elderly émigré gentlemen."

that these were old settlers who at the change of the regime in Yugoslavia adopted the flag as modified by the communists. They did not associate with our people. We wondered how these "*Sokols*" arranged for their inclusion in the rally. In all likelihood the Czechoslovaks were duped by an old communist trick concocted at the *Montreal Sokol Association*, which Vladimir Balta some time ago accused of Marxist leanings. It is interesting that the Czechoslovak Vidlak, one of the most prominent officers of the *Montreal Sokol Association*, sent a postcard from Prague to Dušan Svetlič, editor of the *The Sokol Herald (Sokolski Vestnik)*, right after the Third Rally of the *Czechoslovak Sokol in Emigration* in Vienna at a time when non-communists and even Dubček's supporters were being arrested. It would thus not be the first time that the Czechoslovaks were duped by the communists.

Trifko Vrtikapa said that at the last All-Sokol rally in Prague in 1948, the Yugoslav communists sent a large number of members of the *Partisan association*, which had appropriated *Sokol* halls and posed as the successor of the *Sokol* organization. In the train transporting these partisans to Czechoslovakia no mention of *Sokol* was permitted until the border was reached. At the border, however, a directive was issued that participants were not to address each other as "comrade" but as "brother," according to *Sokol* usage. At the exhibition at the Chicago rally there were photographs from the last All-*Sokol* Rally in Prague (1948) showing a detachment of Tito's "*Sokols*" marching with Titoist hats on their heads and submachine guns slung over their shoulders. By their brotherly reception of the partisans the Czechoslovaks fell for a communist trick. At that time a communist takeover was taking place in Czechoslovakia. One cannot rule out the possibility that to be ready for the prospect, however remote, of a resistance on the part of anti-communists the Czechoslovak communists requested assistance of armed communists from Yugoslavia. In fact there were at that time demonstrations in support of President Beneš and even at the parade the *Sokols* pointedly ignored the communist usurper Klement Gottwald, who angrily left the reviewing stand. The armed assistance of the partisans was not however, needed although they were in all likelihood ready to intervene if requested by the communist Czechoslovak government.

Journal *The Czechoslovak Sokol (Československi Sokol)* published the article: *Behind the Scenes of the Last All-Sokol Rally*.[202]

> *When the Rally was taking place many Sokol activists were already imprisoned by the communists, fired from their jobs, or at least removed from the leadership of the Sokol Associations. Many new ministers in the government were not favorably disposed toward the Sokols and communists officials did whatever they could to impede the preparations for the rally by delaying approval of various permits and making the things difficult in general.*

[202] *The Sokol Herald (Sokolski Vestnik)*, Milwaukee, Wisconsin, 1973, No. 3, p. 126.

Shortly before the rally, however, the pressure diminished somewhat because the communists thought that they will show both to the ravaged homeland as well as to the rest of the world that nothing has changed in the Czechoslovak republic if the Sokols can still hold their rallies. The communists were enraged because the Sokols would not allow them to meddle in their affairs, declined financial support and refused to admit communists to the committees in charge of preparations for the rally. It was with great effort and sacrifice that the Czechoslovak Sokols secured the necessary transportation, housing and boarding arrangements and the construction of the rallying place. Although some Sokol workers wavered under the pressure of the regime, the rally was nevertheless conducted as a strictly Sokol affair, a democratic event without the stigma of communist party domination. The main concern of the communists was ensuring strict police surveillance. The Prague police had been reinforced not only by thousands of policemen from other cities but also by hundreds of reliable party officials and militiamen. No one among the attendees suspected that there was hidden under the western viewing stand a detachment of the SNB (Czechoslovak version of the Yugoslav political police UDB) armed to the teeth and equipped with howitzers ready to go into action in the event of anticommunist demonstrations. No one knew that the communist party of Czechoslovakia, without permission of the Sokols had printed several hundred of forged committeemen identification cards for its informers among the participants and spectators.

Nor did anyone know about this criminal act because all the tickets to the rally have been sold out long before the event the secretariat of the communist party had printed ten thousand forged tickets for distribution among its militiamen and confidants directed that they take up places well before the beginning of the performances, especially on the day when Klement Gottwald was coming to watch the activities. Because of this many of quarrels broke out between those who bought genuine tickets and the communist impostors. The militiamen, of course, arrested those who protested and made them leave the rallying place for "disturbing the peace." The communist party also equipped numerous of its agents with "monitor" badges. Fortunately those were of a different color than the genuine badges issued by the Sokol organization and the Sokols kicked out the communist "monitors." The rally showed to the communists that the Sokol organization was not going to be impressed into their service. A great majority of coaches and officers refused to cooperate with the regime at the cost of their employment. After the rally the communist regime confiscated all Sokol property and Sokol Halls as well as 100 million of annual revenue. It was not until the so-called "velvet revolution" when the communist lost total control of the government in Czechoslovakia — although many still remain in the government — that the Sokols could recover their property. [203]

After the conclusion of the rally Ivanka and I traveled to Gorizia where Ivanka's brother Janko came to visit us and to take Ivanka with him to Yugoslavia. Because of the obvious censorship of our correspondence many advised Ivanka not to travel to Yugoslavia especially as it was undoubtedly known that she was the coach of the *Yugoslav Sokol in the Free World.* However, the Yugoslav consulate in San Francisco issued her a visa without hesitation and in Yugoslavia no one bothered her although the *UDB* (political police) was in the habit of inviting visiting political émigrés for a "chat." Apparently the Yugoslav authorities observed the agreement concluded between the American ambassador George Allen and Edward Kardelj according to which Yugoslavia guaranteed free exit to those United States citizens of Yugoslav origin who come to Yugoslavia for a visit with a Yugoslav visa in their United States passport.[204]

After many years I saw again my good friend Stanko Dimnik, who was allowed to come to Gorizia. Dimnik had spent eight years in prison for his involvement in the resistance movement of Dragoljub Mihajlović. He would not say if they were threatening him because of the meeting with me. In any case the threats did not impress him much. My friend Etbin Bežek, on the other hand, was reminded that he was receiving a pension from the government and that he should think over any plans for meeting certain people from abroad. As earlier mentioned, Bežek's daughter Tatjana Canestri-Trotti was married in Italy and Bežek and his wife were sometimes allowed to visit her, but only one at a time lest they both decide to leave Tito's paradise for good. Following Dimnik's visit Pavle Pestotnik, son of my friend and Sokol ideologue Pavel Pestotnik, arrived and kindly drove me in his car to the interior of Italy, to Florence and to visit my niece, Majda Della Santa in Lucca.

I went to Florence to pay my respects at the grave of my late friend Radmilo Grdjić, the first presiding elder of the *Yugoslav Sokol in the Free World.* His grave is in one of the three cemeteries called Cimitero di Trepiano. The cemetery is on a large steep hill with mausoleums at the top and the graves on the slope below. At the entrance is the registry office where I received directions to Radmilo's grave — square VII, number 19/89 (vertical 19, horizontal 89). When I entered the place

---

[203] Note by Editor: Restitution of Czech and Slovak Sokol Halls is in a glaring contrast with the situation in today's Republic of Slovenia where, nowithstanding the fact that the Law on Restitution Art. 3, 10, expressly provides for the return of property confiscated to organizations, including the *Sokol organization*, the authorities refuse to return the *Sokol* Halls to the *Sokols*. Slovenian authorities cite the convenient fact that the *Sokol* organization has been dissolved and prohibited — already by the Italians — so that the re-established *Sokol Associations* cannot be considered the same as the former *Sokol Associations.*

[204] *US Treaties and other International Documents*, Department of State, US Government Printing Office, Washington, 1952, pp.471-472.

I was overcome by the solemn silence which accompanied me between alleys of tall cypress trees to the final resting place of brother Radmilo. His grave is marked by an attractive stone inscribed with his name and dates of his birth and death and decorated by a lantern and a vase for flowers.

As I reflected on the busy and successful life of the deceased, I felt that the place was indeed suitable for a visitor to turn his thoughts to the memories of the days when Radmilo was striving to advance the *Sokol* idea on its way to the perfection which, although unattainable, is now closer to its goal. The success of the Yugoslav delegation at the *Fourth Czechoslovak Sokol Rally* in Zürich was proudly shared with brother Radmilo.

With such thoughts I parted from my friend Radmilo wishing peace to him and eternal glory to his memory. I was also thinking of my late *Sokol* brethren Janko Jazbec, Franjo Mačus, Dragutin Živanović, and Rudi Žitnik who are no longer with us.

Rally of the American Sokol in Chicago — June 1977

As already mentioned, as early as 1865 the Czech immigrants founded their *American Sokol* which in 1977 celebrated 122 years of existence. Between 22 and 26 June of 1977 the *American Sokol* held its XIV Rally near Chicago in Berwyn which is an independent community adjoining Chicago. The communities of Berwyn and Cicero merge continuously with the other suburbs of Chicago.

Berwyn surprises one with the ambiance of a true Czech town. All names are written in Czech and the mayor who attended the public calisthenics is Czech, too. Participants came from all parts of the United States and Canada as it used to be in the prewar Europe when participants came to the Sokol rallies from all Slavic countries. In 112 years the descendants of Czech immigrants did not lose their national identity; they retained the *Sokol* tradition which nurtured its ideals and desires for the freedom of their nation. By force of circumstances most of the younger generations no longer speaks the language of their parents but they retained all their national traits which they continue to develop through their innate organizational ability. Their progress is in no small way attributable to the positive influence of *Sokol* and its ideas.

When the *Yugoslav Sokol in the Free World* received the invitation to participate in the XIV Rally of the *American Sokol Community* (*Americka Obca Sokolska*) the *Representation* (*Predstavnistvo*) accepted the invitation and requested Ivan Drčar, presiding elder of the *North-American Yugoslav Sokol District* (*Severno-ameriška jugoslovanska župa*), to assume the chairmanship of the *Yugoslav Rally Committee* because he lived in the area. Although members of the *Yugoslav Sokol in the Free World* were scattered all over the United States and Canada, Drčar was able to assemble a respectable group of attendees. The *Representation of the Yugoslav Sokol in the Free World* was thus able to send to the *American Sokol Community* a confirmation that it would attend along with its greetings and congratulations.

At this rally the *American Sokol Community* presented a large number of active members in all categories: children, youth and adult members. The youth, the hope of the future, surprised everybody with its brimming enthusiasm and *Sokol* discipline. Adult members, from the young role models to the graying coaches represented a bridge from the historical past to a bright and promising future. The children and the youth were particularly noteworthy because they were presented by numerous coaches — again evidence of a highly efficient organization.

The public performance on Sunday, June 26, began with a parade of delegations from all the districts and guests with numerous flags attesting to the triumph of the *Sokol* idea. Among the guests was the delegation of the *Yugoslav Sokol in the Free World* with the flag of the *North American Yugoslav district*. The beauty of this pageant was enhanced by groups in national costumes arranged according to the geographic layout of their native country.

Description of the public performance of all categories from the games through the calisthenics is best left to a professional reporter. It demonstrated that the *American Sokol Community* preserved and perfected the irresistible attraction of the *Sokol* in the pursuit of its principle: "A healthy mind in a healthy body."

Our delegation attended the formal banquet on July 24 where our presence was noted. On Saturday, June 25, the Yugoslav delegates held a conference on our task in perfecting the renewal of our *Sokol* organization. In the evening we joined Drčar at dinner in the Old Prague restaurant in Cicero. Twenty two *Sokol* members attended, including the four year old young *Sokol* Steven Stevlić, who showed the success of our development in bringing up our young in the spirit of *Sokol*. After a few words of greeting Drčar invited me to speak. In my address I expressed to all present my sincere pleasure at seeing them on that evening, thanked Ivan Drčar for his hospitality, and mentioned the presence of Dušan Svetlič who as the editor of the *Sokol Herald (Sokolski Vestnik)* advanced the prestige of our paper to the highest level possible and who as deputy secretary general of our organization has won the trust of our members and attracted many new ones. I outlined the problems of our renewal emphasizing the necessity of establishing new basic units — the *Sokol* brotherhoods. The *Sokol* organization devotes all its efforts to the forming of its members in the spirit of the *Sokol* idea: brotherhood, equality, and liberty; and strives to advance the people in their wonderful and happy homeland.

The next item on the program was the formal unfurling of the flag of the *North-American Yugoslav Sokol District.* That flag had already flown at the centennial rally of the *American Sokol Community* (*Američka Obca Sokolska*) in Chicago in 1965 and then on each subsequent rally of this organization as well as at the rallies of the *Czechoslovak Sokol Center Abroad* (*Ustředí Československého Sokolstva v Zahraničí*) in Vienna and in Zürich. The purpose of having a formal unfurling at this time was to have it formally included among the flags of the *Yugoslav Sokol in the Free World.* The Representation requested Milena Svetlič to be the flag's godmother at the unfurling. Milena unfurled the flag and proudly expressed a desire that the flag should lead us at all events where we may present the results of our endeavors and demonstrate the power of the *Sokol* idea. Following that, my wife Ivanka, coach of the *Yugoslav Sokol in the Free World*, appended to the flag a memorial ribbon which she made herself and said: "I am very proud to have the privilege of being present at the formal unfolding of the flag of the *North American Sokol District.* May the words embroidered on this ribbon continue to encourage us in our *Sokol* activities, "Forward without rest following the *Sokol* flag in the fight for our freedom and our home." Jovica Stevlić, a native of Šumadija, then recited in perfect Slovenian Prešeren's ode to Vrba. He explained that he learned the Slovenian language as a student in the intermediate school and later perfected his command of it when he was working near the Yugoslav northwest frontier. Throughout the proceedings recordings of *Sokol* marches were quietly played in the background.

Dušan Svetlič then read the greetings of secretary-general France Vadnal from Paris, deputy presiding elder Rudy Perhinek and his wife from Cleveland, coach Nikola Daškević from Boulogne, deputy coach Savo Barać and Miloš Vauhnik, member of the court of honor, both from Germany, mentor Jovan Nikolić from Zürich, Lojze Povhe from Speicher, Switzerland, and treasurer Dimitrije Kralj and his wife Tatjana from Milwaukee. The latter because of illness could not attend the rally or our meeting.

Then followed speeches by Vojislav Djelević, Dragan Adamović, Peter Urbanc, Dragomir Stevanović, Uroš Zonjić and Jerko Gršković. All of the speeches were overflowing with enthusiasm and satisfaction with the results of our work which show progress toward the renewal of our organization.

We continued with general conversation among the attendees. Dušan Svetlič distributed souvenir pencils inscribed "The *Yugoslav Sokol in the Free World* — Chicago Rally, 1977." He also collected signatures for greeting cards of his own design which were sent to all active *Sokols* worldwide.

Cities of our *North American Sokol District* represented at our conference include: Berwyn, Chicago, and Cicero in Illinois; Detroit, Michigan; Mamaroneck,

New York; Middletown, Ohio; Milwaukee, Wisconsin; Napa, California; and Toronto, Canada.

The principal production of the *American Sokol Community* (*Americka Obca Sokolska*) and the events characterizing the historical significant of the rally have already been discussed above. To this I might add that the main bleachers were overflowing with the spectators. I do not have exact information as to the number of people in attendance exclusive of those performing. It would not be an exaggeration, however, to estimate the crowd at six thousand persons. Those were mainly people who after several generations still retained their cultural heritage.

After the production Jovica Stevlić, our flag bearer, invited all the Yugoslavs to his home only a few blocks away from the stadium. We were all glad to come. In the seven years since his arrival in the United States he has acquired a very nice house. He, his hospitable wife and two sons received us in the garden. We were already acquainted with four year old Steven, since he was sitting with us at the dinner the day before sporting a *Sokol* hat. His younger brother — age two — was acting in every respect like his older brother in extending his welcome to all of us.

It was a very hot day and Jovica had his hands full plying us with refreshing drinks. We hardly noticed that during that time his wife was busy inside. When they invited us to come in, they led us to a small hall with a table set for another festive dinner. All the time the conversation was flowing freely in spirited discussions, greetings and jokes. Pleasant singing gave a special atmosphere to the meeting. Before we left Jovica showed us the rest of the house and I must say that I was full of admiration for his ability to create a home of tremendous value in such a short time. We parted from our kind hosts with warm expressions of appreciation for their kind hospitality.

After we left the Svetličes invited us to spend at least three days as their guests in Milwaukee. Ivan and Paula Drčar, too wanted us to stay with them for the two days before our return home.

Milena and Dušan Svetlič share their lovely home with their daughter Breda Čebašek. Milan, their son in law, is an electrical engineer. The young couple has two sons, Marko and Gregec, aged fifteen and eight, respectively. The Svetličes occupy the upper floor of the house while Breda and Milan live on the ground floor, each family having a separate household, of course. During our stay we were guests of both families. There was no end of conversations. We had not seen Dušan and Milena in 29 years with the exception of a brief encounter in the refugee camp in Italy. Dušan spends eight months of a year editing and printing the *Sokol Herald* (*Sokolski Vestnik*) and on other *Sokol* business. The remaining four months are his free time which he spends with his family.

In Milwaukee we also visited Dimitrije Kralj and his wife Tatjana. Dimitrije, treasurer of the *Yugoslav Sokol in the Free World*, fell ill shortly before the rally and had to be hospitalized. When he was discharged from the hospital it was too late for him to come to the rally. We all hope that now that the cause of his illness had been diagnosed, he would remain in good health and be on guard for any possible relapse. We had met again Lojze Galič and his wife Frances with whom we have met at the rally.

It was difficult to part from our good friends where we felt as if we were at home among our closest relatives. On Wednesday, June 29, Drčar picked us up and drove us to his home in Cicero near Chicago where we were his and his wife Paula's guests for the time we had left until our return to California. We were again guests in a nice home where we had good time with friends. Ivan took us to see the sights of Chicago and the adjoining communities. At the entrance to a large park in Chicago we saw the equestrian sculptures of a cowboy and an Indian brave created by Ivan Meštrovič. These were particularly interesting to me because Meštrovič had shown them to me in his study in Zagreb when he was finishing them. Ivan also showed us many other points of interest and he was our indefatigable guide and interpreter of Chicago's sights.

At the beginning of July we returned home after eight unforgettable days with our *Sokol* brethren and with pleasant of dreams of the *Sokol's* promising future. Ivan Drčar took us to the O'Hare airport where we took our leave. We parted thinking about our dreams and plans for the accomplishments expected by the next *Sokol* rally.

Rally of Czechoslovak Sokol in Vienna, July 1982[205]

Ivanka and I attended the 5th Jubilee Rally of the *Czechoslovak Sokol in Vienna* from July 2nd through 5th , 1982. Vienna is the seat of the *Sokol District* there. France Vadnal, Miloš Vauhnik, Ivo Mavko and Krsto Todorović, leader of the dancing troupe "*Oro*" from Osnabruck and, of course, members of the Yugoslav Sokol Brotherhood in Vienna also attended with us. Our small delegation attended all the productions at the rally and was a visible presence especially at the unveiling of the memorial plaque to Miroslav Tyrš in the gymnasium of the *Czechoslovak Sokol Association*, where Ivanka, following the unveiling of the memorial, delivered a greeting address in the Czech language. France Vadnal gave a lecture at the meeting on the history of the *Sokol* movement. Ivanka was the only Yugoslav who took part in the calisthenics production of members and youth on the principal day of the event.

The flag of the Yugoslav *Sokol* was carried by Ivo Mavko from Canada. Ivanka made memorial ribbons to decorate the *Sokol* flags on the principal day of the event,

---

[205] See also: *V.- Jubilejní slet Československého sokolstva v zahraničíve Vídni 1982*, Uřední pořad, *Ustředí ČS sokolstva v zahraničí*, Tiskem Goldman Druch, 3430 Tulln, Austria.

including the flag of the Czechoslovak Sokol, which was again unfurled, the flag of the *Yugoslav Sokol in the Free World*, and six other flags.

The rally was attended by a great number of Czechoslovak *Sokols* from all over the free world, even from remote countries like Australia and South Africa. The three Yugoslav participants from North America, Ivanka, Ivo Mavko, and myself, also came from distant lands. Our flight from San Francisco to Vienna took fifteen hours. Vadnal came from Paris, Miloš Vauhnik and Krsto Todorović from West Germany, and Ivo Mavko from Edmonton in the Canadian province of Alberta.

The small Yugoslav delegation was kindly welcomed by a three member rally committee of the *Yugoslav Sokol Brotherhood in Vienna*. That committee made arrangements for our lodging, transportation to the remote places of various events

Visiting Tyrš' grave at Oetz; Dr. Miloš Vauhnik and standard bearer Savo Barać (1982)

and even for entertainment. The Representation had sent some money from the general fund and the fund for rallies to the secretary of the rally committee to cover the expenses only. All expenses for travel to Vienna and other personal expenses were borne by the attendees themselves.

On Friday, July 2, a delegation of *Sokols* went to Mauthausen, the site of one of the most horrible Nazi concentration camps where, according to records as of January 15, 1945, over 73,000 persons were interned. How many died there we shall never know. The *Czechoslovak Sokol* under leadership of presiding elder, Jan Waldauf, unveiled on that date a memorial plaque to the *Sokols* who were victims of Nazism. The *Yugoslav Sokol in the Free World* was represented by France Vadnal who published an extensive report of the proceedings in the *Sokol Herald* (*Sokolski Vestnik*).[206]

The same day the President of the Austrian Republic, Rudolf Kirschlager, received the *Sokol* delegation in an audience. I had been selected to represent Yugoslavia. The Czechoslovaks sent a rather large delegation of members from all the places where they have associations so that the palatial hall of Maria Teresa was brimming full. Following the speeches, the President remained almost for an hour in friendly conversation with the attendees. In the evening mayor of the Capital City of Vienna, Leopold Gratz, prepared a get-acquainted evening at the City Hall (*Rathaus Keller*). Both Austrian dignitaries sent official greetings to the *Sokol* rally. After the reception we sat down at a long table for dinner. The rally committee of the Yugoslav Vienna Sokol Brotherhood was host to all of the Yugoslav members in attendance. Official Austria certainly showed itself considerably more attentive and welcoming than the aloof Swiss at the *Sokol* rally in Zürich in 1976.[207]

The next day, July 3, the gymnastic competition took place at the *Schmelz Education Center* and there was a formal unveiling of a memorial plaque to Miroslav Tyrš in the gymnasium of the *Czechoslovak Sokol Association Vienna No. 16*. Before the unveiling the presiding elder of the *Sokol* District addressed the assembled members, the plaque was unveiled at the singing of the Czechoslovak national anthem "*Na Tatarou se blyska*" and then Ivanka recited a short poetic greeting in the Czech language. Ivanka's recitation brought lively applause — people crowded around her and everyone wanted to greet her personally. All the Yugoslavs who attended the rally were present at this event.

The same day there was a symposium on the *Sokol* idea and *Sokol* movement at the *Schmelz Education Center*. France Vadnal gave a talk about the history of *Sokol* leaders. Ivanka, too, gave a talk about her *Sokol* reminiscences of seventy five years ago. In the evening there was a concert in the Volkstheater. Our faithful rally committee obtained admission tickets to all these events for all Yugoslav participants.

---

206 *Sokolski Vestnik* (*Sokol Herald*), Milwaukee, Wisconsin, 1982, No. 3-4, pp. 15–23.

207 Note by editor: In 1986, at the occasion of the 6th Rally of the *Czechoslovak Sokol* in Zürich, the Swiss aloofness was somewhat relaxed and the President of the City of Zürich, Dr. Thomas Wagner, sent official greetings to the rally.

In 1997 the 19th Rally of the *Czechoslovak Sokol* in Chicago — Berwyn was attended also by the ambassador of the Czech Republic. *Yugoslav Sokol in the Free World* was at that time represented by the presiding elder Branko Svirčić, *Sokol I Tabor* from Ljubljana was represented by presiding elder Matjaž Anžur, dressed in the classical *Sokol* uniform. Naturally, there was no trace of any representative of the Republic of Slovenia which to this day holds the confiscated property belonging to the organization.

France Vadnal arranged an exhibition of Yugoslav *Sokol* activities in two places: on July 3 on the grounds of the *Czechoslovak Sokol District Vienna* near Hotel Post with the assistance of Ivanka and Ivo Mavko; then on the next day he moved the exhibition to the stadium.

On July 4, the principal day of the rally, a rehearsal for the afternoon production was held at the Hütteldorf West Stadium (Weststadion). Early in the morning it began to rain. The athletes carried on with their rehearsal and were drenched to the skin by the unrelenting rain. Of the Yugoslavs only Ivanka took part in the production. She has been doing this for almost nine times nine years as she told the

Ivanka Bevc addresses the Sokols — on the left are embroidered memorial ribbons she made herself.

audience in the *Schmeltz Education Center.* She was training for this production at the *American Sokol San Francisco* which has a gymnasium in San Mateo, a drive of two hours from her home. In the afternoon the rain subsided and there were only intermittent showers between which the sun came out. Thus the youths had to exercise in rain but the members even had some sunshine. Although the rain discouraged many spectators, the covered spectator benches were packed. Many took pictures when the sun came out. I too was taking picture from the upper part of the tribune and was surprised by how well they came out. When I got the prints in Vienna more than half of them were very good.

Following the public production on July 4 was a parade of flags whose color guards then assembled in front of the main viewing stand. The leadership of *Czechoslovak Sokol* affixed standard memorial ribbons to the flags. On behalf of the *Yu-*

goslav Sokol in the Free World Ivanka also attached five memorial ribbons which she made herself to the flags of the *Czechoslovak Sokol Abroad,* the *Czechoslovak Sokol District of Vienna,* the *Yugoslav Sokol,* the *Yugoslav West European District in Paris, and* the *Yugoslav Sokol Brotherhood in Vienna.*

The organizers of the Representation of the *Yugoslav Sokol in the Free World* who generously work for its success were secretary general France Vadnal, editor and publisher of the *Sokol Herald (Sokolski Vestnik),* and deputy secretary general Dušan Svetlič, coaches Nikola Dašković and Ivanka Bevc, treasurer Dimitrije Kralj, registrar Milena Svetlič and numerous members of the Executive Council whose successful work has produced results set down in their reports.

Because members of the *Representation* are scattered all over the world it is not possible to hold regular meetings and conferences. Their business must thus be transacted by mail and they depend on the regularity of the postal service to achieve the purposes of their various communications. The expenses of correspondence are thus a considerable and sometimes excessive burden on the treasury and also on the administration of the *Sokol Herald (Sokolski Vestnik).*

The *Sokol Herald (Sokolski Vestnik)* has become an important and irreplaceable publication of the *Yugoslav Sokol.* It belongs among the leading publications of Yugoslav émigrés. When the so-called Allies abandoned the leader of a democratic and independent Yugoslavia Dragoljub Mihajlović, and, by the infamous Yalta agreement, delivered the Slav countries to international communism, they at the same time sealed the fate of democratic fighters for freedom who had to leave a homeland enslaved by the communists. At that time we saw the communists dissolve and ban the *Sokol* organization, the third time in this century. The first and second time *Sokol* was dissolved and banned by the fascists and the Nazis. The communists logically followed in their steps as enemies of liberty and the *Sokol* idea.

At the same time as the Rally in Vienna the *Slovenian Littoral Daily (Primorski dnevnik),* a Trieste paper supported by the Yugoslav government, reported on the celebration of the centennial of the *Trieste Sokol. The Littoral Daily* of course did not mention that the communists, after seizing the power in 1945, dissolved the *Sokol* organization as one of their first official acts. *The Littoral Daily* glossed over this reprehensible fact merely stating that the *Sokol* centennial was celebrated by recreational physical education organizations which "assumed the heritage left to them by the *Sokols.*" Accordingly, no *Sokols* were present at the celebration of the *Sokol* centennial although *Sokols* are still around. It is to those surviving *Sokols* that the *Representation of the Yugoslav Sokol in the Free World* extends its heartfelt congratulations.

Ivanka remembers the situation in the Slovenian Littoral, that part of our land which the Italians "earned" as reward for their treacherous abandonment of their former Austro-Hungarian ally. I have already mentioned that Ivanka took part in the *Sokol* production in the *Sokol* stadium at St. Peter near Gorizia. The *Sokols* marched through the town of Gorizia but only when they were outside of the city limits were they allowed to unfurl their Slovenian flag. Their coach was Bruno Ravnik; his assistant coaches were brothers Bašin, the Kozman family, and Anuša Čigoj.

When at the end of World War I the Sket family returned to the ruins of the Sokol hall, they immediately resumed exercising in the courtyard of the *Slovenian National Loan Association*. Among them were the four Gorjan sisters, the two Primožič sisters, the Sougans, Mozetičes, Lukežičes and many others. Their coach Fon once took them well prepared for a *Sokol* production in Komen and in the Vipava valley. The traditional *Sokol* attire for women was a blue skirt, white blouse and a red kerchief on the head. The Italian occupation authorities could not stand this display of Slovenian national colors and the *carabinieri* ripped the red headgear off their heads. But the girls replaced their red kerchiefs with white ones and proudly gave their presentation at the high Charso plateau of Komen as brave *Sokol* women.

Coach Nikola Daškević wrote the *Manual for Sokol Leaders* published by the *Representation of the Yugoslav Sokol in the Free World*. The manual is an important piece of work and congratulations are in order to Nikola Daškević for making this unique manual available to our members and coaches for use in their exercises. *Sokol* Associations and Brotherhoods have acquired this manual and set up calisthenics exercises for the youths and children. Simple, light exercises can be practiced by every *Sokol* member. With this manual selecting exercises is not a problem and anyone can now begin with at least the simplest exercises.

After the conclusion of the Fifth Jubilee *Sokol* Rally on July 5 we had a conference of the *Federation of Slav Sokol Organizations*, members of which are thus far the Central of the *Czechoslovak Sokol Center Abroad* (*Ustředí Československého Sokolstva v zahraničí*) and the *Yugoslav Sokol in the Free World*. The Yugoslavs were represented by Ivanka Bevc, Ivo Mavko, myself and three members of the Yugoslav Sokol Brotherhood in Vienna. Elder Jan Waldauf presided. In opening the session he called upon us to speak in our native languages and not in foreign languages. I read the reports of the absent General Secretary France Vadnal and Treasurer Dragan Adamović and made the necessary explanations. A motion by elder Waldauf that the membership dues of both member organizations be increased to $100 was approved. On my request Waldauf made up a list of all Czechoslovak *Sokol* associations in the free world and promised that those associations would admit Yugoslav *Sokols* to their gymnasiums in places where they do not have their own gymnasiums.

After the conclusion of the rally some Yugoslav *Sokols* made a pilgrimage to Oetz on the banks of river Ach in Tyrol where during the Hapsburg monarchy Miroslav Tyrš, founder of the *Sokol* movement, was interned. Tyrš drowned in the rapids of the Ach River at Oetz under unexplained circumstances. In 1962 the *Czechoslovak Sokol* placed a memorial plaque at Oetz. Carrying our flag we paid respects to the memory of our mentor. In 1984 at the centennial of Tyrš's death, the *Czechoslovak Sokol Center Abroad* (*Ustředí Československého Sokolstva v zahraničí*) organized a three day symposium on Tyrš, his ideas and his work. At that time the *Yugoslav Sokol in the Free World* was represented by secretary general France Vadnal, who gave two lectures: "Tyrš's influence on the Yugoslav *Sokol*" and "Tyrš's *Sokol* and Marxism."[208]

The Viennese population behaved correctly toward hose participating at the rally. The Fifth Jubilee Rally of *Sokols* was in every respect a significant event not only because of its productions but also because of the fact that it was welcomed in a city where in the not too distant past *Sokol* was considered anathema.

Visit to the American Embassy in Vienna

Vienna is no longer the city I knew in the years 1908 to 1913 when I was studying at the Technological University. In those years a conflict was developing between Austria and Serbia because of Austria's annexation of Bosnia and Herzegovina which Serbia was opposing. Austria declared war on Serbia on July 28, 1914 thereby precipitating World Wart I. In October of 1918 Slavic nations in Austria proclaimed their independence. New sovereign states were created: Yugoslavia, Czechoslovakia and Poland. Austria was reduced to the present rump state of seven million inhabitants. During World War II it was absorbed by the German Reich. When Hitler and his satellites attacked and partitioned Yugoslavia on April 6, 1941, Yugoslav prisoners of war who passed through Vienna were abused and humiliated by Viennese burghers. Today, however, such hatred is a thing of the past. Hitler, who did not survive World War II which he provoked, took to his grave his plans for the annihilation of the Slavs, Jews and Gypsies. After the demise of Hitler's Germany, Austria was restored by the State Treaty of May 15, 1955, unfortunately with the same territory it had prior to the *Anschluss*.

Thus I did not find the Vienna I left after completing my studies in 1913. With perhaps two million inhabitants it may be too large for a state of seven million. The lavish palaces of princes, counts and other nobility still exist in all their former splendor but only as tourist attractions. Multilingual signs in these buildings as

[208] *The Sokol Herald (Sokolski Vestnik)*, Milwaukee, Wisconsin, 1984, No. 3-4, p. 9–18.

well as greetings at the railway stations are in German, French, English and Spanish but an inscription in a Slavic language, if any exists, is hard to find although it was the Slavs who through centuries toiled to help create those luxurious edifices in what at the time was the capital of the Hapsburg Empire.

Today Vienna survives thanks to tourism and support from the country's reduced territory. A Viennese told me that a tourist facility can be found in almost every second house of the city. While that might have been a slight exaggeration it does succinctly sum up the problems of the country's reduced territory. Vienna has for a long time been home to a sizable Czech colony, mostly autochthonous from the times before the war, with the *Sokol* associations which existed before the two world wars and whose membership increased after World War II. The number of immigrants from Czechoslovakia as well as from Yugoslavia is also increasing with the arrival of the so-called "guest workers" (*Gastarbeiter*) some of whom intend to remain in Austria permanently as economic or even political émigrés.

Yugoslavia has tried to influence the status of guest workers with respect to pensions and other benefits. In doing so it may have considered the status of the Slovenians in Carinthia and Croatians in Burgenland and the problem of abridgment of their ethnic rights as small change in these negotiations. Carinthian Slovenians did not want to have contact with political émigrés because their organizations were receiving subsidies from Tito and Yugoslavia. In the role of a protector of Carinthian Slovenians Yugoslavia did make some noises about the unjust treatment of the Slovenians and created tensions in its relations with Austria.

Hence in March 1980 the Yugoslav foreign minister Vrhovec arrived in Vienna and, after concluding various economic agreements, to the satisfaction of the diplomats, advised the Slovenians and Croats in Austria that their best chance to resolve their difficulties was to engage in a dialog with the Austrian authorities. Woodward Romine, political counselor at the American Embassy in Vienna, reported to Washington that the disappointed Carinthian Slovenians were reassured by the Yugoslav ambassador in Vienna, Novak Pribičević, and the Yugoslav consul in Klagenfurt that Yugoslavia would always support their legitimate aspirations. The American diplomat, who obviously knew his Yugoslav colleagues only too well, could not help but place the words "always support" in quotation marks.[209]

Although Carinthian Slovenians had no contacts with the political émigrés, United States Senator Frank Lausche, on the request of professor Ciril Žebot, asked Secretary of State Henry Kissinger to suggest that a representative of Carinthian Slovenians be invited to participate at the celebration of the twenty fifth anniversary of the *Austrian State Treaty* and thus symbolically show an awareness of the

[209] Woodward Romine in State Department telegram *Vienna A-403*, February 15 1980, addressed to the Department of State in Washingtonu.

rights guaranteed to them by the Treaty. Kissinger arrogantly ignored Lausche. However, he was soon replaced by Edmund Muskie, Lausche's former colleague in the Senate, who responded to the effect that no Slovenian representative would be invited to the celebration but that the American Embassy in Vienna was willing to receive representatives of Slovenian organizations that had not up to that time sought to inform the representatives of the Four Powers (United States, Great Britain, France and the Soviet Union) of the problems caused by the intolerance of the German nationalists.

On the occasion of my visit in Vienna I used the opportunity to visit the political counselor Carl T. Clement at the American Embassy to apprise him of the situation of Carinthian Slovenians. Mr. Clement was very courteous and told me that since the Embassy advised Matevž Grilc, chairman of the *Slovenian National Council* in Klagenfurt, that it was willing to see him and other Slovenian leaders, Grilc had visited the Embassy several times and that Clement, too, would make use of the first opportunity to go to Klagenfurt. The counselor concurred with my opinion that it would be a good idea to have an appointed representative of Carinthian Slovenians and Croats from Burgenland in the Austrian parliament as they cannot, because of their small numbers, elect a deputy of their own. Counselor Clement certainly showed himself concerned about the issue under discussion, a considerable change from his predecessor who had flippantly responded to an inquiry from Washington that he had no funds for gasoline to take trips to Carinthia. It is possible that Clement had read the remarks of Congressman Fortney Stark of California who pointed out that no representative of Carinthian Slovenians was invited to the celebration of the State Treaty jubilee.[210] Clement, who is also from California — his home is in Marin County — soon made good on his promise. He attended a Marshal Plan exhibition in Klagenfurt and on that occasion met with several representatives of the Carinthian Slovenians to the consternation of the local Austrian politicians. The Americans may not be aware of the influence of even a small gesture indicating interest in human rights in a country which is, at least to some extent, dependent on America's good will. On the other hand, of course, it may be that they are reluctant to exploit this power as a matter of policy.

[210] Fortney (Pete) Stark of California: *Jubilee of Austrian State Treaty Eclipses Ethnic Rights*, Congressional *Record — Extension of Remarks*, E-3810, August 18, 1980

Friends and Acquaintances

Friends

Rudi Marinčič

Rudi Marinčič was a first class *Sokol* organizer who already as a youth occupied a prominent place in the society. He remained faithful to the *Sokol* idea during the Axis occupation and persevered under the barrage of enemy calumny and communist propaganda. He visited me once and described the campaign conducted against him by the communists. He was in the *Chetnik* detachment at Grčarice where he was captured and taken to Kočevje to be tried before the communist kangaroo court and sentenced to death.

Pavel Pestotnik was among the leading *Sokol* members in Ljubljana. He was prominent in the founding of *Sokol I at Tabor* and in the building of the *Sokol Hall at Tabor.* He wrote several articles on the *Sokol* idea and *Sokol* activities. During the war he was constantly threatened by the communists. To calm his nerves he went for a few months to Italy but on his return the situation did not get any better. When another communist assassination occurred in Ljubljana he had to go to a sanatorium but even there he could not find peace because the communists had at that very time brought in one of their assassins seriously wounded by an Italian policeman who pursued the getaway car after the murder. On May 5, 1945 Pestotnik emigrated to Carinthia where he learned from a refugee that he had been sentenced to death by the communists. It took a long time before he could be safely taken from Carinthia to Trieste and then to Camp Riccione. Because his heart was failing him he at first could not get a visa to come to the United States. He was allowed to come only later when the United States decided to accept a large number of terminally ill persons. He lived in Denver, Colorado, in a retirement home operated by the *Little Sisters of the Poor.* He visited me once in Napa when we were pursuing his lawsuit in San Francisco. He died of a heart attack on the anniversary of his wife's death and is buried in the Mount Olivet cemetery in Denver. I had a modest tombstone placed on his grave which a local flower shop decorates with flowers four times a year. His children never received permission to transport his remains to the mausoleum where his wife is buried. His memories were published in the *Sokol Herald* (*Sokolski Vestnik*) and, in a separate edition titled: *Moja življenska slika* (*My life's portrait*).[211]

[211] Pavel Pestotnik: *Moja življenska slika,* Vlado Bevc, San Ramon, California, 1998.

Janko Mačkovšek was two years ahead of me in the preparatory school (*Realka*). He already had a good reputation because of his work on national minorities. He gained additional expertise on this subject in Prague where he studied at the Technical University and graduated as a civil engineer. Employed as a civil engineer by the Municipality of Ljubljana he was active in many areas of public life; most notable was his work with the *Cyril and Methodius Association* where he applied his scholarly expertise to the national minority questions, particularly to the problems of education. After Franja Tavčar died he succeeded her as the leader of the *Cyril and Methodius Association.* When the Italians occupied Ljubljana they began to track down refugees from the Littoral whom they regarded as Italian subjects and forced them to join their army; in some cases they even prosecuted them for desertion. That was the reason why Andrej Glušič, whose home was in Tolmin, had to go into hiding so that he had no fixed residence. Mačkovšek was of course singled out as leader of the *Cyril and Methodius Association.* He was summarily arrested and it took a long time before he was released. In the *Society's* offices we had meetings about the burning questions of that sad epoch. Following the death of Albert Kramer, chairman of the Slovenian liberals, Mačkovšek was nominated chairman of the seven member directorate of the party. Once he found out that the Germans were preparing a massive raid on the progressives. He asked me to intervene with Rupnik to prevent it. I declined because such intervention would be contrary to all my public activities. Then he asked me to at least speak to Franc Krener, the commander of the *National Guard* who might be able to help. Krener told me that there was no reason for our concerns but as it turned out he was too optimistic. As everybody in an exposed position Mačkovšek, too, sensed the approach of new violence against him. He invited me to his home and explained to me the details of his clandestine activities; one of his secrets was the cipher he used for sending radio messages to London. Soon after that he was arrested and sent to Dachau where he died in an epidemics because of malnutrition, exposure and terrible conditions.

Janko Mačkovšek

Rudolf Žitnik was a sheet metal contractor by trade and owned a thriving business. He was active in the *Yugoslav Nationalist Party.* During the war he was diligently attending to the shipments of supplies to the *Chetnik* detachment at Grčarice. Once on returning by car from one of such missions he was stopped by a truck full of communists who were surprised when they recognized Žitnik whom they wanted to capture. They put him and his companion on the truck and drove away in the direction of the communist command at Ribnica. On the way Žitnik and his companion overpowered the driver, jumped off the truck and ran up the nearby slope, each in his own direction. The communists pursued them and shot at them. Žitnik was wounded in the throat but managed to escape as did his companion. Soon they found friends who seeing that Žitnik's wound that was still bleeding gave him the first aid and bandaged the wound. Later Žitnik noticed that the bullet was moving around under his skin; at Camp Riccione he felt it in his leg but he never had the time to have it extracted.

Rudolf Žitnik

In the camps Žitnik was very restless because he worried about his family and his business. His wife, sister of photographer Pogačnik, remained at home in Yugoslavia with two sons, son Janez, however, went abroad with him. Because of that Žitnik could not stay in one place and he used to travel a great deal. At one time he attempted to go to Austria to visit the refugees there. He went in a car with Colonel Glušič who was in the uniform of the Royal Yugoslav Army. The British who were guarding the border between Italy and Austria, however, allowed no one to cross. Glušič, who had papers issued by the American intelligence, was allowed to pass but Žitnik was not and he had to return without success.

Žitnik and his son emigrated to Argentina where he had a hard time establishing himself. He continued to be active in the *Slovenian Democratic Party* and with Mačus opposed the leading members of the party in Argentina. At times he acted on his own which sometimes embarrassed the party. He was publishing the bulletin *Free Speech (Svobodna Beseda).* Eventually he gave up his opposition and somehow settled down. He died in very modest circumstances.

Janez Žitnik. Of the Grčarice *Sokol Legion* detachment only the seventeen year old Janez Žitnik, sentenced to death by the kangaroo court in Kočevje, was miraculously saved. Janez was already tied to the other victims by a common strand of telephone wire and in the process of being loaded on a truck that would take him to the place of execution when Zdenka Kidrič, president of the kangaroo court, arrived on the scene, called for Žitnik and ordered that he be untied. Kidrič then told the young man that his sentence was commuted to hard labor for life and snarled: "So they won't be able to say we did you in because of your father."

Janez Žitnik

As a prisoner Žitnik and other men had to carry water up a hill under the watchful eye of a communist guard. On a foggy day the prisoners were able to organize an escape, disable the guard and flee, each on his own, into the Kočevje forests and mist. After several days of wandering through the forest Žitnik came out near Prezid, a village on the Charso, where he saw a German patrol near the edge of the forest. He had one of those leaflets the Germans were throwing from the airplanes offering good treatment to those partisans who surrendered and he approached the German officer with it. But the German officer would have none of it and was about to order his men to shoot the young Žitnik. Fortunately a man who knew Janez's father was present and he managed to persuade the German officer that the young man was not a communist. The officer relented and provided Janez with transportation to Kočevje. In Kočevje, by then occupied by the Germans, Janez asked the prison warden to let him see the cell where he had been imprisoned. In the prison cell, under the bench he found the sheaf of banknotes, 30,000 lire, he had hidden there. This was his share of the money the commander Koprivica at Grčarice distributed among his men before he surrendered the fort and shot himself. Janez then went to town and had a good time. Soon he found a secure transportation to Ljubljana but he could not stay put. He went to Trieste and joined the *National Guard* because from there it was easier to get into the *Chetnik* detachment of General Andrej–Prezelj. The latter retreated with his troops to Italy where they were interned at Forli near Bologna. Later the detachment was moved to Eboli in the South of Italy. Rudolf Žitnik could visit Janez there often in the course of his travels. It was his wish that Janez complete the secondary school and it was for this purpose that he had him transferred from Eboli to Riccione.

Etbin Bežek was the secretary of *Sokol I at Tabor* and remained in touch with me throughout the Axis occupation and the concurrent civil war. He also was a member of the *Sokol War Council* and commander of a *Sokol Legion* detachment. When in 1944 the Gestapo arrested me he managed to get to my office through a concealed door and remove a document that listed all members of the *Sokol Legion* and their order of battle. As he and other members of the *Sokol Legion* were likely to be arrested he fled to Postojna where friends hid him in a detachment of the National Guard. In keeping with the Nazi practice of *Sippenhaft* the Gestapo arrested and imprisoned his wife Ljubica. Ljubica did not want to endure the Gestapo prison and she demanded that Etbin give himself up which he did. He was with me in the Gestapo prison where he contracted an inflammation of the pulmonary membrane. With great difficulty we prevailed on the Gestapo to transfer him to the hospital, where he was guarded by armed Gestapo men. At that time the Gestapo interrogator Wurnigg asked his wife: "Has he croaked yet?" Etbin lived to see all of us released. He joined us at the Parliament session at Tabor on May 3,

Etbin Bežek

1945 prior to the communist occupation before we left the country. He intended to leave too, but when he was already on the truck that would take him to freedom he decided to stay at home. The communists arrested him immediately and sentenced him to prison. After he had served his term they arrested him again under an "administrative measure"[212] and sent him to the concentration camp at Strnišče near Ptuj, to do the so-called "socially useful work." There Bežek was abused; once when at "gymnastics" he did not lie down in the mud puddle fast enough a guard jumped on his back with his heavy boots and permanently injured his spine. When President Truman agreed to provide aid to Tito who had a family quarrel with Stalin he demanded that Tito disband the concentration camps and Bežek was released. Today, of course, American diplomats no longer make such demands lest the "leaders" of various foreign dictatorships be annoyed.

Later the communists would allow the Bežeks to visit their daughter in Italy but never together. When Etbin went, Ljubica had to remain behind as a hostage and vice versa. In Italy he read my booklet *On the Ramparts Defending Sokol Ideas* (*Na braniku za Sokolske Ideale*) published in England and wrote me. He identified the man whom Lubej and Rus smuggled to Czechoslovakia at the last Sokol rally. All the suffering and the communist attempts to brainwash him could not undermine his devotion to the Sokol idea.

In 1982 I was traveling in Europe where I met many friends, including some who came to see me from Yugoslavia. Bežek was called in by the political police and reminded that he was receiving a government pension which the government giveth and taketh away at will and that he should think carefully on the advisability of meeting with "dangerous people who these days are agitating around our borders." Thus we never got to see each other again. He died in Kamnik in 1983.

Stanko Dimnik. I knew Stanko since my days in Vienna where we were on the committee of the *Slovenian Engineering Students in Vienna* (*Društvo slovenskih tehnikov na Dunaju*); after my graduation he took over the chairmanship of the society. Together we collected the sum of 200 Krone for the *Cyril and Methodius Association* (*Cirila Metodova Družba*) by virtue of which our association became a founding member of the *Society*. When World War I broke out Stanko had not yet graduated and was immediately conscripted. Only after he had returned from the Italian prisoner of war camps he was able to complete his studies in Vienna. He, too, was a *Sokol*. Until 1945 he was a professor at the Engineering Intermediate School (*Tehniška srednja šola*) and at the same time a lecturer at the Faculty of Engineering of the Technical University in Ljubljana (*Tehniška visoka šola*). He was a

212 "Administrative measure" was an euphemism for arresting a politically suspect person at the discretion of the political police and holding him indefinitely in a concentration camp. No international protests that we know of have ever been voiced against such practices by the otherwise vocal community of the civil rights activists.

consultant for structural engineering problems at numerous construction projects such as Hotel Slon and the Ljubljana's skyscraper. He studied the earthquake problems as applicable to the construction of buildings and sent me his paper on the subject. Stanko made all the structural calculations for the *Sokol Hall at Tabor.* Professionally he was very successful to which attest numerous buildings and the suspension bridge across Sava River at Medno. Today no one in Medno knows who designed that bridge. As a professor at the Engineering Intermediate School he was entitled to maintain a private practice by virtue of which he was also a member of the Engineering Chamber. Stanko was a born professor who educated a large number of technicians.

His wife Edit and he had two daughters: Edita, married Sovinc, and Staša, married Gašperšič. I was honored to be the godfather to their first-born. His wife was very supportive and stood by him when they were saving and working hard to build their own house in Ziljskega Street (now *Tobačna ulica)* in Ljubljana. If he could find spare time Stanko liked to hike in the mountains. As a lonely wanderer he was making rounds of the mountains, he walked by the moonlight, too, or rested on a rock wrapped in his cape.

It was natural that we collaborated in the resistance organizations from the very beginning of the Axis occupation when the communists were still sitting on the fence and with malicious glee watched the depredations of the occupying forces. Communist Polak-Stjenka, one of his neighbors wrote on the portal of Stanko's house:"Death to Mihajlovic's men!" At Christmastime in 1944 Stanko was arrested by the Gestapo shortly after they arrested me. We were together in prison until April of the next year. He also was a member of the Parliament convened by the *National Committee* on May 3, 1945 in the small hall at Tabor. The night following that session I slept at Dimnik's' house as we expected the Germans to react with repressive measures. I discussed the option of leaving with friends, most of whom recommended it for the sake of personal safety but Stanko felt that he could remain. He thought he could hide with his relatives in Jarše north of Ljubljana during the first days of the communist occupation as he thought that the fact that he was imprisoned by the Gestapo was sufficient evidence that he had done nothing against the national interests of Yugoslavia.

The communists arrested Stanko soon after the onset of the communist reign of terror. Around Christmas of 1945 the communists staged a trial against a group of people whom they accused of having worked against the so-called *Liberation Front.* The state prosecutor accused Stanko, among other things, of being ideologically responsible for unspecified murders and demanded death penalty. Among us, in exile, was Judge Fran Orožen who had been Stanko's friend from his early years. Orožen characterized accusations leveled against Dimnik as an excessive abuse of power and a vendetta against ideological opponents. He ruled out any possibility that Stanko could be responsible for murders; he knew his character thoroughly. Stanko was never aggressive, on the contrary: he saw and respected human dignity in every man. Even when a sadistic prison guard at Novo Mesto humiliated and abused him in presence of his wife and daughter Staša at one of their rare and brief visits he explained to his wife that the guard could be basically a good man but that he had bad luck with those who brought him up. Judge Orožen categorically rejected any doubts of Stanko's integrity in absence of an objective and impartial trial. Stanko–murderer? Absurd! The "trial" in presence of a howling mob which was abusing the attending relatives of the accused ended with Stanko being sentenced to twelve years' imprisonment.

In prison the communists exploited his professional expertise and used to send him, under guard, to *Litostroj*, a new industrial complex they were building in Šiska, to get his professional advice free off charge. He used to jest about that saying that there he could even take a walk toward Šmarna Gora — up to the fence, to be sure. When under the weight of snow some roofs in the Litostroj plant collapsed the communists found errors and culprits everywhere — only the calculations made by Dimnik were flawless and withstood every scrutiny. After seven years of imprisonment the communists found out that he was wrongfully convicted because they took him for a forestry engineer Dimnik who was in no way related to him and they released him. Still he was not allowed to resume his teaching at the Engineering Intermediate School where only the most fervent supporters of communism could be enrolled. The communists did continue to exploit his professional expertise but only in very low and poorly paid positions.

Stanko was unchanged by all the suffering to which he was subjected. He used to describe to my son the atrocities the communists were committing in prison vividly and in gory detail. His younger daughter who was present at one of such narratives once nearly fainted from horror. But Stanko justified such narratives by explaining to her that my son would once get to the free world and would be able to report these deeds. In fact there had been no paucity of such testimonials. It was some disappointment to my son when he found out that in the free world no one cared.

Once, at the restaurant Čad, Dimnik was accosted by the warden of the Ig prison, and Dimnik responded addressing him as "Mister warden" rather than "comrade Warden," as communists naturally cannot allow prisoners to address their jailers as "comrade." The response, perfectly correct under the communist system, still caused some public embarrassment to the warden for at that time the people already began to feel a certain revulsion toward the guardians of the communist paradise. On another occasion, in the alpine lodge at the Kamnik Pass (*Kamniško Sedlo*), conversation between the tourists turned to the topic of the partisans and their times. As is customary everybody participated in the conversation including Stanko. Seeing that he was somewhat older than most of the partisans were at the time but still seemed to know all about them one of the tourists, who apparently was a former partisan, asked him if he, too, had been with the partisans to which Stanko responded in the affirmative. Naturally he was then asked where he was with them to which he retorted: "In Jail," causing a somewhat strained situation.

Stanko's sentence also included loss of his accrued pension rights about which he was jesting that he needed only 35 years more of service to earn a full pension. After a few years, however, the communists gave him a decree stating that his conviction was expunged and that he was considered as "having no record." That means that they have officially admitted that all the accusations against him were false, that he has done nothing against the Slovenian people and had committed no crime. I last saw him in 1982 when I was traveling in Europe and Stanko came to see me in Gorizia where I was waiting for my wife to return from her visit to Yugoslavia. In fact, he visited me twice, ignoring the veiled threats of the communists. We had long talks about everything under the sun. After that I never saw him again.

Premature death of his wife Edit devastated Stanko so much that he sank into a profound depression. To compound the misfortune his older daughter Edita, a respected civil engineer also died from an insidious illness. That aggravated Stanko's sadness still further. He became so helpless that he could sign the letters Staša wrote for him only with difficulty and he could not read because he could no longer connect words. He became totally dependent on others so that he could no longer move around on his own. He went to a very nice home for assisted living in Tržič where he died on the third day after his arrival.

Ivan Vurnik, world class architect, was visited by misfortune as soon as the World War II started. His son was shot by the Italian political police in a tragic incident involving an attempted assassination. He probably had forebodings of the impending tragedy and tried to get in touch with me at the time but we somehow did not get together. That might have been the reason for his somewhat detached attitude toward me during the war. After the war, however, when I was already abroad, he resumed friendly contacts with my family who got to know him when we were guests in his mountain home on Pokljuka above Bled. Later on, when I was in the United States, he wrote me often and in a friendly manner as before.

The communists were making unreasonable demands on him; they wanted him to sponsor as prospective architectural faculty members persons who had no professional qualifications. In 1946 nearly all faculty positions in the Department of Architecture at the Technical University in Ljubljana were occupied by the political friends of various ministers. If those had any higher education it was exclusively from Plečnik's school of which neither Vurnik nor I thought much. Once

Vurnik was summoned to see the Minister of Education Juš Kozak, a mediocre writer and communist. In the minister's office were already the dean of the technical faculty Alojzij Hrovat, a prospective faculty member and Marjan Tepina, a mediocre architect from Plečnik's school who once admitted to Vurnik that he had no moral scruples. As Vurnik sat down the prospective faculty member began to reel off the names of other candidates he expected Vurnik to sponsor. Vurnik was ill at ease because he had other people in mind for the newly created positions and gave a noncommittal response just to get out of the unpleasant situation. In a few days one of the candidates came to see Vurnik and requested that he propose him for a position as a professor to which Vurnik responded that he could not do that. His principled and outright refusal to go along with what at that time was a customary practice brought demands from the ministry of education that he take an early retirement. The faculty, however, was courageous enough to resist that and arrangements were made for Vurnik to take a year's sabbatical whereupon he was allowed to return to teaching.

In the summer of 1957 friends of Plečnik, whom the regime was promoting, tried to open an exhibition of Plečnik's his life work at the Vienna Academy of Arts; the event was to be in connection with the life work of his mentor Otto Wagner. Vurnik wrote me that his greatest sin committed against the Slovenian culture was that he had in 1921, because he was not familiar with Plečnik's character and education, helped Plečnik to become a full professor of architecture at the University in Ljubljana. Plečnik's creations were narrowly conceived and generally had the appearance of an artisans' work of a past era. Vurnik wrote to Vienna that he, too, wanted to present his life work at an exhibition concurrent with Plečnik's. In response the dean of the faculty inquired why Vurnik imagined that something created in Ljubljana could possibly be of interest to the Viennese. Faced with such an attitude Vurnik simply filed a candidacy for a doctorate in his field for which he was well qualified and technically eligible as a graduate of the Technical University in Vienna (*Technische Hochschule*). The doctorate required an examination, the *Rigoroz*. The dean had to accede to that and Vurnik's exhibit was scheduled for the end of June 1957. At the opening of the exhibit all distrust and misgivings of the Viennese professors toward Vurnik evaporated and following his lecture they became very friendly. In recognition of his work he received a gold medal of the founder of the Technical University with citation: *To the Friend of the University and the Educator of the Youth.* Plečnik's exhibition, however, was never held.

Vurnik's wife Helena née Kottler, an artist whom he met in Vienna, died in 1962. In 1966 he was proposed for a membership in the Slovenian Academy of Arts and Sciences but he declined because two others whom Vurnik considered not qualified were proposed at the same time. He obviously did not care to be associated with "academicians" like Josip Broz, Edward Kardelj[213], Srečko Brodar, Jovan Hadži and the like whose political positions earned them honorary doctorates from the University in Ljubljana. The Academy took a deep umbrage at his refusal and from that time on used every opportunity to snub him. He did, however, accept the nomination as an honorary citizen of Radovljica.

Although justifiably embittered he remained professionally active and was working on his last project, a monumental building in Rome, eight hours a day in his father's house in Radovljica.

Lado Kham, architect with the Ljubljana Municipal Public Works Department, *Sokol* and a good friend, was likewise sentenced to several years of imprisonment. Among other things the public prosecutor accused him that after all the brainwashing to which he was subjected he still could not find a good word for the people's liberation movement. That accusation reveals the perverted mentality of the communist players at that trial and in general. It is incomprehensible to any reasonable man how could anyone who was stalked by communist assassins, whom the communists imprisoned, and treated him and his family with unheard of cruelty find a good word for a movement which represents denial of all human rights, repression of liberty and violence over the mind as well as the body. Kham was once a healthy athlete, tourist, mountaineer, and swimmer but the mistreatment he endured in communist prisons left him ill with asthma and with impaired mobility. In his last letter to me in April of 1978 he wrote that he was determined to make every day of those few he felt he had left count. He wrote about his family and how his daughters succeeded in overcoming the obstacles to their professional aspirations raised by the communists and eventually achieved professional success and recognition. He wrote of the happiness he enjoyed in the circle of his family. He wrote about the Slovenians, their ancient history and how as a nation they are now rushing down the path of destruction. It was clear that he no longer cared if the censors read his letters. The prisons did not break him. He remained a free man.

Lado Kham

[213] Note by editor: Strangely enough, the University in Ljubljana never conferred a doctorate on Milan Kučan, the last publicly known communist party boss and former President of the Republic of Slovenia. The privilege of conferring a doctorate on that communist was reserved to the sycophants at the Cleveland State University, apparently on recommendation of Senator Voinovich.

Promoters of the Revolution

Now and then my friends or acquaintances from Yugoslavia travel abroad. Although they can spend only a short time in the free world they send me uncensored news from Yugoslavia. Of course I am glad to receive news from the old country as I am interested in what is going on there. The communist censorship is still strong; not so long ago I received a letter from Yugoslavia on which the censor brazenly wrote in red ink:" We shall see about that!" With incredible arrogance and cynicism they thumb their noses at the basic human rights and fundamental freedoms.

Following are brief notes on how various persons who helped the communists to attain absolute power fared under the communist system they helped impose on the Yugoslav nation.

Jože Rus — Although Milovan Djilas,[214] at one time Vice President of Yugoslavia, considered Jože Rus to be a politicaster without talent and prospects, Rus nevertheless found a sinecure for which he was well suited as it required no analysis of legal matters nor even an interpretation of the statutes. He became the vice president of the federal senate, a position in which he did not have to suffer deprivations such as were the lot of the masses that could only hope for improvement if the increasingly more unrealistic five-year-plans ever came to fruition. How Rus lived after the war was described to me by *Sokol* Silvo Bernobić of Sao Gaetano Del Sul in Brazil as follows:

> *"I used to know well his grace super comrade Jože Andrej Rus, the vice president of the Senate chamber in Belgrade. I met him one Sunday morning in Belgrade at the State Institute for Physical Culture. There I saw an elderly man who was still exercising at his age. I addressed him as follows: "Comrade if you are still exercising at your age you must have been a Sokol and I can be on the first name basis with you." He was quite agreeable. We talked, he admired my expert gymnastics on the equipment and invited me to lunch. Meanwhile his chauffeur was freezing in the limousine waiting outside while the super comrade Rus exercised. When he saw my appetite he understood that I had nothing to eat, being an engineer who was paid lower wages than a street sweeper. After that I frequently simply "invited" myself to his house to mitigate my hunger. Whenever I came on my own he would open a tin of sardines and poured a glass of wine with which he was well supplied. The super comrade Rus had a well stocked wine cellar. He lived not far from me in a confiscated villa at Dedinje, one of the best districts in Belgrade, at Malgajska Street 22. I found out that there were no guards at his house. Rus talked incessantly and there was no end of his prattle which was full of platitudes, clichés and incomprehensible "philosophical" terms, exactly as you described his speech in the Sokol Herald (Sokolski Vestnik). I listened to him patiently but with all my wit I could not comprehend whether the sound with which he was shaking the air had a head or a tail. I looked at the ceiling or at the floor and never contradicted him because I did not want*

[214] Milovan Djilas, *Wartime*, Harcourt, Brace, Jovanovich, New York, 1977,,p335.

to lose the food he gave me. I did not lack the education to understand him and from his meanderings I soon figured out that the man was not quite ruight in his head or as the Serbs use to say: whacked with a wet sock. After almost thirty years I read in your article how people began leaving the meetings when Rus was recognized to speak and spouted a logorrhea of incomprehensible words. It is simply uncanny how our observations agree. He was flailing the empty straw as we say in Slovenia or as the Portuguese would say: pounding water in a mortar. I sized him up as a born sycophant, feeding at the public through. He gave me his telephone number. My landlords who by government decree had to sublet a room to me because I was needed at the so-called Automobile command — a glorified name for the auto repair shop — tried to help me find his name in the telephone directory but he was not listed. When I asked the super comrade why he was not in the telephone directory he explained that diplomatic listings, such as his, were not published. I did not notice any animus toward Sokol in him, though. At that time I did not know that Rus was one of the gravediggers of the Sokol organization, as you describe him in your book. I treated him as a Sokol and he did not object. It seemed to me that he was still a Sokol in his heart but that he was dissembling because of his job. I think that in his heart he had never forgotten the Sokols. Ivo Pustišek was getting along with him very well. The latter was leading the gymnasts to the XI All-Sokol rally in Prague in 1948 (the last held in Czechoslovakia). It was truly a sad spectacle to see brother Ivko Pustišek wear for the occasion a Sokol hat with the rim turned down just as the Croatian soldiers who were returning from the front were told to do to be distinguished from the Serbs.

Apotheosis of Jože Rus and Franjo Lubej

In 1983 the Presidency of the Socialist Federative Republic Yugoslavia decorated Franjo Lubej and Josip Rus for bravery, the former on his eighty fifth birthday and the latter on his ninetieth. From the daily press it is not apparent for which specific deeds of bravery these two were in the twilight of their life elevated among the brave and even among the heroes. From Polič's writings we know that each of them had brought some of his cohorts in the fold of communism thus forming a new cell for the spread of communism among the *Sokols* under the tutelage of Edward Kardelj. At one time they tried to take the *Sokol* organization to court for having hinted at that. Today they no longer need to sue for defamation. Both were co-founders of the so-called *Liberation Front.* At the inception of the latter, Rus was instrumental in promulgating that infamous decree according to which anyone who failed to place himself under the direction of the so-called *Liberation Front* was to be summarily sentenced to death. Later he distinguished himself as a judge advocate with the partisans who meted out only two kinds of sentences: death and forced labor for life. At the session of *AVNOJ*[215] at Jajce it was his privilege to have the honor of proposing that Josip Broz — Tito be elevated to the rank of the Marshal of Yugoslavia, an unknown title until then. That reportedly enraged Marshal

[215] *AVNOJ — Antifasističko veće narodnog oslobodjenja Jugoslavije — Anti Fascist Council of People's Liberation of Yugoslavia* — Tito's wartime rubber stamp parliament

Stalin so that he had his title changed into *Generalissimo,* presumably meaning something like a super general, one notch up from the *Marshal* in any case. In his later years Rus reportedly felt overlooked, disappointed and embittered.

Franjo Lubej shared in the heroism of his mentor Rus as they both remain inseparable as the Siamese twins. One of Lubej's specialties used to be fulminating against the *Yugoslav Nationalist Party* (*Jugoslovenska nacionalistična stranka — JNS*) to which he belonged as a teacher in Ljutomer where he said he was instrumental in causing Anton Korošec to lose at the elections in his own birthplace. Lubej who paraded barechested singing the *Sokol* anthem at the *Sokol* rallies now roams over Rožnik singing something appropriate to the banning of the *Sokol.*

Lubej was for a time director of the spa at Dobrna where he managed to achieve a loss of 350 million Dinars and was eventually removed because of incompetence.

The above attests to the duo's merits for the communist party. The president of the Republic Conference of the Socialist Alliance of the Working People (*Republiška konferenca Socialistične zveze delovnega ljudstva*) Franc Šetinc abbreviated his formal speech to a simple affirmation of heroism of the communist duo Lubej-Rus and added that Rus always rides a bicycle while Lubej takes walks on Rožnik and sings. He then handed to each of them the Order of Bravery and hurriedly ended the tedious ceremony.

People's Heroes in General

In this connection it is appropriate to examine what the word *hero* means in contemporary Yugoslavia. It is in vain to seek the interpretation of the word in the Slovenian dictionary[216] although the term has been adopted as a word in literary use. In the introduction to the dictionary one finds: "If a word is not included here, it is not necessarily unacceptable or bad." *Hero* must then be a good word; the only information about it in the dictionary is that it is of male gender. The new dictionary has obviously considerably widened the scope of literary words. So much foreign clutter was added that the dictionary makes for a quite good reading, particularly if one is conversant in French. To the new Slovenian words now belongs "*partija*" even though we have a *congenial* expression for it. The *hero,* however, is not such a simple concept as the *party.*

Bear this in mind if you happen to come across some of the Ljubljana newsprint. This year a series of articles brought a good description of "*Life and work of the People's heroes.*" If you read that you will be edified. A typical representative of a *hero* is **Edo Brajnik**, lieutenant colonel of the Yugoslav People's Army, born August 16, 1922 in Kamnik, secondary school student in Ljubljana at the outbreak of World WarII. He soon became commander of the *Security Intelligence Service* (*Varnostna*

[216] *Slovenski Pravopis*, Slovenska Akademija Znanosti in Umetnosti, DZS, Ljubljana, 1962.

obveščevalna služba — VOS) in which capacity he showed an exceptional ability. He personally *liquidated,*[217] i.e., murdered, Fanouš Emmer, Ivan Polak, and Avgust Praprotnik, to mention only the better known names from the list of his victims. For all this he now has the title of People's *hero.*

This, then, would clarify the meaning of that new Slovenian word. But we already have an ancient term for it: it is *"executioner"* and we really should not use the erroneous form imported from the Greek, Latin, and French or Russian language. On its long journey the word has been corrupted, until at the last but one station only its form suffered, but at the last the meaning, too, was changed.

Before the war Ljubljana used two official executioners. Messrs. Seifert and Hart were called upon only about every seven years to perform their services. Brajnik, however, had the opportunity to get plenty of experience in a very short time and on top of that he even got a purloined title. Seifert and Hart no longer dressed in red according to the medieval custom, they went about their business dressed in black with a top hat on their head. When the executed subject died in front of the officially appointed witnesses Brajnik's predecessor took off his gloves and announced that justice had been done.

Brajnik thoroughly revolutionized all that. He used go to work on foot, riding a bicycle or even in an automobile. The most convenient way for him was to liquidate the victim from behind in accordance with the communist practice; a shot in the nape was the standard procedure in Moscow's *Lubyanka* prison. A considerable innovation was also that he would announce the judgment *ex post facto*, after the execution, by proclaiming the liquidated person to be "an agent of the Gestapo" or a "traitor to the people." That could not be appealed. If somebody should inadvertently express doubt as to the guilt of the deceased, Brajnik would be readily behind him — on a bicycle, perhaps.

Today, in the free world, we can say aloud that the Slovenian word for an *executioner* is not *hero.* Likewise we can maintain that judgments declared *post mortem* reflect the criminal depravity of the society that uses such executioners.

Milan Lemež, old communist from before World War I, at one time the chairman of the communist party in Slovenia — but probably not approved by the *Comintern* — reportedly became disillusioned because of the repressive methods of the Yugoslav communists and became a great opponent of the current communist regime. He was totally excluded from any public political activity. Only because of his advanced age he was not imprisoned. When he died in 1973, his death rated nothing more than a plain notice in the Ljubljana *"Delo"* — no high flown farewells or eulogies from his former party comrades.

---

217 Prešeren, the greatest Slovenian poet, observed: "Thieves, thugs, and robbers speak a language unknown to others."

Dragotin Gustinčič

Dragotin Gustinčič. — I already reported how Gustinčič fared at the hands of the understanding comrades in *OZNA*'s prisons. He had no friends and upon his release everyone avoided him as he was limping forlornly around Ljubljana. His second wife died and his sons are apparently still in Russia. Curiously enough his first wife, Anica Lokar, advanced claims that it was he who wrote *The Development of Slovenian National Question* (*Razvoj slovenskega narodnostnega vprašanja*) and not Edward Kardelj alias *Sperans* to whom the work is attributed.[218] Knowing him from as far back as my Vienna years I do not think that Gustinčič had the ability for producing such a work or any other work for that matter. But it is certainly interesting to find out that in addition to private property intellectual property also is subject to appropriation by the communists.

Marjan Tepina, the young communist whom Mayor Hribar foisted upon the *Committee for the Memorial of King Alexander* had, as a less than mediocre architect, a career only after he had joined the partisans where he became a political commissar. Until 1967 he was some kind of a Grand Mayor of Ljubljana and following that a Yugoslav consul in Trieste. The United States government fawningly invited him on a tour of American engineering achievements.

Josip Jeras, a Yugoslav volunteer during World War I and a member of the first communist parliament became entrenched in several lucrative positions where he lasted until 1967 when he died of a heart attack. His monthly income amounted to half a million Dinars while an average Yugoslav subject earned 50,000 to 200,000 dinars at best. For that price Jeras was hanging onto his various sinecures to the bitter end. None of his former friends would have anything to do with him; they did not even look at him.

Stane Vidmar, brother of the mighty Josip Vidmar, silent partner and front of the communist infiltrators at Tabor, who presented his factory to the communists, died disappointed and embittered in 1957.

Milan Vidmar, brother of Josip Vidmar, who at the arrival of the Italians went to Rome and paid an obsequious visit to Mussolini in the name of the Slovenian Academy of Arts and Sciences, had no problems with the new regime; he was only pushed aside a little. But he did have his new Institute for Electrical Economy and Industry to play with. He was kind enough to offer a job to my son before Vlado "flew the coop."

---

218 Note by editor: See for instance: A. Puhar, Delo-Znanost, Ljubljana, October 3, 2001; Gustinčič's life, personality, contributions to the advancement of communism in Yugoslavia, and imprisonment are subject of a dissertation by Dejan Kac, *Gustinčič – Kominterna – Komunizem – Nacionalno vprašanje*, University of Maribor, 2007

Ivan Jelačin, merchant and industrialist, vigorous opponent of taxes prior to World War II, a member of the *Association of Friends of the Soviet Union, a* generous supporter of the so-called *Liberation Front,* robbed by the Soviet General Bogomolov, died in 1952 of paralysis, completely abandoned, totally disappointed and, of course, liberated of all his property.

Vinko Vrhunec — one time director of the Trbovlje Coal Mining Company, the director of the Yugoslav Cement Cartel, did not attain the cabinet post which he craved. The communists confiscated his precious villa on Mirje and gave it to Edward Kardelj. Now he is in a banking enterprise in Koper and dreams about the unattainable cabinet post for which he would abandon any and all ideals if he ever had any. He abandoned them too soon, before the coveted position slipped away from him. His son, a partisan, declared in the woods that he was ashamed of his illustrious father.

Miha Kambič, minister in the first communist government who succeeded Gustinčič in our apartment after Gustinčič was arrested. Prior to his moving in our apartment all our furniture was removed by OZNA. Edita Dimnik who happened to be passing by inquired of the movers where they were taking the furniture which under the order of the communist court was to be returned to my wife only to be told to move on or be arrested. At that time the portrait of my wife painted by Tine Gorjup was tossed out. Stanko Dimnik's son in law, architect Marjan Gašperšič wrote me that the janitor picked the painting up and took it to the Dimnik's home a few houses away where it was stored for better days.[219]

Kambič became sorely disillusioned with communism. People were afraid to encounter him on the street because he was so loudly heaping abuse on the regime that it was dangerous for those who were listening to him.

[219] Note by editor: In 2006 the painting was retrieved from the Gašperšič's cellar, restored by artist Marij Vrenko and shipped to the author's family in California.

Rebuttal of Communist Indictment

Communists Stage Trial Against Sokols

The late Josip Rustja,[220] one of my friends in the Sokol Legion, who was among the victims of the communist trial against Sokols, sent me a copy of the indictment against Sokols filed on January 15, 1946 by Anton Zupan, the public prosecutor for the Ljubljana region. Rustja survived the trial and imprisonment and was able to return to Gorizia after the Italian government intervened on his behalf. He gave a photocopy of the indictment to Avgust Sfiligoj with a request that he transmit it to me and that I, as a person familiar with the contemporary situation, publish the truth about it and refute the communist accusations.

The indictment put together by the public prosecutor Zupan is replete with lies and false unsupported statements to the effect that the Sokol Legion was, with the assistance of the accused, resorting to dishonorable means in its fight against the so-called national liberation movement, that is, against the communists. It is appropriate that I make the following observations.

As already narrated above, shortly after I returned from the front on April 18, 1941, Engelbert Gangl, deputy presiding elder of the Sokol Federation of Kingdom of Yugoslavia came to see me and urged me to organize the Sokol for resistance against the invaders in accordance with the resolutions of the Sokol Federation. I convened a meeting of the representatives of all five Slovenian Sokol Districts for consultation. We unanimously decided to establish the Sokol War Council which would with the Sokol Legion conduct resistance against the occupying powers. The communists, on the other hand, established on April 27, 1941 their Anti-Imperialist Front which, in view of the fact that the Soviet Union was at that time an ally of Hitler, was directed against France and England and certainly not against the Nazis whose socialist ideology was akin to that of the communists; the two systems varied only by slight variations in their phraseology. On June 6, 1941 the Serbian communists published a declaration exhorting their adherents to assist the German invaders as much as possible because the British war effort was directed against Germany, an ally of the Soviet Union, fatherland of the communists. It is well known that the so-called Liberation Front was established only after June 21, 1941 when Germany attacked the Soviet Union.[221] When the Germans arrived in

[220] When this article was written, Etbin Bežek and Lado Kham, too, were also deceased.

[221] The communist even sung about that fact in their marching song: "Readily we grabbed our guns/and took it to the hills/ when the Soviet *heroes*/ called us to join the battle."

Slovenia the communists received them as their allies with joyful cheering and manifestations, particularly in the coal mining town of Trbovlje known as a den of communists. To those who naively advanced the opinion that the Soviet Union should come to the rescue of Yugoslavia they responded that until a short time ago Yugoslavia did not even recognize the Soviet Union — or rather its repressive regime — and that consequently there was no reason for the Soviets to help Yugoslavia. Stalin made this clear to the Czechoslovak president Eduard Beneš when the latter during a visit in Moscow tried to appeal to Stalin in the name of Slavic solidarity. To Beneš' overtures Stalin retorted: "What Slavs? We are the USSR."

The Sokol War Council was established two months before the so-called Liberation Front with the sole purpose to organize resistance against the invaders at a time when that communists still considered Hitler to be their ally.[222] The Sokol War Council never did any favors to the occupying authorities nor did it ever engage in spying of any kind. The organization of Sokols was dedicated solely to the work inspired by the Sokol idea. The communists, on the other hand, sent to the commanding officer of the Sokol Legion, Edi Antosiewicz, a spy who acted as his courier and carried messages to the various members of the Legion. That man then, in 1944, handed over to the Germans Antosiewicz's circular in which Antosiewicz advised members of the Legion to be prepared as the time was coming to expel the invader from our country. The Germans sent Antosiewicz to Dachau. In 1945, when we were in the Gestapo prison, we found out that the German unmasked and shot the spy.

The indictment put together by communist Anton Zupan reveals his mentality and the way in which he thinks. It spews his hatred of any opponents of communism. In keeping with the precepts of Marxism-Leninism, according to which lies are a cardinal virtue of communism, he heaps lies and calumny upon the accused, lies not supported by any material evidence but constituted solely of communist assertions. The trial staged by Zupan is in no way different from the infamous kangaroo court in Kočevje at the time of Italian surrender in 1943 when the communists with the assistance of Italian artillery and personnel succeeded to capture a battalion of the Yugoslav Army in the Homeland at Grčarice and sentenced all the prisoners to death. Present at the Kočevje trial were representatives of the so-called "Allied"[223] military mission at the headquarters of the so-called People's Liberation Army who declared that the trial was legally impeccable and later broadcast their observation over Radio London.[224] After the war the communists killed over 350,000 noncommunists, who were Yugoslavs patriots.

[222] That the communists truly wanted to be allies of the Nazis and did not conclude the pact only for tactical reasons was clear from the lame question put to the German ambassador Schulenburg by Vjacheslav Molotov when the ambassador informed him of German invasion. Molotov asked him plaintively: "Do you think we deserved that?"

[223] Those were allies of Stalin and not of Yugoslavia, as the events had shown.

[224] Franček Saje: *Belogardizem*, Slovenski knjižni zavod, Ljubljana, 1951, p. 790

From the text quoted in the following it is clear what kind of mentality is still prevailing in the Yugoslav judiciary today.[225]

Public prosecutor for the district of Ljubljana
Criminal Division
Case No. K/1654/45-e
Criminal case: Dr. Ernest Kalan and co-defendants
Under Article 3/Section 3 and 4/

Indictment

I, Public prosecutor for the city of Ljubljana, am filing at the People's court in Ljubljana against;

1. Dr. Ernest Kalan, son of Franc and Ernestina née Franzi, born 25 November 1883 in Kranj, married, residing at: Aškerčeva Street 5, Ljubljana, Slovenian national, Roman Catholic, Citizen of Yugoslavia, not imprisoned;
2. Ing. Bogdan Ferlinc, son Franc and Zofija, née Skaza, born 23 January 1982 in Šmarje pri Jelšah, residing at: Škrabčeva Street 6, Ljubljana, married, agricultural engineer, Slovenian national, Roman Catholic, citizen of Yugoslavia, currently in prison;
3. Dr. Jože Pučnik, son of Anton and Terezija, née Žoh, born 19 May 1886 at Dobrova pri Slovenskih Konjicah, married, residence: Ljubljana, Aškerčeva 15, citizen of Yugoslavia, not imprisoned;
4. Rajko Lederhas, son of Ludvik and Ivana née Ferjan, born 27 October 1893 in Ljubljana, residing at Sv. Petra c. 14, Ljubljana, single, Roman Catholic, Slovenian national, citizen of Yugoslavia, judge of the district court, currently in prison.
5. Etbin Bežek, son of Rihard and Ivanka née Richter, born 25.September 1895 in Kamnik, residing at Groharjeva 7, Ljubljana, married, inspector at Pivovarna Union, Slovenian national, currently in prison;
6. Stane Pirc, son of Ferdinand and Anica née Jana, born 30 April 1905 in Ljubljana, residing at Na peči 19, Ljubljana, married, civil engineering technician, Slovenian national, Roman Catholic, citizen of Yugoslavia, currently in prison;
7. Pavel Borštnik, son of Pavel and Jozefina, born 12 January, 1985 in Ljubljana, residing at Splitska Street 7, Ljubljana, married, official of Slovenec, Slovenian national, Roman Catholic, citizen of Yugoslavia, currently in prison;
8. Ing. Lado Kham, son of Franc and Franciska née Teleminus, born 6 May 1901 in Ljubljana, civil engineer, married, residing in Ljubljana, Puharjeva 5/III, citizen of Yugoslavia, currently in prison;

[225] Note by Editor: The late author who died on November 29, 1988, did not live to see the disintegration of Yugoslavia and accordingly cannot comment on the current state of affairs in Yugoslavia or in the Republic of Slovenia. In view of contemporary judicial practice in Slovenia, however, we can conclude that fifty years of communism resulted in such decline of legal standards and judicial ethics that it will undoubtedly take several generations before Slovenia may again see the rule of law.

9. Karl Benič , son of Mavricij and Viktorija née Doljak, born 11 october 1918 in Bruck an der Leita, residing in Ljubljana Murnova 6, akademician, single, Slovenian national, citizen of Yugoslavia, currently in prison;
10. Zdravko Deleja, son of Zdravko and Dragotine née Sarc, born 14 February 1892 in Rečici pri Gornjem Gradu, commercial employee, residing in Ljubljana Vodnikova 31, married, Slovenian national, citizen of Yugoslavia, without property, currently in prison;
11. Oton Frölich, son of Filip and Antonija Petrovič, born 15 March 1897 in Ribnica na Pohorju, commercial assistant, residing in Ljubljana Dolenjska c. 48; Slovenian national, citizen of Yugoslavia, married, without property, currently in prison'
12. Rudolf Zorc, son of Rudolf and Pavla née Lotrič, born 19 May 1914 in Ljubljana, single, attorney, residing in Ljubljana Rožna Street 9, Slovenian national, citizen of Yugoslavia, without property, currently in prison;
13. Božo Borštnik, son of Pavel and Barbara née Malič, born 27 December 1889 at Hinje pri Žužemberku, journalist, residing in Ljubljana, Gestrinova 5, married, Slovenian national, citizen of Yugoslavia, without property, currently in prison;
14. Filip Kosec, son of Franc and Marija née Srčnik, born 29 April 1895 in Ljubljana, residing at Bleiweisova Road.13, married, warehouse superintendent of the national theatre, Slovenian national, citizen of Yugoslavia, currently in prison;
15. ing. Josip Rustja son of Pavel and Marija née Bajc, born 14 April 1894 at Škrilje pri Ajdovščini, residing in Gorizia, Via Pollgini 6, married, merchant, Slovenian national, citizen of Italy, currently in prison;
16. Oktavij Cek, son of Ivan and Josipina née. Koleša, born 16 February 1898 in Trieste, residing at Vita Mazzini 11 Trieste, single, journalist, Slovenian national, Roman Catholic, citizen of Italy, currently in prison;

An Indictment

I. That they were at the time of enemy occupation ideological instigators of Mihajlović's treason in Slovenia and organizer of treasonable Sokol or National Legion which, joined with the Slovenian Legion in the so-called regular Yugoslav army under the command of the invader's serf and war criminal Draža Mihajlović collaborated with the occupying powers in the fighting against the liberation movement so that they with false propaganda destroyed the unity of the Slovenian people and created treason to and desertion from the movement of national liberation.

   Sent their members to the white guard, chetnik and later national guard armies which represented the cadres of the so-called Royal Army in the Homeland.

   Directed their members to the secret intelligence service (Tajna obveščevalna služba — TOS) of the occupying powers which was incorporated with the enemy police and the Gestapo.

   Organized within each legion a widely spread espionage network joined within the framework of the treasonable regular Yugoslav army into the so-called government intelligence service (Državna Obveščevalna Služba — DOS), whose purpose was to betray the adherents of the liberation movement, reported to the occupation authorities' police and the Gestapo sympathizers of the liberation front, ordered massacres of activists of the liberation front which were carried out by special detachment of the Slovenian National Guard under the name of "Black Hand."

During wartime and enemy occupation acted as instigators and organizers of the above mentioned terrorist organizations which perpetrated assassinations, death sentences and their executions, imprisonments, torture, forced deportations of the Yugoslav population to concentration camps, internment or forced labor and were ideological instigators and organizers of military formations consisting of Yugoslav citizens with the purpose of supporting the enemy and fight against the homeland with enemy supplied weapons.

II. Dr. Ernest Kalan, Pavel Borštnik, Karel Benič, Zdravko Deleja and Oto Frölich were, in addition to the above, either within the framework of the above mentioned espionage organizations or directly, reporting to the occupation authorities or to domestic traitors the adherents of the liberation front which resulted in the already mentioned arrests, internments, torture and murders perpetrated on the Slovenian population during the time of enemy occupation.

II. Dr. Ernest Kalan, ing. Bogdan Ferlinc, Dr. Jože Pučnik, Edvin Bežek, Stane Pirc and Pavel Borštnik, as members of the so-called National Committee of [2]9 October 1944 collaborated at the proclamation of the one-day national government of 3 May 1945 at Tabor the purpose of which was to start a civil war and disrupt the already established government structure of the democratic federative Yugoslavia, endanger its international security and destroy the basic democratic, political, national, and economic achievements of the war of liberation.

IV. Rudolf Zorc collaborated as sergeant of the Slovenian National Guard in the investigations of the National Guard's military tribunals against those members of the National Guard who clandestinely worked for the liberation front, among others in the matter of Černe whom the above mentioned tribunal sentenced to death and subsequently executed, he had accordingly assisted in carrying out acts of repression and murder.

V. Dr. Jože Pučnik had furthermore politically and with propaganda collaborated with the occupation authorities by attending the weekly meetings of "Jutro" thereby assisting the notorious propaganda against the national liberation movement which resulted in acts of repression against the Yugoslav population.

With this all the defendants under I committed a crime under Article 3, Section 3/ Zknd.

Dr. Ernest Kalan, Pavel Borštnik, Karel Benič, Zdravko Deleja and Oto Frölich under II, a crime under Article 3, Section 3/Zknd.

Dr. Ernest Kalan, ing. Bogdan Ferlinc, Dr. Jože Pučnik, Edvin Bežek, Stane Pirc and Pavel Borštnik under III, a crime under Article 3, Section 2/Zknd

Rudolf Zorc under IV, a crime under Article 3, Section 3/Zknd

Dr. Jože Pučnik under V, in addition a crime under Article 3, Section 6/Zknd and should be punished according to Article 4/Zknd

Explanation

All defendants are representatives of the old corrupt political parties of the former Yugoslavia. Treason against the nation which they were carrying out during the time of enemy occupation was a continuation of the treasonous anti-people policies carried out by the political entities already several years prior o the breakup of Yugoslavia.

The politicians of the pre-April[1941] Yugoslavia were pursuing only their own class interests. Therefore they were building a type of government which would be most amenable to them and the pursuit of the advantages of their class. Thus they created an absolutist and centralist Yugoslavia which in its anti-democratic character assumed a definitively pro-fascist character. To ensure the security of such kind of a state the former summit politicians of the old Yugoslavia necessarily looked for support from the Nazi Germany and fascist Italy, that is, to the historical enemies of Slavic nations.

By this treason the ruling circles have foregone any right to represent the national interests of the peoples of Yugoslavia. In addition they were responsible for facilitating the invasion of our country by the Nazi and fascist hordes.[226] The army of former Yugoslavia had because of the treachery of the fifth column and reactionary ruling clique miserably and dishonorably succumbed under the onslaught of the invaders.[227]

In the situation that ensued and which was for our people the most dreadful experience of its entire history[228] new forces arose which alone were capable of protecting the national interests of the peoples of Yugoslavia.[229]. These were new forces of the people that represented a huge majority among the Yugoslav peoples and which alone were capable of achieving the unity of everything that was truly patriotic in Yugoslavia.

The working people have through their progressive majority forces became obligated to champion the goal of national unity in the struggle against the occupying powers.

Throughout the entire Yugoslavia a national liberation movement arouse spontaneously, an integral part of this movement was the Slovenian liberation front.

The popular masses have in great majority joined this movement because they understood that only that movement with its sound democratic foundation would be able to unite all of the Yugoslav peoples on a firm foundation and thus successfully achieve national freedom.

---

[226] In April of 1941 the communists urged the Yugoslavs not to respond to the mobilization and were greeting the Nazi hordes as allies of the Soviet Union and comrade Stalin.

[227] Under the onslaught of Nazi aggression succumbed the armies of Norway, Belgium, The Netherlands, Denmark, Poland and France while the British army barely managed to flee from Dunkuerque back to its island. In each and every instance the communists were urging governments of those countries not to resist Hitler.

[228] Today it is known that the most dreadful experience of any nation anywhere is communism.

[229] Communists, who constitute a criminal international conspiracy and a terrorist organization, never represented the national interest of any nation. This was clearly demonstrated in the case of Yugoslav territorial demands for Trieste and Slovenian Carinthia. In both cases the Soviet Union abandoned Yugoslavia. In case of Trieste, it was satisfied that it was able to support the powerful communist party of Italy which could get in trouble if the Soviets supported the Yugoslav demands, as for Carinthia, it was traded for the right to haul out of Austria as much of industrial equipment as could be moved so that the Austrian State Treaty reads more like a bill of lading than a political document. The Soviets consoled the Yugoslavs by telling them that one day the entire world would be communist and then no borders would be necessary.

The reactionary and anti-people clique did not want to join this movement because of its class interests, on the contrary, it was wrecking the national unity and floundered into treason. That treason was exacerbated to such extent that it reached for the weapons supplied by the occupying power and stabbed in the back its own people during its heroic struggle for liberty. The reactionary politicians had established contact with the occupying power already during the first days of occupation. At that time already they urged our people to assume a loyal stance toward the occupying authorities and placed the entire government apparat of the old Yugoslavia in the service of the occupying power.[230]

The liaison between the occupying authorities and the clique of traitors was particularly strong on the territory occupied by Italy. The latter found itself in unpleasant military and political situations because of the decidedly contrary attitude of our people. For this reason the Italians were in dire need of allies which they found in the domestic traitors. These tried during the first days of the Italian occupation to create some kind of a puppet state in the service of Germany and Italy particularly because they expected the attack on the Soviet Union. Therefore they were early in establishing the Slovenian, Sokol and National League which were expected to constitute the army of that puppet state similarly as was done in Croatia and Slovakia. Feverishly they sought an opportunity to bring that army of theirs to life. They succeeded with the assistance of the Italian occupying authorities. In concordance they established Milizia volontaria ant icommunistica which our people dubbed White Guard[231] and which became a horrifying symbol of inhuman suffering and persecution of our people during the four years of occupation.

While the Slovenian Legion is a product of the clerical camp the Sokol Legion and National Legion represent the liberal camp of Slovenian reaction. The liberals, too, saw in the people's liberation movement an end of its expressly anti-people politics therefore it directed all its strength to combating the people's liberation movement. Mutual anti-people interests have thus brought about a rapprochement between the formerly irreconcilable antagonists, the so-called clerical and liberal camp. In pursuing mutual interests their former antagonism stemming from the power struggle was forgotten.

The moral and propagandist background of the Sokol and National Legion thus represents the liberal camp of Slovenian reaction represented by the former Yugoslav Nationalist Party (Jugoslovenska Nacionalna Stranka — JNS). All the attributes the representatives of that party had in the former Yugoslavia they retained during the occupation. Like the corrupt politicians of the clerical camp the so-called liberal politicians quite soon entered in combat against the liberation movement, covertly at first and openly after some time and placed themselves at the disposal of the occupying authorities with all the organization they possessed.

Characteristic representatives of the liberal reaction were the Sokol and National League. In these two groups were organized members of all nuances of the liberal reaction, both the old JNS headed by Dr. Pirkmajer, Dr. Zajc, Žitnik, and Ing. Dimnik, the Young Yugoslavia or OJNS (Youth of the Yugoslav Nationalist Party — Omladina JNS) headed by Dr. Drnovšek and others. Other members of this groups were former radi-

230 It is a well known provision of international law that an occupying power, prior to the conclusion of a peace treaty, is required to maintain the existing civil service of the occupied country.

231 *White Guard* is a Bolshevik term for their opponents, as is well known.

cals, the representative of which is Dr. Köstl, members of the ORJUNA (Organizacija Jugoslovenskih Nacionalista — Organization of Yugoslav Nationalists), and finally the Stara Pravda[232] and part of the Independent Democratic Party. The entire liberal reaction had already in 1941 united in the so-called progressive working alliance in which participated all of the above enumerated parties. That progressive working alliance published its paper "Zarja svobode"(Dawn of Liberty) which became the most heinous propaganda for the treasonous movement of Draža Mihaljović.

The purpose of that literature was transparent. It wanted to create desertions in the national liberation movement and thus spread treason to the widest extent possible."Zarja svobode" therefore began to smear at its very inception the essence of the patriotic war of the Yugoslav people while at the same time promoting the war criminal and traitor of his people Draža Mihajlović[233] That Mihajlović, who was already at that time known in Slovenia as a traitor the anti-people reaction selected as its leader. They knew quite well that Mihajlović was agent of the fugitive government in London, hence an instrument of anti-people autocrats who charged him with the task of suppressing the freedom loving movement with all means possible — including the assistance of the Nazi and fascist conquerors. By order of the treasonable émigré government in London Mihajlović already in 1941 passed a death sentence upon former lieutenant colonel, now lieutenant general Jaka Avšič because the latter solemnly stated that the detachment of partisans and organizations of people's security[234] as mass military formations are the only Yugoslav people's army on the territory of Slovenia and that the Executive Council of the Liberation Front is the only legitimate political leadership of the Slovenian people and people's liberation movement. By promoting the criminal Mihajlović the liberal progressive working alliance became the fundamental ideological instigator of Mihajlović's treason in Slovenia and with this through all its representatives morally responsible for all crimes resulting from that treason over the Slovenian people in time of its heroic struggle for liberty.

Progressive working alliance had trough its paper "Zarja svobode" took care to induce the as yet not informed people into the so-called regular Yugoslav army consisting of the Slovenian, Sokol, and National Leagues thereby becoming the ideological instigator and executor of terrorist organizations of the Sokol and National Leagues.

Liberal reactionary politicians under the guise of the abovementioned progressive working alliance united with their former political opponents, that is, with the clerical reactionary camp. They established Slovenska zaveza (Slovenian confederation) which

---

232 *Stara Pravda — Ancient Rights* was the slogan under which once relatively free Slovenian peasants rebelled against the erosion of their recorded rights because of arbitrary acts of the feudal landlords. The political group adopted this as its name possibly hinting at advocacy of some social reforms.

233 Characteristic for the communist judiciary is that in January 1946, when the trial against *Sokols* began, Mihajlović had not yet been captured much less convicted of any treason in any communist court. Mihajlović was, unfortunately, captured in March of 1946.

234 Organization of People's Security *(Organizacija Narodne Zaščite — OZNA)*, during the war communist intelligence service with the principal task of murdering people who opposed the communists. After the war *OZNA* was elevated to the status of political police at the level and reputation of *KGB*, the *Gestapo* and *OVRA*, it was later renamed *Adminstration of State Security — UDV (Uprava Državne Varnosti)*.

became the symbol of united reaction and moral support of Mihajlović's treason in Slovenia.

Analogous with political liaison of ideological instigators, that is leading politicians of former Yugoslavia, the three legions likewise joined into the so-called regular Yugoslav Army in the Homeland. That army, as was shown at the recent trial against Dr. Köstl and his co-defendants, was not created to fight the occupying powers but had quite a different purpose. It was created to suppress the freedom loving movement of the Yugoslav people and secure the anti-people positions of exploiters for the event of Nazi Germany's defeat.

Such movement of course could elicit no response from our popular masses. These were confronted with the fateful decision: live or perish, with which the accursed occupying power was threatening them. They opted to fight for their existence against the hated Nazi conqueror. Through their progressive majority forces, workers, farmers, and working intelligentsia they joined the heroic combat for their national liberation thus starting the finest period of their history.

Because of that spontaneous resistance stemming from the popular masses themselves and having from its very beginning the character of the people, the character of people's power the reactionary clique reached for the weapons supplied by the occupying authorities. Its army under the ringing name of some kind of Royal Army in the Homeland thus found its true form, to wit: the already mentioned MVAC or White Guard. With the assistance of savage fascist hordes that army attempted to suppress everything that was truly patriotic in our people

Detachments of MVAC were first deployed in the big offensive of 1942 when the Italian army with overwhelming force mounted an onslaught on the Slovenian partisan army. The intended annihilation of our people's army did not succeed but the offensive inflicted the most dreadful atrocities upon the Slovenian civilian population. Italian troops were setting our villages on fire, killing our men, women and children and hauled thousands away to concentration camps.[235] The worst part of it was that the occupying forces were assisted in this by the detachments of the White Guards. With their advice and particularly with informing that offal of the Slovenian people that offal assisted in the horrific persecution of Slovenians.

All three legions which within the framework of the White Guard carried out the above mentioned atrocities in Dolenjska did not leave Ljubljana at peace, either. Here, too, all three legions, in accord with each other and in agreement with the occupation authorities established a special MVAC police which was organized under direction of General Gallo the domestic traitor and war criminal lieutenant colonel Peterlin. This ill-famed police amounted to approximately 150 men. These were members of the said legions each of witch contributed fifty men with corresponding officer cadres. That police collaborated at all mass arrests; clad in Italian uniforms they were deciding in the Italian barracks on the fate of the arrested. The well known Christmas raids in 1942 that exacted so many victims were likewise the work of that criminally minded police, which carried out those raids independently with the assistance of the Italian carabinieri.

[235] The Italian concentration camps are never mentioned in connection with the German concentration camps when the latter are periodically discussed at length in Western press.

At that time all three legions organized a widespread spying network for detection and reporting of the adherents of the liberation front. These intelligence services were integrated within the framework of the treasonable Mihajlović army under the name of Government Intelligence Service (Državna obveščevalna služba).

The Slovenian people viewed those criminal acts of traitors with utmost revulsion and they repudiated the criminal activities of the MVAC detachments. For the same reason moral deterioration began within the ranks of the MVAC which no amount of propaganda could arrest. Members of the White Guard began to desert and join the people's army with increasing frequency. In this predicament the reactionary clique thought up of exploiting the uncompromising anti-occupation stance of the Slovenian people by deluding it with reports that an illegal anti-occupation chetnik army exists on the terrain. This was the time of the most intensive activity of all three legions which were competing with each other which one would send more members in those chetnik detachments. The well known political charade of old politicians reappeared again, namely the power struggle. In their myopia they imagined that those chetnik detachments would control the situation in Slovenia and thus in fact represented all authority. Consequently the political groups were interested to make as big a contribution as possible so that they would eventually have a corresponding share of power.

The same people who had organized the White Guard now appeared on the terrain as organizers of a new chetnik movement. The same major Novak who had earlier been organizing the criminal detachments of the MVAC appeared as commander of the Yugoslav Army in a new guise. The same persons who as officers and enlisted men in the ranks of the MVAC units carried out crimes and atrocities.

The same people who earlier organized the white guard appear now on the terrain as organizers of a new chetnik movement. The same major Novak who on orders of Draža Mihajlović organized the criminal detachment of the MVAC is now appearing as commander of the Yugoslav Army in a new guise. The same people who as officers and enlisted men in the units of the MVAC carried out crimes and atrocities appear now as chetniks. Naturally the methods of these new chetniks remained the same: murders and violence perpetrated on partisan fighters as witnessed by the lamentable chronicle of the so-called promotional march in Dolenjska in 1943. Not a single shot against the occupying authorities, combat directed solely against the people's liberation army.

The Sokol and National Legions were the most avid participants in all those criminal undertakings. They knew how to exploit the patriotic feelings of the youth and managed to deceive it that the chetnik detachments were fighting at the side of the people's liberation army. That cruel and loathsome lie was unmasked by the Italian capitulation in 1943 following which those chetnik detachments following the orders of the reactionary politicians in Ljubljana joined the Slovenian National Guard. In this way they became protagonists of the most abject disgrace and evil our people had to sustain.

While the Slovenian National Guard, which took an oath of loyalty to Hitler[236] himself, was being modeled along the lines of the Waffen-SS the leading reactionary clique assiduously continued its behind the scenes policies. The principal organizers and ideological

---

236 That the *Slovenian National Guard* took an oath of loyalty to Hitler personally is simply a communist lie. Hitler's name was not even mentioned in the oath taken by the *National Guard.*

instigators of treason kept themselves in the background. Their policy had the following task. The German invaders were receiving blow after blow from the ever victorious Red Army and was inexorably driven to his eventual and final defeat. It was necessary to plan for the contingency of German defeat when the Slovenian people would demand a well deserved retribution to be visited upon all those who collaborated with the enemy. For such contingency the united reaction wanted to created a so-called unblemished group, a group with clean hands, which could at the right moment step forward and again establish its anti-people government. However that could not be done before it could with the assistance of the occupying power extirpate everything that was progressive, democratic and truly patriotic, in brief, until the liberation front with the partisan detachments still existed.

For this reason a truly violent combat against the partisan detachments was joined. The criminal activity of the National Guard units was receiving most of its support from the camouflaged criminal reaction in Ljubljana.

That reaction escalated ten fold its espionage network of all three legions and at the same time extended it to the country in the form of the so-called information centers.

The occupying authorities established a special political police in which many members of the Sokol and National legion were active. Among them was the notorious criminal Rus, brothers Kobi, brothers Adamič, Dušan Lajovic and others. In addition the Sokol and National legions sent their members to the so-called Secret Intelligence Service (Tajna obveščevalna služba — TOS) which was directly incorporated into the police and in direct liaison with the Gestapo.

Through its representatives in this political police and Secret Intelligence Service the legions were reporting to the occupying authorities adherents of the people's liberation struggle. This, unfortunately successful activity was made possible to them by the intelligence network of DOS (Državna obveščevalna služba — State Intelligence Service) which was a joint effort of all three legions. All these legions were exhorting their members to the greatest effort possible in combating the liberation front which was to consist of discovering and reporting members of the liberation front. Thus all their members were organized in the informers' activities to which can be attributed the fact that arrests were following one after another so that it seemed that the liberation movement in Ljubljana might indeed be suppressed. That delusion encouraged those reactionary and criminal elements to openly show their true inhuman and murderous essence. That was the beginning of the most dreadful terror at which human reason shudders questioning whether human beings are capable of such conduct.

All the reaction at that time emphasized the principle of total war against the democratic goals of the Slovenian people. All that was democratic in the Slovenian people had to be extirpated, roots and all.

A most intense campaign against the liberation movement began in the so-called anti-communist action. The essence of our patriotic war was smeared which led the as yet uninformed popular masses into combat against a fratricidal war.

All that horrible and criminal campaign against the people's liberation movement is responsible for the extent to which criminally depraved persons could be found among our people who in the ranks of Slovenian National Guard comported themselves with utmost cruelty against the partisan detachment. Captured partisans were tortured in the most atrocious manner, gouging out their eyes, cutting off and out their ears and tongues, peeling the skin off their back, nailing them alive to the trees, breaking their bones and murdering them after such horrendous ordeal.

Just as the National Guard detachment were committing inconceivable crimes in the countryside members of all three legions carried out no lesser atrocities in Ljubljana. With their widely spread network they had thousands of Slovenians shipped off to Dachau and other concentration camps throughout Germany. Because of the criminal propaganda many informers were found among Slovenians who were sending their own brothers to death in German concentration camps. In this respect, too, all the blame for the suffering of proud Slovenians is attributable to the criminal reaction. Still all this was not enough. In their fanatical hatred of the forces of liberation they went so far that their savagery exceeded even that of the occupying forces. They began to order massacres of activists of the liberation front under the name of the "Black hand" which were carried out by the detachments of the Slovenian National Guard.

In this way they attempted to kill the freedom loving movement of the Slovenian people which would supposedly enable them to regain their anti-people government with the help of the so-called Regular Yugoslav Arm. Politicasters who are to blame for all the atrocities committed against the Slovenian people wanted to achieve their aspirations by proclaiming their one-day government of May 3, 1945. At a time when Yugoslavia, with the exception of a few regions of Slovenia was already liberated[237] they attempted to tear out of it the not yet liberated part and incite a civil war.

At this attempt collaborated all the representatives of former political parties who were preparing for this act since October 29, 1944 when they established some kind of a National Committee. While they used to justify their treasonable politics by invoking the authority of a legitimate émigré government in London they now came out showing their true nature. Because Yugoslavia was liberated by democratic forces which accordingly gave it a new democratic federative foundation based on people's government they tried to cut out of it at least Slovenia and in this way destroy the desires of the Slovenian people for a unified Yugoslavia which would according to the above mentioned people's essence offer a reward for the unequal struggle against the occupation powers. Representatives of the so-called one-day government called in special telegrams upon the western allies and the traitorous king Peter[238] demanding an armed intervention.

They were unsuccessful in that attempt to incite a civil war. In view of the Tito-Šubašič agreement[239] it was also fairly illusory, its failure was ensured by the genius of the leadership of the Yugoslav People's Army.[240]

---

237 The communists did not arrive to Ljubljana one hour sooner than Germany signed the act of unconditional surrender at Reims. As shown elsewhere in these memoirs the Germans, when faced with the defeat sought out the appropriate representatives of the Yugoslav government and would have done so whether any partisans existed or not.

238 Communist Zupan does not tell whom did King Peter betray and it is difficult to see why should a constitutional monarch owe allegiance to an alien ideology.

239 The British dictated agreement between Josip Broz and representative of the Yugoslav émigré government in London former Croatian governor Šubašič appointed Josip Broz as minister of war and Šubašič as foreign minister. The communists, of course, never intended to respect that

For the contingency of their failure the reactionary politicians were preparing yet another treason They were perpetrating treason in the Slovenian Littoral, that is, on the Slovenian territory beyond the confines of the former Yugoslavia which the liberation front had in its initial program annexed to Yugoslavia for ethnic reasons.[241]

In the Slovenian littoral they were packing in the so-called Royal Army which was an integral part of the occupation forces in the form of Nedić's formations, Djevdjelević's chetinks and Slovenian Security Guards in the Littoral.

The greatest abettors of treachery in the Littoral were the Sokol and National Legions. These two were sending from Ljubljana emissaries who were expected to incite dissatisfaction over the people's liberation movement. By creating such treason and particularly by amassing military formations the Sokol and National Leagues attempted to detach from Slovenia at least the littoral which would thus remain a separate buffer state and a preserve for the reaction.

Activities of united reaction in the Littoral had the gravest consequences. Ten thousands of traitors as cooperating because of the propaganda efforts of said legions from Ljubljana in the dreadful German offensive in the Spring of 1945 directed against the partisans. The intended destruction of the people's army did not succeed. The criminal traitor did, however, succeed in that the peoples of Yugoslavia must despite their magnificent political victory still fight for the territory which they themselves liberated with the heaviest casualties and to which belongs to them by every ethnic and historical right.

Such is a brief outline of the treason of united Slovenian reaction. It is with respect to this that the guilt of the defendants listed below must be judged.

1. Dr. Ernest Kalan
   is a representative of the former JNS (Jugoslovenska nacionalna stranka — Yugoslav Nationalist Party). After 1941 he was even member of the directory of that party. He was thus one of those leading politicians of the liberal-reactionary[242] camp which

... agreement, as the British fully expected they would not, and they arrested Šubašič as soon as it was practicable as their foreign minister always was Edward Kardelj, an agent of the Comintern.

240 What constituted the genius of the partisan's army is still not known because the Yugoslav partisans contributed nothing to the successful military conclusion of World War II in Europe. The Soviet Red Army occupied Belgrade because that was necessary for its conquest of Vienna and East Austria, the Germans did not withdraw from the other regions of Yugoslavia before they surrendered unconditionally to the Allies.

241 The communist army had to leave Trieste and the Littoral as soon as President Truman telephoned Marshal Tito and sent him packing. The communists continued to carry on their reign of terror in the Slovenian littoral under the military government of the so-called Allies and have kidnapped several democrats, dragged them to Yugoslavia and murdered them. Defendants Jože Rustja and Oktavij Cek were among those brought from the littoral to Yugoslavia. The consequence of communist revolution in Yugoslavia was that the western powers would not want to hear anything about returning the Slovenian territory to Yugoslavia because they did not want to expand the territory controlled, — in everyone's opinion forever-- by the communists. Thus the communists forever forfeited the opportunity for the Slovenians to live in one nation state.

242 Liberal-reactionary epithet is of course an oxymoron as are all too many other communist characterizations.

joined with its former adversaries in a joint uncompromising fight against the basic human rights of the Slovenian people. In this fight Dr. Kalan was one of the most active members although he knew how to remain carefully hidden in the background. He attended all important meetings of the former JNS at which such pernicious decisions were made as was for instance the joining of the Slovenian Confederation (Slovenska Zaveza) and the like.

Toward the liberation movement he had already at is inception assumed a negative stance. He was the source of all the mendacious propaganda which as early as in 1941 attempted to suppress the liberation movement by various assertions such as for instance that the liberation front was pernicious because it involves members of the communist party, that armed resistance against the occupying powers is nonsense, that liquidations of traitors is a crime,[243] etc. On the other hand he fervently defended the criminal activities and rampages of the so-called Village Guards claiming that it was a spontaneous resistance against communism and the liberation front. He was thus defending that offal of the Slovenian people who accepted weapons from the occupation authorities and stabbed the Slovenian people fighting for liberty in the back.

For such criminal propaganda he naturally found justification in the traitor Draža Mihajlović who was supposedly representative of the old authorities of pre-April [1941]l Yugoslavia to whom everybody should be subordinated and whom everybody should follow even in the treason against his own people.

In his influential position which he enjoyed because of his position in the party he caused indescribable damage to the liberation movement.

Defendant Kalan is the type of clever politician who knew how to conceal his treasonable activities before the general public. He knew how to use connections with notorious traitors such as are Dr. Marjan Zajc, Lah and Hacin which he used even to help some adherents of the liberation front. But we should not be fooled by that obvious political maneuver. Thus his interventions were only a palpable attempt to create disinclination toward the liberation movement including with the adherents of the liberation front themselves whom he attempted to induce with his "broad minded" assistance to his tactics of waiting and speculation and thus pry them away from the ranks of the liberation movement. Numerous interventions which he was making eloquently demonstrate the degree of trust and respect he enjoyed with the treasonable elements. Those assuredly would not be accommodating him if they did not see in him one of their most faithful supporters and moreover if they did not think that such activities of Defendant Kalan was not detrimental to the liberation front.

Defendant Dr. Kalan who tried in this way to preserve to the end his false appearance of non-compromised politician is of course again at work in the establishment of the National Committee and subsequent to it of the so-called National Government which was in power for full six hours. He took part in the subject meetings and by that very action demonstrated his behind the scenes politics which he revealed with his collaboration in the last attempt of the criminal reaction which at

[243] Slovenian greatest poet Prešeren wrote: "A thief and bandit speaks a vernacular unknown to others." The communists have their own jargon: liquidation here means murder.

the last moment attempted to wrestle away from the Slovenian people the arduously attained people's democratic government.

Defendant only partially and evasively admits his activities in the ranks of the JNS politicians during the time of occupation. The proof of his guilt, however, is in the treasonable activity of his political group,[244]as well as in the testimony of major Marko Vrhunec which describes the Defendant in his true light as a traitor of the most clever and insidious kind. What a depraved informer was Dr. Kalan is further shown by the cases of Majda Treo and Drago Kralj whom he had sent to the internment as is clearly shown by witnesses Drago Kralj, Srečko Kobi and Majda Treo.

2. Ing. Bogdan Ferlinc:

   was together with Ing. Mačkovšek, Dr. Pirkmajer, Dr. Ernest Kalan and others member of the directorate of the JNS. Within this political party he helped the Mihajlović movement, to wit, spreading the treasonable thinking by distributing the subject literature and oral propaganda.[245] In addition he was connected with the notorious traitor Ing. Martinjak who had in the meantime already received his punishment.[246]

   He attended the larger meetings of the Sokol Legion at which were adopted decisions such as for instance collaboration in the treasonous regular Yugoslav Army and finally the disgraceful decision that members of the Sokol Legion should join the Slovenian National Guard. Characteristic also are his liaisons with traitor Marjan Strniša who had likewise already been punished.[247]

   At the organizing of the Blue Guard movement he participated as quartermaster. He admits that he was supplying in association with Ing. Martinjak the chetnik detachment with potatoes and linen. In addition-he was disbursing monetary aid[248] to the families of the chetnik traitors, the funds for which he received from Ing. Mačkovšek and which he was distributing through Verk, one of the commanders in the Sokol Legion.

   As organizer the accused Dr. Ferlinc is responsible for all the atrocities committed by the united Slovenian reaction through creating and supporting the criminal legions and eventually the Slovenian National Guard.

   How important, although carefully camouflaged was Defendant Ing Bogdan Ferlinc in the ranks of Mihajlović's traitors is evident from the fact that he was jointly with co-defendants Dr. Kalan and Dr. Pučmik invited to Tabor to the proclamation of the one-day government. From the declaration proclaimed at that meeting it is

[244] Communist justice holds that association with a political group is proof of criminal activity of an individual if criminal actions are imputed to the group. This shows how far the Slovenian judiciary is from the judicial systems of the civilized world.

[245] As foreseen in Orwell's novel *1984* the communists hold thinking to be a criminal conduct — *crimethink*.

[246] Martinjak was kidnapped by the communists in Trieste and dragged to Yugoslavia where he was imprisoned for many years.

[247] The communists captured Captain Marjan Strniša at Grčarice in 1943, took him to their kangaroo court in Kočevje, sentenced him to death, and executed him.

[248] Between December 1944 and April 1945, when the author was imprisoned by the Gestapo, his wife received aid from Milan Lenarčič and Rudolf Žitnik, probably from their own funds.

apparent that the attempt to form such government was based on the declaration of some kind of National Committee of October 29, 1944 signed by 300 persons in public life. One of those was certainly Dr. Ferlinc which proves that he was invited to the session at Tabor.

3. Dr. Jože Pučnik:
   was likewise member of the directorate of JNS. As such he was helping his group in promoting the Mihajlović's movement. During the occupation he, as member of the JNS directorate collaborated as delegate of his party jointly with Verk, Antosiewicz, and Ing. Ferlinc participated in all meetings of the Sokol legion and thus became one of the essential ideological instigators of the Blue Guard movement. Defendant himself admits that he always had reservations against the liberation front and that he adhered only to Mihajlović. Characteristic for the Defendant's liaisons with some treasonable politicians is the fact that the Defendant was able to successfully intervene on behalf of some who were arrested. This fact indicates that Pučnik was a political speculator to the very end. A solid and inconvertible evidence of that is the fact, corroborated by his co-Defendant Dr. Ernest Kalan, that Pučnik was invited to attend the forming of the one-day government at Tabor. From this follows that he was member of the National Committee and that he thus assisted at the apex of treason which is represented by the initially outlined attempt of starting a civil war.[249]

   From the statements of co-Defendant Dr. Kalan it furthermore follows that Pučnik was always taking an active part at the weekly meetings of the owners of "Jutro." With this he supported the said newspaper in its notorious campaign against the people's liberation movement.[250] Concerning Dr. Kalan, who also attended those meeting, it should be stated that he had already been convicted for this crime of his by the Court of Slovenian national honor but that he was given an amnesty.[251]

4. Rajko Lederhas:
   is one of the most obdurate opponents of people's liberation movement who was especially active in personally carrying out propaganda activities with which he was trying to recruit as many supporters as possible for the treasonable Mihqajlović. As such he was constantly seen in company of the most visible representatives of treason, to wit: Žitnik, Dr. Šmajd, Zajec and others. Most intimate were his relations with the fugitive judge Orožen,[252] a notorious organizer of the Blue Guard. With his personal propagandizing Defendant Lederhas damaged the people's liberation

---

[249] After the communists, who were unable to offer any real resistance to the Germans, have been carrying on a civil war for four years they are accusing the *Sokols* of having tried to start one in 1945.

[250] As a matter of fact *Jutro* was rather reserved with respect to the communists and, in contrast with *Slovenec* and *Slovenski dom*, published only what the authorities required.

[251] Translated from the communist jargon this means that they intended to assassinate Dr. Kalan during the war already but their attempt to murder him was not successful.

[252] Fran Orožen had timely removed himself to Trieste. Among the members of *OZNA* and partisans who in May 1945 came to arrest him was Jorgi Tavčar, at the time only fifteeen years old grandson of the famous writer and liberal politician Ivan Tavčar, who would a few years later meet a tragic end in a shootout with his younger brother Igor over a dispute concerning the Cominform. At that heroic undertaking Jorgi Tavčar pounded the floor with his rifle butt and shouted at judge's wife: "Where is Fran Orožen?"

movement because he ideologically instigated treason thereby assisting in the organizing of the so called regular Yugoslav Army. His activities in this respect were so visible that adherents of the liberation front were already watching[253] him already during the occupation and found out that Lederhas had been attending organizational meetings of the Blue Guard at the Kregar inn in Štepanja vas together with Dr. Zajec, Dr. Šmajd, Bajič and Žitnik. There they conducted very intensive propaganda campaign and recruited about twenty farmers for the treasonable movement. Proof of this is report included in the material of the trial in Kočevje [in 1943].

From the said it is clear that Defendant Lederhas, too, was an ideological instigator of treason[254] and as such shares the responsibility for all the evil perpetrated by traitors of the people.

5. Etbin Bežek:

   was according to the statement of Franc Mikuž one of the most alert organizers of the Sokol Legion. He was recruited for the legion by Antosiewicz. Because of his active and successful operation he was soon made commander of a battalion or the so-called regular Yugoslav Army. He was particularly active in spreading of the treasonable propaganda and was as such even accepted in the anticommunist committee at the Provincial Administration. There he jointly with Milan Finc and others ensured the dissemination of criminal propaganda against the liberation movement under the guise of anticommunist action. That propaganda which urged annihilation of the people's liberation movement by all means possible and especially by informing caused that the ranks of people's traitors increased to such lamentable proportions. Etbin Bežek as ideological instigator shares the responsibility for the crimes of all the traitors as a whole.

   As member of the treasonable National Committee of October 1944 Etbin Bežek, too, participated in the proclamation of the one-day government which proves his treasonable activity in full.

6. Stane Pirc:

   was a very active organizer of the Sokol Legion. He participated in the meetings jointly with Antosiewicz, Verk, Marinčič, and Dušan Verbič. He was entrusted with the position of liaison officer, that is reporter, among the representatives of the illegal organization of "Blue Guard," that is those who hypocritically remained in the background and the legal organization of the "Blue Guard" which existed in the National Guard itself, that is in the persons of Glušič and Pleško. Stane Pirc and Etbin Bežek also collaborated in the forming of the National Committee or the so-called National Government at Tabor. In addition Stane Pirc was commander of the Sokol Legion at Kodeljevo which fact fully confirms his complicity with respect to the ideological instigation of the said treasonable organization.

7. Pavel Borštnik:

   comes from the ranks of the most cruel opponents of democratic liberties of the Slovenian people. Since as early as 1925 he was a member of ORJUNA and subse-

[253] From that statement one can clearly see who was spying on whom. The communist prosecutor states that communists were sending their spies to the meetings of Mihajlović's legions. He cited no instance of the legions ever having sent a spy to the meetings of the *Liberation Front*.

[254] Another communist oxymoron: public prosecutor speaks about the "ideology of treason!" By definition, however, treason is devoid of ideology.

quently member of the JNS. As such he became one of the most active organizers of the Sokol and National League. He participated in the most important meetings which were in connection with the treasonable movement taking place in Zvezda. There used to come also the notorious traitors Žitnik, Bevc,[255] Bežek, and Kalan.

Pavel Borštnik is moreover guilty of the arrests of activists of the liberation front at Poštna Hranilnica (Postal Savings bank). He was in close contact with Ivan Tonja and knew, as he himself admits, about all adherents of the liberation movement at that institution. In this way it came to the deplorable treason because of which so many Slovenians of good character had to be marched off to prison. Always a clever manipulator of intrigue we naturally find him together with Ivan Tonja at the forming of the one-day government at Tabor on May 3, 1945.

8. Ing. Lado Kham:
is a type of one of the most carefully concealed and therefore also most active instigators of Mihajlović's treason in Slovenia. He was the military organizer of the so-called "Blue Guard," member of the National Legion and had close contact with major Novak as well as with other prominent representatives of the "Blue Guard."

While the Defendant denies membership in any of the legions he does admit that he had already at its inception declined to collaborate with the liberation front acting together with the other adherents of Nagode's[256] group, the so-called Stara Pravda.

Although he denies any kind of activity in the ranks of the Mihajlović's movement he admits having contact with Ing. Martinjak, Ing. Dimnik and others who however were only coming to see him about the news from allied radio broadcasts.

Although he even today cannot find a good word for the national liberation movement, he characterizes the formation of the so called village guards as a spontaneous resistance of the Slovenian people against the partisans.[257] He further admits that Ing. Dimnik and the known organizer Glušič were telling him about the organizing of the so-called regular Yugoslav Army but maintains that he himself was member of neither the Sokol nor National League.

---

[255] Note by editor: In 1995 it was possible to gain partial access to the author' *dossier* in the Archive of Slovenia where it is kept under No. 4235. The dossier was prepared by communist Svetozar Frantar. "Treason" supposedly committed by the author consisted, according to that dossier, of his failure to observe the Decree on formation of the so-called *Slovenian people's army*, that is, he did not subordinate himself to the so-called *Liberation Front*, that he on October 29, 1944 signed the *National declaration* and on May 3, 1945 *Proclamation to the Slovenian People*. The dossier was transferred to the Archive by the *Administration for the Execution of Penal Sanctions* of the Republic of Slovenia which presumably means that the present Republic of Slovenia does not intend to prosecute the said accusations.

[256] Črtomir Nagode, one time sympathizer of the Soviet Union, naively attempted to organize a democratic opposition to the communist regime. He was tried in a show trial managed by the Moscow trained political police boss Mitja Ribičič, sentenced to death and executed in a manner that involved torture. The Slovenian government to this day refuses to disclose where he is buried.

[257] Brainwashing attempts on Lado Kham were clearly unsuccessful. It is truly inconceivable how the communists can expect that someone could find "a good word" for a movement which deprived him, as well as the entire Slovenian nation, of liberty, treated him and his entire family in a cruel and inhuman manner, ruined his health and made out of him who once was a healthy athlete, tourist, swimmer and mountaineer, a living skeleton.

The entire defense of Defendant Kham is clearly shown to be false by the statement of Franc Juvan who stated that the office of Ing. Kham in the skyscraper was the command post of the regular Yugoslav army where used to convene lieutenant colonel Krener, commanders of all three legions, and other traitors. From this it is quite clear that Ing. Kham was one of the first, albeit carefully concealed, organizers of treason.

9. Karel Benič:

was member of the Sokol Legion and could always be seen in the company of its most active organizers, particularly with Joško Jurman. He participated at the meetings where within the framework of the intelligence service they reported adherents of the liberation front to the authorities. At one of those meetings into which a member of the Liberation Front managed to slip in undetected[258] Benič stated that Jože Valentinčič was a communist, member of the communist party and that he should be put out of the way. Indeed the said Valentinčič was subsequently arrested. That, by witness Valentinčič confirmed event, clearly presents the purpose and character of the meetings of the Sokol and other legions.[259] Proof that Karel Benič was an informer in the service of occupation authorities is also Defendant's statement confirmed by Janko Pirkovič who at the arrest of Kremžar and Berdajs stated that they could be released if they became supporters of his cause.[260]

Very significant is furthermore the statement of witness Božena Kremžar from which can be undoubtedly established that Defendant Benič was responsible for the arrest of her husband who was subsequently killed in Barje. That case at the same time proves that activities of the so-called Black Hand were the intended result of the working of all three legions rather than isolated incidents perpetrated by a small criminal clique as the Slovenian people in its innocence suspected during the occupation. At the mobilization of the National Guard Defendant Benič naturally joined its ranks but he did not stay long in that uniform. He soon donned the mufti again because in this way he could easier carry on his criminal business of the informer.

10. Zdravko Deleja

was an active organizer of the Sokol Legion which is particularly apparent from the statements of Defendant Anton Lajovic whom Deleja recruited in the Sokol Legion. The same follows from the statements of Janko Tavzes whom Deleja attempted to recruit for the Sokol Legion and to whom he was even bringing the Blue Guard literature.

Zdravko Deleja was particularly active in distributing the literature. He also had direct connections with the German Gestapo. Although he maintains that those

[258] From the above it is clear who was spying on whom. The communist prosecutor states that the communists were sending their spies to the meetings of Mihajlović's legions but he cited no instance where the legions had sent a spy to the meetings of the so-called Liberation Front.

[259] No judiciary system of any government based on laws considers sequence of two events to as evidence that the prior event caused the later. Such sequence could at best be considered an indication of possible connection, but causal relationship between the two must be demonstrated by additional evidence. In this case such evidence would consist of records on the information provided to the police which the communists did not produce because they did not exist.

[260] It is not credible that democratic legions were willing to admit former communists to their ranks because such converts would be unreliable.

connections were only such as required by his business it is of course clear that he, as an organizer of the Sokol Legion, could only have been an informer of the Gestapo.

Defendant Zdravko Deleja is accordingly one of those through whom the Sokol Legion informed on the adherents of the liberation movement to enemy police.

11. Oto Frölich:

organized the Sokol Legion in the Dolenjska region. He was leader of the Sokol Legion as is apparent from the statements of Franc Juvan. He was constantly in the company of other organizers of the Sokol Legion, to wit: Juvan, Starin, Vrančič and others. He was in touch even with the notorious Black Hand member Berlot.

He was particularly active in the intelligence service and spied on the adherents of the Liberation Front. He himself admits that he was tracking Francka Janežič whose photograph was furnished to him by Franc Juvan.

Significant are frequent meetings in the shop of Defendant Frölich in which participated among others Juvan, the well known leader of DOS. How close were the ties of that company with the occupation authorities is evident from the statement of Štefan Pokovec to the effect that Juvan was promising to the arrested that he would arrange for his release as soon as he would state that he would collaborate with the Blue Guard. Witness Štefan Pokovec further made well founded accusation against Frölich to the effect that he was responsible for the murder of comrade Bolha. The latter once mentioned to him that he spoke with Frölich about the Liberation Front whereupon the witness warned him to be wary of Frölich. It was, however, too late as Frölich had already taken steps the consequence of which was murder of comrade Bolha.[261]

The Frölich case proves that the operation "Black Hand" was a result of a widely spread espionage network of all three legions. Frölich was clearly member of the informer group which through Juvan, a member of DOS, reported the names of the Liberation Front activists.

12. Rudolf Zorc

was already at the time of Yugoslavia member of the general committee of JNS. Already at the time of occupation he had jointly with the most prominent traitors such as Andrej Uršič,[262] Dr. Vrčon, Ing. Rus, Ing. Sancin, Franc Juvan, Albert Hertič and others took part in six or seven meetings of OJNS (Omladina Jugoslovenske Nacionalne Stranke — Youth of the Yugoslav Nationalist Party) where it was discussed how to combat the Liberation Front.

As member of OJNS he was automatically member of the Sokol Legion and as such he following the call of the said legion joined the Slovenian National Guard.

Defendant Rudolf Zorc is thus one of the first ideological instigators and organizers of the treasonable Sokol Legion.

---

261 The indictment contains neither a description nor the record of the steps taken by Frölich.

262 Andrej Uršič, lawyer, politician, member of *JNS* and editorial board of the daily *Jutro*, editor of the paper *Demokracija*. Kidnapped in 1948 in the Slovenian Littoral, taken to Yugoslavia where he was murdered. *Dosje Mačkovšek*, published in 1994 by *Arhivsko društvo Slovenije* in Ljubljana states that he was interrogated by *UDB* whereupon he disappeared if we are to use the quaint parlance of Slovenian "historians."

13. Božo Borštnik
had initially cooperated with the Liberation Front but had, as representative of Stara Pravda, dropped out of it together with Nagode. With his political group he then joined the progressive alliance and later also the Slovenska zaveza (Slovenian confederation) which constituted the ideological background of united Slovenian treasonable reaction.

Defendant Božo Borštnik is thus ideological instigator of treason which is clearly seen from the fact that he was publisher and editor of "Stara Pravda" which opposed the goals of the people's liberation struggle by trying to instill in the people distrust of the Liberation Front thereby exhorting the people to treason.

14. Filip Kosec:
comes from the ranks of the Yugoslav ORJUNA (Organizacija Jugoslovenskih Nacionalista — Organization of Yugoslav Nationalists). As such he already at the beginning declined to cooperate with the Liberation Front although he was given the opportunity do so by the most prominent representatives of the liberation movement. Despite that, Defendant Kosec strayed on the path of treason. He became one of the most active organizers and promoters of the National League working with Dr. Potokar who collaborated in organizing treason in the Gorizia region and with the assistance of Ing. Rustja and Gojko Vujošević. In this way Filip Kosec is an accomplice of the treason in the Littoral.

15. Ing. Jože Rustja:
was from the very beginning a sympathizer of the Mihajlović movement. As he himself says he was not active in politics prior to December 1944. After that he got in touch with Vujošević with whom they began to discuss the situation in the Littoral and planning how to organize chetnik detachment there. They held several meetings at which they resolved that they would fight for the Karadjordjes and the old Yugoslavia.[263]

At one of those meetings they divided the Littoral into three sectors, viz., Gorizia, Trieste and Istria. They established the general committee of the National Legion in Trieste and a regional committee in Gorizia. Defendant Ing. Rustja was very active in the latter writing articles for "Narodna edinost" (National Unity), a paper that was making every effort to induce the Slovenians of the Littoral to treason. In addition Defendant Rustja collaborated in the intelligence service in the Littoral and constantly informed on the adherents of the Liberation Front to Vujošević. Around Ing. Rustja gathered all the reaction of Gorizia which fact contributed to the great suffering of the People of the Littoral in their final phase of combat.[264]

The greatest guilt of Defendant Rustja consist of the fact that he was through his assiduous propaganda attempted to weaken the trust of the people of the Littoral

[263] Communist dialectics glibly calls loyalty to the king and country "treason." In the works of Marx, Engels, Lenin and Stalin an explanation may perhaps be found how can one betray something he never supported.

[264] "In the last phase of its combat" here of, course, does not mean combat against the Germans but only the civil war carried out by the communists knowing full well that the Germans would leave Yugoslavia without their help after surrendering to the Allies.

in the Liberation Front. Especially damaging for him is that he collaborated with Vujoševič although he knew that the latter was collaborating with the notorious criminal Jevdjević who is responsible for many atrocities in the Littoral.

16. Oktavij Cek:
is a type of opportunist who would not join the liberation front of the Slovenian people because it would be inconvenient as he himself admits. He is one of those who for a long time remained in the background and waited that the Slovenian people exhaust themselves in their unequal combat against the occupation forces and then appeared in the forefront. According to the plans of Dr. Vrčon, however, Oktavij Cek, too, became active and joined the ranks of the Sokol Legion. He was entrusted with the task of organizing treason among the Slovenian of the Littoral. To this end he went to Trieste where he began to organize some kind of center collecting adherents mainly from the intelligentsia. That center, of course, was a clandestine Sokol Legion whose basic purpose was suppressing the partisan movement which became apparent when it, in the Littoral as well, established liaison with the chetniks and thus take an active part in combating our people's army.
Oktavij Cek is, accordingly, the organizer and ideological instigator of the treasonable movement in the Littoral which constitutes the gravamen of his offense.

All the defendants admit the above described acts in their entirety with only rare exceptions. The guilt of the defendants is thus completely proven by their own admissions[265] and finally by testimony of witnesses[266] and other evidence.

Death to fascism — liberty to the people[267]

Public prosecutor for the Ljubljana region
(signed) Anton Zupan

[265] A basic tenet of every judicial system in countries where the rule of law exists is that admission by itself and in absence of corroborating evidence does not suffice for conviction. If this were not so it would be possible to convict innocent people — such as is the case in Slovenia — or it would be even possible to execute persons who because of suicidal tendencies admitted to crimes they did not commit. The purpose of communist judiciary, however, is not the establishment of guilt or innocence but annihilation of class enemies.

[266] The quality of "testimony" of witnesses in this connection is apparent from the cross examination which Stanko Dimnik was allowed to make of one such "witnesses." When the "witness" concluded her tirade as to how the defendants promoted criminal ideology, Dimnik inquired first if the "witness" knew him. The "witness" said she did not, therefore Dimnik introduced himself whereupon the "witness" said she had *never heard of him*. Dimnik then inquired how the "witness" can then assert that he had been carrying out the said crimes. The "witness" repeated her speech as she was coached to the effect that Dimnik was one of the ideological promoters of the Mihajlović movement. Such "evidence" would not be admissible under a government of laws.

[267] Communist greeting equivalent to the Nazi "*Heil Hitler*!"

Rebuttal of the Communist Indictment

The public prosecutor states that, with rare exceptions, all the accused had confessed to all his accusations in their entirety. He does not, however, tell what they did not confess and what he was unable to prove with his court proceedings. Zupan's evidence consists of crass invective which serve him instead of the analysis of statements of credible witnesses. He dreams of ghosts and so he sees them. Zupan trembles before ghosts, swoons over mendacious claims of espionage and informing allegedly carried on by the Sokol and National Legion, over murders supposedly committed by members of the Legions against adherents of the so-called Liberation Front, and asserts that the entire detachment of the Yugoslav Army in the Homeland joined the Slovenian National Guard. The accused had, in Zupan's opinion, committed an unpardonable offense by attending the meeting in the Sokol Hall at Tabor where the "one-day government" was constituted which still exist as the Slovenian National Committee. There are too many of such fabricated and mendacious claims in the Zupan's indictment to be enumerated here.

I will limit myself to the following observations. In April of 1941 Germany and Italy occupied Yugoslavia. As narrated earlier, Deputy Presiding Elder Engelbert Gangl[268] called on me to organize resistance of the Sokols along the lines envisioned by the Sokol Federation. I convened a meeting of representatives of all five Sokol Districts at which we established the Sokol War Council with me as chairman. The Sokol War Council decided to entrust the resistance against the occupation authorities to the Sokol Legion. The purpose of the Sokol War Council was conducting resistance against the occupying powers and eventually expelling them from our country. The Sokol War Council never abandoned that goal and never had any connection with the occupation authorities much less much less conducted any espionage. Who was spying on whom was told by prosecutor Zupan himself among his accusations directed against Karl Benič: he revealed that members of the so-called Liberation Front were sending their spies to the meetings of the Sokol Legion.

When the Sokol War Council was established in the spring of 1941 there was as yet no trace of the so-called Liberation Front. The communists, who later became members of the so called Liberation Front followed Stalin as an ally of Hitler's at the time and were quite indifferent to the suffering our people had to endure under Hitler's and Mussolini's occupation. But when Hitler in June 1941 attacked the Soviet Union, the true homeland of all communists, the Slovenian communists naturally opted for Stalin and immediately, in July of 1941, established the so-called Liberation Front. The communists decreed that everyone who should refuse to join

[268] Although he had taken this initiative Engelbert Gangl soon withdrew from the *Sokol* resistance movement and played a rather strange role ill befitting a Deputy Presiding Elder of the *Sokol Federation of the Kingdom of Yugoslavia.*

the so-called Liberation Front was ipso facto sentenced to death. This meant civil war during which the communists carried out their war against their own people. At the inception of the so-called Liberation Front we noted numerous deportations to concentration camps, among the interned was the first commander of the Sokol Legion who was murdered in the concentration camp at Mauthausen at a time where there was as yet no trace of any Liberation Front. Anton Zupan, if he wanted to be consistent, would have to say that the communists who at the time were allies of Hitler were responsible for the death of those victims.

The Sokol Legion never engaged in murder nor did it carry on any war against the domestic people even though they might be of a different political persuasion nor did it have resources for something like that. It was preoccupied with the education of its membership along the guidelines set forth by Tyrš. Its war aim was to become prepared, with countrymen of the same persuasion, for the eventual expulsion of the enemy from our country. That already sufficed so that the Sokol Legion Legion felt no need to join the Slovenian National Guard. We were integrated into the Yugoslav Army in the Homeland under the command of General Draža Mihajlović which defended the homeland against all enemies, foreign and domestic, including the communists whose objective was to seize power in Yugoslavia.

In Yalta the allies of Stalin partitioned the sphere of interests in Yugoslavia but the end result of that agreement was that Stalin seized all power in Yugoslavia for the benefit of international communism, an agreement that will live in infamy.

The so-called Western Allies[269] directed the Italians after they surrendered to turn over all their weapons to the Yugoslav communists who used that armament for the annihilation of the resistance of Mihajlović's troops.

Commander of the Sokol Legion Antosiewicz needed a courier who carried messages to members of the Legion. The communists placed one of their confidantes into that position who betrayed to the Gestapo Antosiewicz's circular alerting members of the Legion to be prepared for the approaching expulsion of Germans. The Gestapo arrested Antosiewicz and sent him to the concentration camp at Dachau.

The British who occupied southern Carinthia conveniently got rid of Tito's partisans who were coming to the Slovenian part of Carinthia by making an agreement with Tito according to which Tito would withdraw all his partisans from Slovenian Carinthia in exchange for witch the British promised to turn over to the communists all Yugoslav refugees. In accordance with that agreement the British turned over to Tito 13,000 Slovenians, Croatians and Serbs who left their homeland before

[269] The term "Western Allies" of Yugoslavia raises the question whether such designation was appropriate for the British and the Americans in view of which they acted toward Yugoslavia and its nations. Of course they were not allies of Yugoslavia: they were allies of Stalin and only in this respect they can be properly called allies.

the advent of the communists and took refuge in the Slovenian part of Carinthia claiming asylum under international law and agreements.[270] Southern Carinthia was at that time occupied by the British. Nikolaj Tolstoy, grandson of the famed Russian novelist, exposed that invidious British act in the journal Encounter[271] and later in a book titled: "The Minister and the Massacres."[272] Tolstoy was able to gain access to the British wartime documents on which he based his documentaries.[273] Later, however, when he needed the originals do defend himself against a libel suit in court those documents were conveniently lost by the government agencies where they were kept.

In its broadcast of January 3, 1984, the British Television BBC 2 presented a highly condensed account of this British war crime exposed by Tolstoy. The broadcast was announced in the leading London paper The Times. The Yugoslav government protested but could not prevent the broadcast. The sinister forces that control the policy of the British empire, however, found a way to visit a revenge upon Tolstoy in the form of a suit for defamation of character leveled at him by Lord Aldington, formerly Toby Low, who was in charge of the British Expeditionary Force in Southern Europe where the now retired British Prime Minister Harold Macmillan was diplomatic counselor. The court imposed multimillion damages upon the writer which, however will not clear the British reputation. The only good thing about the outcome of that suit was that Tolstoy did not have that kind of money and the British lord could not collect. The British were probably concerned about the possibility that Yugoslav victims of this British war crime sued the British empire for damages which would undoubtedly amount to tens of billions of dollars. The damages would be only for the support of the families of those murdered who lost their providers when they placed themselves unarmed under the British protection; nothing, of course, can pay for their lives. However, the British empire need not lose sleep over that — the Yugoslavs will not bother them with such suit.

---

270 Nicholas Bethell: *The Last Secret*, Basic Books, Inc., New York, 1974.

271 Nikolai Tolstoy: "The Klagenfurt Conspiracy, War Crimes and Diplomatic Secrets," *Encounter*, v. 40, No. 5, p.. 24-37, May, 1983.

272 Nicholas Tolstoy: *The Minister and the Massacres*, Century Hutchinson Ltd, London, 1986. On publication of this book Lord Aldington alias Toby Low sued Tolstoy for defamation of character and the British courts, subservient to the British nobility, imposed a multi million fine on Tolstoy and banished the book from all libraries in Britain, including even the Bodleian library at Oxford university, repository of everything printed in England. The book, however is available from second hand book dealers as well as in university libraries in the United States where the British writ does not run. The European Court of Human rights in Strasburg had subsequent to that voided the fine imposed on Tolstoy as unreasonably high administering a well deserved, if rare, rebuke to the British judiciary.

273 Considering the manner in which Anglo-Saxon countries keep documentation on delicate affairs authenticity of documentation made available to the public is always somewhat suspect.

During the war two leaders of the democratic party died: Albert Kramer of natural death, Ivan Pucelj in the concentration camp at Dachau. The party's affairs were taken over by a directorate of several members. For a while the chairman was Janko Mačkovšek whom the German Gestapo arrested and deported to Dachau where he died. He was succeeded by Otmar Pirkmajer whom the Gestapo held in prison in Ljubljana for some time and then took him to Dachau where he survived the epidemics and lived to see liberation by United States Army. After the war Pirkmajer remained in Germany where he was employed by the International Refugee Organization.

I, too, was arrested by the Gestapo the day before Christmas evening. I was in prison until April 1945 when the Germans already began to retreat. Zupan's imputations that the JNS had any connections with the occupying authorities is accordingly absurd. Nevertheless Zupan was talking at the trial that my arrest and stay in prison was staged for the appearances sake. What is clear is that someone had to suggest to the occupation authorities whom they should get rid of. In view of the case of Antosiewicz's courier who was shot on the way to Dachau it is clear that the communists knew how to get rid of persons they considered dangerous. An evil person thinks that others will act as he would. Anton Zupan would have to turn around his lies, change them to truth and haul his masters to court because it is they who were savagely murdering their countrymen who opposed communism. The worst of all communist crimes is the Slovenian Katyn of Kočevje. All the communist crimes have for a long time been crying for a Nűremberg trial equal to that mounted against Hitler's war criminals. At that process Zupan's mendacity should also be condemned and the honor of all sixteen accused smeared by Zupan in the unjust communist court vindicated.

My friend Stanko Dimnik was at his trial condemned to twelve years' imprisonment. He was incarcerated in the prison at Ig. His wife and daughter were only seldom allowed to visit him and on those occasions his communist jailers took special sadistic pleasure to humiliate him in any way they could. The communists were also exploiting his professional expertise of a structural engineer and demanded that he work for a pittance on the construction plans for the factory Litostroj in Šiška. After eight years of imprisonment he was released and the communists gave him a decree stating that his conviction was expunged and that he is considered as "having no record." That meant that they have officially admitted that all the accusations against him were false, that he has done nothing against the Slovenian people and had committed no crime.

My friend Etbin Bežek was also condemned at that trial to several years' imprisonment. Concerning Zupan's lies that the Sokols were spying on members of the so-called Liberation Front it is noteworthy that he was during the war once called by an acquaintance who was a police employee and warned him that the police found in a seized communist archive his photograph made by a lurking communist when Bežek stepped out of a house where the Sokols were meeting. After five years Bežek was released but only after a year he was arrested again under an "administrative measure"[274] and sent him to the concentration camp at Strnišče near Ptuj where the communists were finishing a huge aluminum factory started by Herman Göring during the war. After several years Bežek was allowed to visit his daughter Tatjana in Italy where he had the opportunity to read my booklet on the communist infiltration attempt at Sokol I at Tabor[275] and write to me identifying person whom the communists smuggled to Czechoslovakia at the time of the Sokol rally in Prague in 1938.

Jože Rustja returned to Italy after he had served his sentence. No doubt he was able to do that only because of the intervention of the Italian government. Before his death he expressed a desire that I publish a rebuttal to the communist accusations. This I have done.

---

[274] Administrative punishment" for usually unspecified charges was a repressive measure whereby the communists would every so often imprison a whole series of people about which they knew that they are against them. The purpose of those raids was strengthening of the communist reign of terror. When Tito and Stalin quarreled in 1948 and Tito sought American aid President Truman set as the only condition that Tito disband the concentration camps. Tito agreed to that and at that occasion Bežek was released from Strnišče. That was the only instance we know that an American president demanded that Tito show at least a modicum of respect for human rights. All other presidents competed in fawning over that repulsive Balkan dictator. All concentration camps were not disbanded, however. The camp *Goli Otok* (*Barren Island*) in the Adriatic continued as a special re-education facility for the adherents of the *Cominform*.

[275] Lado (Ladislav) Bevc: *Na braniku za sokolske ideale*, Slovenska Pravda, Enfield, England 1965; Second edition, V. Bevc, San Ramon, California, 1988; English translation: *On the Ramparts*, V. Bevc, San Ramon, California 1990. See also: L.Bevc: *Defense of Sokol Ideas*, The South Slav Journal, v. 14 (No. 53-54) pp 28-45, 1991; v. 15 (No. 55-56) pp 44-64, 1992.

Conclusion

In 1986 the author retired as the active chairman of the *Slovenian Democratic Party* and became its chairman emeritus as Peter Urbanc was unanimously elected to take his place. The author died in 1988, two years before Yugoslavia for which he worked all his life disintegrated. He is buried in Santa Paula, California, not far from the Pacific Ocean.

The president of the Slovenian National Committee Miha Krek (1897-1969) died in Cleveland in 1969. He was succeeded by Miloš Stare (1905-1984), publisher and editor of *Svobodna Slovenia* (*Eslovenia Libre*) in Buenos Aires who was in turn succeeded by Rudolf Smersu (1905-1998).

It gradually became clear that the Yugoslavia we knew was no longer a viable political entity. On May 20, 1991 Slovenia and Croatia declared their independence and in autumn of the same year the Yugoslav People's Army withdrew from Slovenia tacitly acquiescing to its secession. On April 7, 1992 the United States, still mesmerized by the memory of Marshal Tito in whom it had invested so heavily, reluctantly recognized Slovenia as an independent nation. Slovenians opted to cast their lot with the faceless bureaucracies of Brussels and Strasburg after the endless frustrations endured when one nation with their Slav brethren.

The *Slovenian National Committee* headquartered in Argentina assisted the democratic coalition Demos in Slovenia to achieve political plurality. When *Demos* won the election it was clear that that political activity was again possible in Slovenia and the *Slovenian National Committee* voted unanimously to dissolve itself. The grip of the communist party on the country had weakened somewhat, but the communists who remained in control of most administrative and judicial offices soon regained control of the parliament. The reorganized communist party is now called *The Liberal Democracy of Slovenia* and remains true to the principles of Marxism-Leninism. A reconstituted *Slovenian People's Party* elected a significant number of its members to the new parliament and became part of the communist-dominated coalition government. The *Slovenian Democratic Party* is home to those with social democratic views but, except for the name and a general attachment to the democratic way of life, the new *SDS* had nothing in common with the *Slovenian Democratic Party* that once existed in exile. Other parties directed to addressing present conditions have also come into being. In 2002 the communists finally lost their majority in parliament. The new government seems to be slowly but steadily making as much progress as it can given holdovers and remnants of the ruinous communist era.

The sentences of some convicted as enemies of the state were rescinded by the courts and in token compliance with a *Resolution of Council of Europe 1096* calling for the dismantling of the heritage of totalitarian communist systems, parliament hesitantly and reluctantly enacted a restitution law under which properties confiscated from class enemies could be returned. However, with the United States, the avowed champion of human rights, taking the position that property rights are not human rights, no effective international pressure was brought to bear neither on Slovenian communists nor on those in any other reorganized communist country to restore property to those few survivors of communist terror who claimed it. Bevc property seized by the communists has not been returned.

The voluminous archives of Bogumil Vošnjak, the doyen of Slovenian democrats in exile and one of the founders of Yugoslavia, were delivered to Janez Arnež for his collections of Slovenian émigré activities which he maintains at *Studia Slovenica* in Šentvid near Ljubljana. The *Hoover Institution of War Revolution and Peace* at Stanford University has graciously consented to accept the part of the archive of the *Slovenian Democratic Party* in exile that was in the author's possession.

The editor of these memoirs has received from the Slovenian Ambassador in Washington a scroll dated April 7, 1992, bearing a letter of thanks from the government of Slovenia "*in appreciation of your concern, help and irreplaceable contribution to our common endeavors for international recognition of the Republic of Slovenia by the United States of America*" signed by Milan Kučan, communist party boss and President of Slovenia; France Bučar, President of the Assembly (parliament); Lojze Peterle, President of the Government (prime minister), the only non-communist; and Dimitrij Rupel, Minister for Foreign Affairs. The contribution presumably consisted of letters to the United States Senate Foreign Relations Committee and participation at a demonstration in San Francisco.

Tahoma, California
February 2007

Vladislav Bevc

Index

About the Author

Ladislav Bevc, was born in Škocijan in Lower Carniola — Slovenia. He was graduated as civil engineer from the Technical University in Vienna and entered civil service in the Slovenian Littoral. In World War I he served as a member of the Austrian military engineering corps on the Eastern front in Russia and on the Western front in France.

In Ljubljana he was the senior engineering counselor at the regional administration for construction of roads, bridges and hydraulics and concurrently taught civil engineering subjects at the Technical University in Ljubljana. He was the chairman of the City's Engineering Committee and City Planning Commission, member of the City Council, on the committee of the Reserve Officers'Association, Presiding Elder of Sokol Association I at Tabor, a member of the committee of the Sokol Federation of Kingdom of Yugoslavia in Belgrade, president of the Aeronautical Club and vice chairman of the Association of Democratic Academicians. In recognition of his civic and professional work he was decorated with the order of Sveti Sava, Beli Orao, Jugoslovenska Kruna and with the Bulgarian order for Civic Merits.

During World War II he was a member of the clandestine Sokol War Council which organized resistance against the Italian and German occupation within the framework of the resistance movement of General Dragoljub Mihajlović and member of the directorate of the Democratic Party. In 1944 he was elected to the Slovenian National Committee whose member he was throughout his life. As member of the National Committee he left his homeland before it was occupied by communists.

After spending four years in refugee camps in Italy he in1949 arrived to the United States where he was invited by relatives and lived in Napa, California. He was a Registered Civil Engineer and from 1959 on a Fellow of the American Society of Civil Engineers. In 1963 his alma mater, the Technical University in Vienna, conferred upon him the Gold Diploma in recognition of fifty years of professional activity. Since 1950 he was Chairman and later from 1988 Honorary Chairman of the Slovenian Democratic Party in exile. In 1974 he was elected Presiding Elder of the Yugoslav Sokol in the Free World. He died on November 29, 1988.

Part of author's memoirs was published by the *Sokol Herald* (*Sokolski Vestnik*) of Milwaukee, Wisconsin and the South Slav Journal in London. He also wrote: *Brick and Timber Construction* (1920), *Regulation on Engineering Honoraria of the Yugoslav Engineering Association* in Belgrade (1920), *Ethics in Sokol Associations* (1931), *General Plan for Public Buildings in Slovenia* (1936) and numerous other professional and political papers published in reviews and journals.

Studies in Modern European History

The monographs in this series focus upon aspects of the political, social, economic, cultural, and religious history of Europe from the Renaissance to the present. Emphasis is placed on the states of Western Europe, especially Great Britain, France, Italy, and Germany. While some of the volumes treat internal developments, others deal with movements such as liberalism, socialism, and industrialization, which transcend a particular country.

The series editor is:

Frank J. Coppa
Director, Doctor of Arts Program
in Modern World History
Department of History
St. John's University
Jamaica, New York 11439

To order other books in this series, please contact our Customer Service Department: